T0298972

AN INTRODUCTION TO RINGS AND MODULES

Also available in this series

An Introduction to Rings and Modules with K-theory in view

A. J. Berrick & M. E. Keating

CAMBRIDGE
UNIVERSITY PRESS

CAMBRIDGE UNIVERSITY PRESS
Cambridge, New York, Melbourne, Madrid, Cape Town,
Singapore, São Paulo, Delhi, Tokyo, Mexico City

Cambridge University Press
The Edinburgh Building, Cambridge CB2 8RU, UK

Published in the United States of America by
Cambridge University Press, New York

www.cambridge.org
Information on this title: www.cambridge.org/9780521632744

First published 2000

A catalogue record for this publication is available from the British Library

ISBN 978-0-521-63274-4 Hardback

To the memory
of
Hamilton Berrick
(6 March 1987 – 8 September 1994)
who in his short life
brought others so much joy;
with the thought that this book
might one day have given him some pleasure.

Contents

PREFACE

This text is an introduction to the theory of rings and modules for a new
graduate student. What gives this book its special flavour is that it is particularly aimed at someone who intends to move on to study algebraic K-theory.
This aim influences both the choice of rings and modules that we discuss, and
the manner in which we discuss them.

Starting from a knowledge of undergraduate linear algebra together with
some elementary properties of the integers, polynomials and matrices, we provide the basic definitions and methods of construction for rings and modules,
and then we develop the structure theory for modules over various kinds of
ring.

These classes of ring reflect the historical roots of algebraic K-theory in geometry and topology on the one hand, and representation theory and number
theory on the other. Thus the rings that interest us are Noetherian rings, in
particular skew polynomial rings, Artinian rings, and Dedekind domains.

The text pursues ring and module theory up to the point where the aspiring
K-theorist needs category theory. The necessary category theory is dealt with
in the companion volume [BK: CM], both in the abstract and in relation to
more advanced topics in the ring and module theory. Our division of the
subject matter in this way means that the present volume can be regarded
simply as an introduction to some fundamental topics in the theory of rings
and modules.

Here is a more detailed survey of the material in this text.

The first chapter gives the definitions of rings and modules, together with
some examples and constructions. The second chapter develops the ground-
work of the theory of modules, with a particular emphasis on the basic ways
of constructing and comparing modules. We see how to combine modules to
make new ones, using direct sums and, more generally, short exact sequences.
The use of short exact sequences to describe the relationships between mod-

ules lies at the core of algebraic K-theory, and so the notion of a short exact
sequence plays a central role both in this text and in [BK: CM]. Also crucial
to K-theory are the free modules, which are assembled by taking direct sums
of copies of a ring, and the projective modules, which arise when free modules
are in turn split up into direct sums. A projective module over a ring is the
appropriate generalization of a vector space over a field.

Other topics that we look at in some depth in the second chapter are the
representation of homomorphisms between free modules in terms of matrices,
and the use of matrices to relate the different bases of a given free module.
This leads to a discussion of the circumstances under which a free module has
bases with differing numbers of elements, that is, when the coefficient ring
fails to have invariant basis number.

The primary aim of the remainder of this book is to acquaint the reader with
some important classes of rings and their module theory, with an emphasis
on the description of the projective modules. The prominence of these rings
in this text is in part due to their importance in applications, and in part due
to the fact that their modules are amenable to calculation. Thus, in Chapter
3 we look at skew polynomial rings and, more generally, noncommutative
Euclidean domains. The module theory of such rings is fairly transparent,
and the construction of a skew polynomial ring can be iterated to provide
some interesting examples of K-groups. Next, in Chapter 4, we investigate
the structure of Artinian semisimple rings. These can be characterized by
the property that any module is projective. We classify the finitely generated
projective modules over such rings. We also look at the structure of Artinian
rings in general.

The motivation for our final two chapters is that some of the most interesting results and questions in K-theory originate in number theory. In Chapter
5 we show that a ring of algebraic integers is a Dedekind domain, and we
introduce the ideal class group. The theory is illustrated with some explicit
computations in quadratic fields. Finally, we show how the projective modules over a Dedekind domain depend on the ideal class group, and we classify
all the finitely generated modules over a Dedekind domain.

Since this book is an introduction, we have kept a fairly leisurely pace
throughout. There are nearly two hundred exercises, some of which give brief
introductions to topics that are not covered in the body of the text.

Our approach takes full advantage of the powerful abstract methods that
were introduced in the early part of the 20th century by Emmy Noether
and her contemporaries. Thus our treatment of the subject matter does not
reflect its historical development. To counterbalance this, we have thrown in
the occasional comment on the origins of definitions and results. Many of our

comments are based on information obtained from [Bourbaki 1991], [Dedekind 1996], [Srinivasan & Sally 1983] and [van der Waerden 1980], which the reader should consult for fuller details.

This text is divided into chapters and sections. Within each section, all subsections, theorems, propositions, and lemmas are numbered consecutively. Exercises are to be found at the end of each section and have their own separate numbering.

The symbol □ indicates either the end of a proof <u>or</u> that none is needed. Once or twice we use the symbol ◯ after the statement of a result to indicate that the result is not a consequence of the arguments of this text; in this event, a reference is given.

We thank our departments, and Cambridge University Press, for their patience and encouragement during the lengthy gestation period of this text and its companion [BK: CM]. Particular thanks are due to Oliver Pretzel at Imperial whose advice, provided over a long timespan, enabled us to produce this text in LaTeX. We also thank several referees for their comments on drafts of this work; we have made some modifications to take account of their views.

We also thank SERC (Visiting Fellowship GR/D/79586), the London Mathematical Centre, the British Council in Singapore, NUS (Research Grant RP950645), and the Lee Kong Chian Centre for Mathematical Research for providing travel and subsistence expenses so that the authors were able to meet occasionally.

Especially, we thank our families for their tolerance in allowing us to vanish from sight for sporadic and inconvenient periods of very variable lengths, ranging from the odd month to the odd 'five minutes' when something just had to be finished before supper.

1
BASICS

This first chapter introduces the fundamental definitions and properties of rings and modules. Our starting point is that the reader knows something of arithmetic and of linear algebra, and our explanations and examples will often invoke such knowledge.

In arithmetic, we use the fact that unique factorization holds in the ring of integers, and we also use the division algorithm and the elementary properties of residue classes. In linear algebra, we call upon the standard results on finite-dimensional vector spaces and matrices. We also take for granted that the reader is acquainted with the basic language of set theory and group theory, and that he or she is happy to carry out 'routine' verifications to confirm that an object does possess some properties as claimed.

All these topics are met in a standard undergraduate mathematics course and in many expository texts, such as [Allenby 1991] and [Higgins 1975].

1.1 RINGS

In this section, we introduce rings, ideals, residue rings and homomorphisms of rings, and we discuss the relationships between these objects. We show how to construct two types of ring: one is a field of fractions, the other, a noncommutative polynomial ring in several variables. Our illustrations and examples are provided by the ring of integers, and by matrix rings and polynomial rings (in one variable), which we assume the reader has met before.

In this text we usually prefer to work with rings that have an identity element, but we sometimes make an excursion to examine rings that do not, which we call nonunital rings.

The abstract definition of a ring was first formulated by Fraenkel in 1914 [Kleiner 1996], although the term 'ring' had been introduced previously by Hilbert. Before then, the various types of ring that we encounter later –

polynomial rings, noncommutative algebras, rings of algebraic integers – had each been considered separately. Perhaps surprisingly, the idea of an ideal is much older, since it originates in number theory, as we shall see in Chapter 5. However, the explicit distinction between left and right ideals and the formal construction of a residue ring modulo a twosided ideal first occur in the work of Emmy Noether in the 1920s.

1.1.1 The definition

Informally, a ring is a set of elements which can be added and multiplied in such a way that most of the expected rules of arithmetic are obeyed. Familiar examples of rings are the integers $\mathbb{Z}$, the rational numbers $\mathbb{Q}$, the real numbers $\mathbb{R}$ and the complex numbers $\mathbb{C}$.

However, we also wish to work with rings such as the ring $M_n(\mathbb{R})$ of $n \times n$ real matrices, so we cannot assume that multiplication is commutative, that is, that $xy = yx$ always.

The formal definition is as follows. A *ring* is a nonempty set R on which there are two operations, addition and multiplication. Under addition, R must be an *abelian group*, which means that the following axioms hold.

(A1) Closure:
   if $r, s \in R$, then the sum $r + s \in R$.

(A2) Associativity:
   $(r + s) + t = r + (s + t)$ for all r, s and $t \in R$.

(A3) Commutativity:
   $r + s = s + r$ for all $r, s \in R$.

(A4) Zero:
   there is a zero element 0 in R with $r + 0 = r$ for all $r \in R$.

(A5) Negatives:
   if $r \in R$, then there is a negative $-r$ with $r + (-r) = 0$.

(As usual, we write $r + (-s) = r - s$ and $(-r) + s = -r + s$.)

Under multiplication, we require that R is a *monoid*, which means that the following axioms must hold.

(M1) Closure:
   if $r, s \in R$, then the product $rs \in R$.

(M2) Associativity:
   $(rs)t = r(st)$ for all r, s and $t \in R$.

(M3) Identity:
   there is an identity element 1 in R with $r1 = r = 1r$ for all r in R.

The addition and multiplication in a ring are related by

(D) Distributivity:
      for all r, s, t and $u \in R$,

$$(r + s)t = rt + st \quad \text{and} \quad r(t + u) = rt + ru.$$

If there are several rings under consideration, we sometimes indicate the zero and identity elements of R by 0_R and 1_R respectively.

We allow the possibility that $0 = 1$. In that event we have the *trivial* or *zero* ring 0 that has only one element.

Given any ring R, the set $M_n(R)$ of all $n \times n$ matrices over R is again a ring under the usual matrix addition and multiplication. Similarly, the set $R[T]$ of polynomials $f(T) = f_0 + f_1 T + \ldots + f_n T^n$, with $f_0, f_1, \ldots, f_n \in R$ and T an indeterminate, is a ring under the standard addition and multiplication of polynomials.

We assume that the reader is familiar with these constructions, at least when the coefficient ring R is $\mathbb{R}$, $\mathbb{C}$ or $\mathbb{Z}$. They are considered in more detail in sections 2.2 and 3.2.

In everday arithmetic, one takes for granted that products can be computed in any order, that is, $rs = sr$ always. However, this property does not hold for many of the rings that we wish to consider in this text, so we make a formal definition that distinguishes the rings that do have this property.

A ring R is *commutative* if the following condition holds.

(CR) Commutativity:
      $rs = sr$ for all r and s in R.

It is well known and easy to verify that the polynomial ring $R[T]$ is commutative precisely when R is commutative; for $n > 1$, the matrix ring $M_n(R)$ is not commutative except in the trivial case $R = 0$. Naturally enough, a ring is said to be *noncommutative* if it is not commutative.

1.1.2 Nonunital rings

If the axiom of the identity, (M3) above, is omitted, R is called a *nonunital ring* or a *pseudoring*. (In this case, R is a *semigroup* under multiplication.) Thus every ring is necessarily a nonunital ring. Many of the definitions and constructions that we make for rings have evident counterparts for nonunital rings, which we usually do not state explicitly. A systematic way of passing from nonunital rings to rings is indicated in Exercise 1.1.5.

Some authors extend the definition of a ring to include nonunital rings as rings, so that a 'ring with identity' becomes a special type of ring. However, for the purposes of algebraic K-theory our definition is the more convenient.

1.1.3 Subrings

A *subring* of R is a subset S of R with the following properties:

(SR1) $0, 1 \in S$;
(SR2) if $r, s \in S$, then $r + s, -r$ and rs are also in S.

Clearly, a subring of a ring is itself a ring with the same operations of addition and multiplication.

An important example is the *centre* $Z(R)$ of R:

$$Z(R) = \{z \in R \mid zr = rz \text{ for all } r \in R\}.$$

Note that $Z(R)$ is commutative. A method for the computation of the centre of a matrix ring is indicated in Exercise 1.1.4.

Sometimes it is more natural to focus attention on the subring S, for instance when the ring R is constructed from S in some way. We then say that R is an *extension* of S.

1.1.4 Ideals

A *right ideal* of R is a subset $\mathfrak{a}$ of R which satisfies the following requirements:

(Id1) $0 \in \mathfrak{a}$;
(Id2) if $a, b \in \mathfrak{a}$, then $a + b \in \mathfrak{a}$ and $-a \in \mathfrak{a}$ also;
(Id^{r}3) if $a \in \mathfrak{a}$ and $r \in R$, then $ar \in \mathfrak{a}$ also.

A *left ideal* has instead

(Id$^\ell$3) if $a \in \mathfrak{a}$ and $r \in R$, then $ra \in \mathfrak{a}$ also.

If $\mathfrak{a}$ is at the same time a left and a right ideal, we call it simply an *ideal* of R, although we sometimes refer to it as a *twosided ideal* if we need to avoid ambiguity.

The ring R is always an ideal of itself. We say that the ideal $\mathfrak{a}$ is *proper* if $\mathfrak{a} \subset R$, where we use the symbol $\subset$ to denote strict inclusion, that is, $\mathfrak{a} \subseteq R$ and $\mathfrak{a} \neq R$. At the other extreme, $\{0\}$ is always an ideal, the *zero ideal* of R, which we usually denote by 0. Of course, $0 \subseteq \mathfrak{a}$ for any ideal $\mathfrak{a}$.

Observe that any left or right ideal $\mathfrak{a}$ of R is a nonunital ring – one might call it a sub-nonunital-ring or nonunital subring of R – but that $\mathfrak{a}$ will not be a subring of R unless $\mathfrak{a} = R$.

If $\mathfrak{a}$ and $\mathfrak{b}$ are each right ideals of R, then evidently their *sum*

$$\mathfrak{a} + \mathfrak{b} = \{a + b \mid a \in \mathfrak{a}, \, b \in \mathfrak{b}\}$$

is also a right ideal of R. Similarly, if $\mathfrak{a}$ and $\mathfrak{b}$ are both left ideals, then so too is $\mathfrak{a} + \mathfrak{b}$. Moreover, the procedure can obviously be iterated (by associativity (A2)) to give the definition of a finite sum of ideals $\mathfrak{a}_1 + \mathfrak{a}_2 + \cdots + \mathfrak{a}_n$.

1.1.5 Generators

A convenient method of defining an ideal is to specify a set of generators. Given a subset $\{r_i \mid i \in I\}$ of R, where I is some index set, possibly infinite, we say that $r_i = 0$ for *almost all i*, or for *all except a finite set of indices*, if the set of indices i with $r_i \neq 0$ is finite.

Let $X = \{x_i \mid i \in I\}$ be a subset of R. Then the right ideal $\mathfrak{a}$ *generated* by X is the set of all expressions

$$\sum_{i \in I} x_i r_i,$$

where $r_i = 0$ for all except a finite set of indices; X is then called a set of *generators* for $\mathfrak{a}$.

When $X = \{x_1, \dots, x_n\}$ is finite, we write

$$\mathfrak{a} = x_1 R + \cdots + x_n R;$$

if $X = \{x\}$, then $\mathfrak{a} = xR$ is the *principal right* ideal generated by x. The left ideal generated by X is defined in a similar way.

When R is commutative, we use the notation $(x_1, \dots, x_n)$ for the ideal generated by $\{x_1, \dots, x_n\}$. Although this notation is the same as that for a sequence, the context should prevent any confusion.

Suppose that both $\mathfrak{a}$ and $\mathfrak{b}$ are right ideals of R. The *product* $\mathfrak{a}\mathfrak{b}$ is defined to be the right ideal generated by all products ab with $a \in \mathfrak{a}$ and $b \in \mathfrak{b}$. The product of a pair of left ideals or a pair of twosided ideals is defined similarly.

Suppose that $\mathfrak{a}$ is a twosided ideal. If $\mathfrak{b}$ is a right ideal, then $\mathfrak{a}\mathfrak{b}$ is a twosided ideal; on the other hand, if $\mathfrak{b}$ is a left ideal, then $\mathfrak{a}\mathfrak{b}$ is a left ideal which need not be twosided.

1.1.6 Homomorphisms

Let R and S be rings with zero elements 0_R and 0_S and identity elements 1_R and 1_S respectively. A *ring homomorphism* $f : R \to S$ is a map from R to S with the properties that

(RH1) for all r and s in R,

$$f(r + s) = fr + fs \text{ and } f(rs) = f(r)f(s),$$

(RH2) $f1_R = 1_S$.

The facts that $f(-r) = -f(r)$ and $f(0_R) = 0_S$ are consequences of (RH1), but (RH2) is not. This follows from the observation that the obvious map from the zero ring 0 to R is not a ring homomorphism (unless R itself happens to be 0).

Note that there is a unique homomorphism from any ring R to the zero ring 0. In the other direction, there is, for any ring, a unique ring homomorphism, the *characteristic homomorphism*

$$\chi : \mathbb{Z} \to R,$$

given by $\chi(a) = a1_R$, where 1_R is the identity element of R. Since an integer a need not be a member of the ring R, we should say what we mean by $a1_R$. We take $0_{\mathbb{Z}}1_R = 0_R$ and $1_{\mathbb{Z}}1_R = 1_R$. Then, for $a > 1$, we make the inductive definition $a1_R = (a-1)1_R + 1_R$, and for $a < 0$ we put $a1_R = -(-a)1_R$. The fact that χ is a homomorphism can now be verified by an induction argument.

Given a homomorphism f, we associate with it its *kernel*

$$\text{Ker } f = \{r \in R \mid fr = 0\}$$

and its *image*

$$\text{Im } f = \{s \in S \mid s = fr \text{ for some } r \in R\}.$$

Then Ker f is an ideal in R and Im f is a subring of S.

An argument familiar from elementary linear algebra shows that f is injective precisely when Ker $f = 0$; it is a tautology to say that f is surjective if and only if Im $f = S$.

A homomorphism $f : R \to S$ of rings is an *isomorphism* if has an *inverse*, that is, there is a ring homomorphism $g : S \to R$ such that $fg = id_S$ and $gf = id_R$, where id_R is the identity map on R and id_S is the identity map on S. It is not hard to see that f is an isomorphism precisely when it is both injective and surjective. We then write $R \cong S$.

Here is a result, used in [BK: CM], which shows how a homomorphism into a ring may be used to promote an abelian group to a ring. We give the proof in some detail as it is our first.

1.1.7 Lemma

Let $(S, +)$ be an additive group, let T be a ring, and suppose that $\theta : S \to T$ is an injective group homomorphism whose image is a subring of T. Then

*there is a unique multiplication · on S which makes $(S, +, \cdot)$ a ring and θ a
ring homomorphism.*

Proof

Since $\operatorname{Im}\theta$ is a subring of T, for any s_1, s_2 in S we have $\theta(s_1)\theta(s_2) \in \operatorname{Im}\theta$.
Because θ is injective, we may define

$$s_1 \cdot s_2 = \theta^{-1}(\theta(s_1)\theta(s_2)),$$

since this element is uniquely determined. So Axiom (M1) holds. Evidently,

$$\theta((s_1 \cdot s_2) \cdot s_3) = \quad \theta(s_1 \cdot s_2)\theta(s_3) \quad = (\theta(s_1)\theta(s_2))\theta(s_3)$$
$$= \theta(s_1)(\theta(s_2)\theta(s_3)) = \quad \theta(s_1)\theta(s_2 \cdot s_3) \quad = \quad \theta(s_1 \cdot (s_2 \cdot s_3)),$$

so that Axiom (M2) holds in S because θ is injective. Likewise Axiom (D)
holds in S. There remains Axiom (M3). Now $1_T \in \operatorname{Im}\theta$, so let $1_S = \theta^{-1}(1_T)$.
Then again (M3) in S may be deduced from its counterpart in T. Hence
$(S, +, \cdot)$ is a ring. By construction, θ is a ring homomorphism. If $*$ is another
ring multiplication making θ a ring homomorphism, then

$$\theta(s_1 * s_2) = \theta(s_1)\theta(s_2) = \theta(s_1 \cdot s_2),$$

so that, by injectivity again, $s_1 * s_2 = s_1 \cdot s_2$.   □

1.1.8 Residue rings

Given an ideal $\mathfrak{a}$ of a ring R, we can construct the *residue ring* $R/\mathfrak{a}$ of R
modulo $\mathfrak{a}$. (The residue ring is also called the *quotient* or *factor* ring in some
texts.) The definition goes as follows.

We say that $r, s \in R$ are *congruent modulo* $\mathfrak{a}$ if $r - s \in \mathfrak{a}$; this is sometimes
denoted by $r \equiv s \pmod{\mathfrak{a}}$. Congruence is easily seen to be an equivalence
relation on R, and the equivalence class of an element $r \in R$ is called its
residue class or *congruence class*.

The residue class of r is denoted $\bar{r}$, so that $\bar{r} = \{r + x \mid x \in \mathfrak{a}\}$, and the
residue ring $R/\mathfrak{a}$ is defined to be the set of all such classes. Addition and
multiplication in $R/\mathfrak{a}$ are given by

$$\bar{r} + \bar{s} = \overline{r + s} \quad \text{and} \quad \bar{r} \cdot \bar{s} = \overline{rs};$$

then $R/\mathfrak{a}$ is a ring with zero $\bar{0}$ and identity $\bar{1}$.

The *canonical* or *standard* ring homomorphism $\pi : R \to R/\mathfrak{a}$ is defined
simply by $\pi r = \bar{r}$. It is not hard to verify that π is a surjective ring homomorphism, with $\operatorname{Ker}\pi = \mathfrak{a}$.

The basic example of the construction of a residue ring, which also explains the name, occurs when we take R to be the ring of integers $\mathbb{Z}$ and $\mathfrak{a} = n\mathbb{Z}$ for some $n > 0$.

It is wellknown that the *division algorithm* holds in $\mathbb{Z}$: any integer a can be written in the form $a = qn + r$ with $0 \leq r < n$. The integer r is the *residue* (or *remainder*) and q the *quotient*. Thus the residue ring $\mathbb{Z}/n\mathbb{Z}$ consists of the residue classes

$$\bar{0}, \bar{1}, \ldots, \overline{n-1}.$$

We note the following result, which is typical of a class of very useful lemmas that we often use without comment.

1.1.9 The Induced Mapping Theorem for Rings

Let $f : R \to S$ be a ring homomorphism. Then there is a unique injective ring homomorphism $\bar{f} : R/\operatorname{Ker} f \to S$ so that $\bar{f}\pi = f$.

Proof

It must be that $\bar{f}(\bar{r}) = fr$. Note that $\bar{f}$ is often called the *induced* homomorphism.   □

1.1.10 The characteristic

As an illustration, we show how to define the characteristic of a ring R and the prime subring of R.

A familiar argument based on the division algorithm shows that any ideal $\mathfrak{a}$ of $\mathbb{Z}$ is principal, of the form $\mathfrak{a} = a\mathbb{Z}$, where a is uniquely defined as the least positive integer belonging to $\mathfrak{a}$ if $\mathfrak{a} \neq 0$, and $a = 0$ if $\mathfrak{a} = 0$. (This argument is given in a much more general context in (3.2.10) below.)

Now, for an arbitrary ring R, let $\chi : \mathbb{Z} \to R$, $\chi(a) = a1_R$, be the unique homomorphism from $\mathbb{Z}$ to R, and write $\operatorname{Ker} \chi = c\mathbb{Z}$ with $c \geq 0$. Then c is called the *characteristic* of R.

The *prime ring* of R is the subring $\chi\mathbb{Z}$ of R. If R has characteristic 0, then χ is an injective ring homomorphism and $\chi\mathbb{Z} \cong \mathbb{Z}$. If R has characteristic $c > 0$, then the Induced Mapping Theorem shows that $\chi\mathbb{Z}$ is isomorphic to the residue ring $\mathbb{Z}/c\mathbb{Z}$.

1.1.11 Units

An element u of the ring R is said to be a *unit* of R or an *invertible element* of R if

$$uv = 1 = vu$$

for some v in R. The element v is then the unique *inverse* of u.

The set of all units in R will be denoted by $U(R)$. Clearly, $U(R)$ is a group under multiplication, called the *unit group*. In particular, $U(0)$ is the trivial group.

If $U(R) = \{r \in R \mid r \neq 0\}$, then R is a *division ring* or *skew field*; if also R is commutative, then R is a *field*. Familiar examples of fields are $\mathbb{Q}$, $\mathbb{R}$ and $\mathbb{C}$, and the reader should have no difficulty in verifying that the residue ring $\mathbb{Z}/n\mathbb{Z}$, $n > 0$, is a field precisely when the integer n is a prime number. It is also straightforward to show that if the ring R is a field or division ring, then it must have characteristic either 0 or a prime. An example of a division ring is given in Exercise 2.2.3 below.

One particular type of unit group plays a fundamental role in K-theory. For any ring R and natural number n, the unit group $U(M_n(R))$ of the matrix ring $M_n(R)$ is called the *general linear group* of *degree* n, and is written $\mathrm{GL}_n(R)$.

1.1.12 Constructing the field of fractions

It is a very useful fact that a certain type of commutative ring, namely one that is a domain, can be embedded in a field. First, we make a formal definition. A nontrivial ring $\mathcal{O}$ (not necessarily commutative) is called a *domain* if the following holds.

(Dom) If $r, s \in \mathcal{O}$ and $rs = 0$, then either $r = 0$ or $s = 0$.

The terms *integral domain* and *entire ring* are sometimes used instead.

Suppose now that $\mathcal{O}$ is a commutative domain. We construct a field $\mathcal{K}$ in which every nonzero element r of $\mathcal{O}$ has an inverse $1/r$, and further any element of $\mathcal{K}$ can be written in the form r/s for $r, s \in \mathcal{O}$. Naturally enough, $\mathcal{K}$ is called the *field of fractions* of $\mathcal{O}$. The technique is exactly the same as that used to manufacture the rational numbers $\mathbb{Q}$ from the ring of integers $\mathbb{Z}$.

Let $\Sigma = \mathcal{O} \backslash \{0\}$ be the set of nonzero elements in $\mathcal{O}$. We introduce a relation $\sim$ on the set of pairs $(a, s) \in \mathcal{O} \times \Sigma$ by stipulating that $(a, s) \sim (a', s')$ if and only if there are elements u and u' in Σ with $au = a'u'$ and $su = s'u'$. It is easy to verify that this relation is an equivalence relation.

The fraction a/s is defined to be the equivalence class of (a, s) under this relation and $\mathcal{K}$ is the set of equivalence classes; thus $a/s = a'/s'$ if and only if $ax = a'x'$ and $sx = s'x' \in \Sigma$ for some x and x' in Σ.

We define addition by

$$a/s + b/t = (at + bs)/st,$$

and multiplication by

$$(a/s)(b/t) = ab/st.$$

Another routine check shows that these rules are well-defined and make $\mathcal{K}$ into a ring with zero element $0/1$ and identity $1/1$.

Furthermore, $a/1 = 0$ only if $a = 0$, so that we can identify $\mathcal{O}$ as the subring of $\mathcal{K}$ consisting of all elements of the form $a/1$.

Then the identity $r/r = 1/1$ holds for all nonzero r in $\mathcal{O}$, which confirms that r has an inverse in $\mathcal{K}$, and it is easy to see that $\mathcal{K}$ is a field.

1.1.13 Noncommutative polynomials

Many examples of rings arise from noncommutative rings of polynomials in several variables, so it will be helpful to have a brief outline of their construction. (A more detailed construction of rings of polynomials in one variable is given in section 3.2.)

Let A be a ring, which is referred to in this setting as the *coefficient ring*, and let $\{X_1, \ldots, X_k\}$ be a set of 'variables'. A *monomial* in $X_1, \ldots, X_k$ is any expression of the form

$$X_{h(1)}^{n(1)} \cdots X_{h(\ell)}^{n(\ell)}, \quad \ell \geq 0,$$

where

$$h(1), \ldots, h(\ell) \in \{1, \ldots, k\}$$

and

$$n(1), \ldots, n(\ell) \geq 1.$$

The case $\ell = 0$ is to be interpreted as giving an identity element 1. Note that two monomials are the same only if they have the same factors to the same exponents and in the same order, that is,

$$X_{h(1)}^{n(1)} \cdots X_{h(\ell)}^{n(\ell)} = X_{g(1)}^{p(1)} \cdots X_{g(m)}^{p(m)}$$

if and only if

$$\ell = m, \quad \text{and} \quad n(i) = p(i) \text{ and } h(i) = g(i) \text{ for all } i.$$

For example, $X_1^2 X_2$, $X_1 X_2 X_1$ and $X_2 X_1^2$ are all different.

The *total degree* of a monomial is the sum

$$n(1) + \cdots + n(\ell).$$

Multiplication of monomials is defined by writing them consecutively and collecting the powers of consecutive like terms, so that, for instance,

$$X_1 X_2 X_1 \cdot X_1^2 X_2 = X_1^2 X_2 X_1^3 X_2.$$

The *noncommutative polynomial ring* (also known as the *free associative algebra*) $A\langle X_1, \ldots, X_k \rangle$ is defined to be the set of all sums of the form $\sum_m a_m m$, where the summation is over all monomials m, all the coefficients a_m belong to A, and only a finite number of coefficients can be nonzero. We stipulate that

$$\sum_m a_m m = \sum_m b_m m \iff a_m = b_m \text{ for each } m.$$

Addition in $A\langle X_1, \ldots, X_k \rangle$ is given by the rule

$$\sum_m a_m m + \sum_m b_m m = \sum_m (a_m + b_m) m.$$

The multiplication is inherited from the multiplication of monomials using the associative and distributive laws, together with the rule that

$$am = ma \text{ for any } a \in A \text{ and any monomial } m.$$

It is then not too hard to see that $A\langle X_1, \ldots, X_k \rangle$ is a ring. If we identify the vacuous monomial 1 with the identity element 1_A of A, then we can regard A as a subring of $A\langle X_1, \ldots, X_k \rangle$.

Exercises

1.1.1   Let S be the centre of the ring R. Show that the centre of the polynomial ring $R[T]$ is $S[T]$.

1.1.2   Let R be any ring. Prove that $M_n(R)$ and $R[T]$ have the same characteristic as R.

Suppose that there is a ring homomorphism from R to S. Show that the characteristic of S divides that of R.

1.1.3   **Simple rings**

A ring S is said to be *simple* if it has no (twosided) ideals except 0 and S itself.

(i) Show that a commutative ring R is simple if and only if R is a field.

(ii) Let R be a nontrivial ring which has no left or right ideals except 0 and R. Show that R is a division ring. (In the next exercise there is an example of a simple ring with a proper right ideal.)

(iii) A proper twosided (or left or right) ideal $\mathfrak{m}$ of a ring R is said to be *maximal* if there is no twosided ideal (or left or right ideal, respectively) $\mathfrak{a}$ with $\mathfrak{m} \subset \mathfrak{a} \subset R$.

Show that the residue ring $R/\mathfrak{m}$ is simple if and only if $\mathfrak{m}$ is a maximal twosided ideal.

1.1.4 Matrix rings

The rings $M_n(R)$ of $n \times n$ matrices over various rings R are important objects of study in this text. In this exercise, we indicate how some properties of matrix rings can be found by elementary calculation.

(i) For each h and i with $1 \le h, i \le n$, let e_{hi} be the matrix with (h, i)th entry 1 and all other entries 0. Verify the relations

$$e_{hi}e_{jk} = \begin{cases} e_{hk} & \text{for } i = j \\ 0 & \text{for } i \ne j \end{cases}$$

and

$$I = e_{11} + \cdots + e_{nn},$$

where $I = I_n$ is the $n \times n$ identity matrix.

(The set $\{e_{hi}\}$ is the *standard set of matrix units* in $M_n(R)$.)

(ii) Using the fact that a matrix $A = (a_{hi})$ can be written $A = \sum_{h,i} a_{hi}e_{hi}$, show that $A \in Z(M_n(R))$, the centre of the matrix ring, if and only if $A = aI$ with $a \in Z(R)$.

(iii) Let $\mathfrak{a}$ be a twosided ideal of R and let $M_n(\mathfrak{a})$ be the set of matrices $A = (a_{hi})$ with $a_{hi} \in \mathfrak{a}$ for all h, i. Show that $M_n(\mathfrak{a})$ is a twosided ideal of $M_n(R)$, and that there is a ring isomorphism

$$M_n(R)/M_n(\mathfrak{a}) \cong M_n(R/\mathfrak{a}).$$

(iv) Let $\mathfrak{A}$ be a twosided ideal in $M_n(R)$. Show that, for any $A = (a_{hi})$ in $\mathfrak{A}$, the matrix $a_{hi}e_{11}$ is in $\mathfrak{A}$ for all h, i. Deduce that $\mathfrak{A} = M_n(\mathfrak{a})$ for a twosided ideal $\mathfrak{a}$ of R.

(v) Show that R is simple if and only if $M_n(R)$ is simple.

(vi) Show that the 'first row' of $M_n(R)$, that is, the set of all $A = (a_{hi})$ with $a_{hi} = 0$ for $h \ge 2$, and a_{1i} arbitrary for $i = 1, \ldots, n$, is a right ideal of $M_n(R)$.

(vii) Give an example of a simple ring with a nontrivial proper right ideal.

1.1.5 Embedding nonunital rings in rings

Let R be a nonunital ring. Let

$$\overline{R} = \{(r, a) \mid r \in R, a \in \mathbb{Z}\},$$

and define addition in $\overline{R}$ by

$$(r, a) + (s, b) = (r + s, a + b)$$

and multiplication by

$$(r, a) \cdot (s, b) = (rs + br + as, ab).$$

(Additively, $\overline{R} = R \oplus \mathbb{Z}$, the direct sum of abelian groups, which we meet more formally in section (2.2).)

Show that $\overline{R}$ is a ring with zero $(0, 0)$ and identity $(0, 1)$; $\overline{R}$ is called the *enveloping ring* or *unitalization* of R.

The *standard* embedding of a nonunital ring R in $\overline{R}$ is the map $\iota : r \mapsto (r, 0)$. Verify that ιR is an ideal of $\overline{R}$, with residue ring $\overline{R}/\iota R \cong \mathbb{Z}$.

1.1.6 Units and nonunital rings

Let R be a nonunital ring. Define an operation $\dagger$ on R by

$$r \dagger s = r + s + rs.$$

An element is called *quasi-invertible* if $r \dagger s = 0 = s \dagger r$ for some s in R. Let $QU(R)$ be the set of quasi-invertible elements in R. Show that $QU(R)$ is a group under $\dagger$.

Show that if R is actually a ring, then there is a group isomorphism $\epsilon_R : QU(R) \to U(R)$ given by $\epsilon_R(r) = 1 + r$.

Verify that $U(\overline{R})/\epsilon_{\overline{R}}(\iota QU(R))$ has order 2 for any nonunital ring R.

1.1.7 A domain that cannot be embedded in a division ring

In (1.1.12) we proved that a commutative domain $\mathcal{O}$ can be embedded in a field of fractions. One might hope that the method could be extended to show that if R is a noncommutative domain, then R can be embedded in a division ring of fractions. However, the following example shows that this is not always the case. The general problem of embedding a noncommutative domain in a division ring of fractions is investigated in depth in [Cohn 1995]. A special type of noncommutative domain for which embedding is possible is considered in Chapter 6 of [BK: CM]. Our example is due to [Mal'cev 1937]; we follow [Rowen 1988], §3.2.49.

Let $S = \mathbb{Q}\langle X_1, \ldots, X_4, Y_1, \ldots, Y_4 \rangle$ be the noncommutative polynomial ring in eight variables as indicated, and let $\mathfrak{a}$ be the twosided ideal of S generated by the three elements

$$X_1Y_1 + X_2Y_3, \quad X_3Y_1 + X_4Y_3 \quad \text{and} \quad X_3Y_2 + X_4Y_4.$$

We can then write the residue ring $R = S/\mathfrak{a}$ in the form $R = \mathbb{Q}[x_1, \ldots, x_4, y_1, \ldots, y_4]$ where x_i is the image of X_i, etc. Clearly, $x_1, \ldots, y_4$ satisfy the relations

$$x_1y_1 + x_2y_3 = x_3y_1 + x_4y_3 = x_3y_2 + x_4y_4 = 0.$$

Show that $a = x_1y_2 + x_2y_4$ is nonzero. (*Hint.* Choose suitable rational values of the variables with x_3 and y_4 both 0.)

An element f of R can be written as a sum of monomials (which are images of monomials in S), and we say that f is homogeneous if all these monomials have the same total degree (1.1.13).

To verify that R is a domain, use brute force (and much hard work): to check whether or not a product $f(x,y)g(x,y)$ is 0, it is enough to consider the case that f and g are homogeneous. We can further assume that each monomial term in f ends in some x_i and each monomial in g starts with some y_j. Checking cases, we find that fg is nonzero unless one of f and g is 0.

The relations on R give the matrix equation

$$\begin{pmatrix} x_1 & x_2 \\ x_3 & x_4 \end{pmatrix} \cdot \begin{pmatrix} y_1 & y_2 \\ y_3 & y_4 \end{pmatrix} = \begin{pmatrix} 0 & a \\ 0 & 0 \end{pmatrix}.$$

Suppose that R is a subring of some division ring $\mathcal{D}$. Multiplying the equation on the left and on the right by

$$\begin{pmatrix} 1 & -x_1x_3^{-1} \\ 0 & 1 \end{pmatrix} \quad \text{and} \quad \begin{pmatrix} 1 & y_3^{-1}y_4 \\ 0 & 1 \end{pmatrix}$$

respectively, we obtain the equation

$$\begin{pmatrix} 0 & b \\ x_3 & x_4 \end{pmatrix} \cdot \begin{pmatrix} y_1 & c \\ y_3 & 0 \end{pmatrix} = \begin{pmatrix} 0 & a \\ 0 & 0 \end{pmatrix},$$

for some b and c in $\mathcal{D}$, a contradiction.

1.2 MODULES

The concept of a module is one of the most fundamental in modern algebra, since many diverse problems can be recast in terms of the determination of the modules over a particular ring. One such problem, that of finding normal forms for matrices, and its alternative formulation using polynomial rings, will be the subject of a series of exercises, commencing with Exercise 1.2.8 and culminating in section 3.3. Some others, such as the representation theory of groups, will be considered from time to time in the text.

In this section, we give the basic definitions of modules and their homomorphisms, together with our first examples. Modules come in two flavours, left modules and right modules, both of which will be used in this text. We therefore set up two systems of notation for homomorphisms, one suited to left modules, the other to right modules. Although this attention to detail might seem unnecessarily pedantic at first glance, it repays us by making many bimodule structures appear in a very natural form.

We also describe some methods for constructing new modules from old, and we discuss the opposite of a ring, which is a tool for exchanging left modules with right modules.

A more leisurely introduction to module theory can be found in [Keating 1998].

Modules made their first appearance in Dedekind's work on number theory in the 1870s; [Dedekind 1996] is a recent translation. The modern usage originates in [Noether & Schmeider 1920] and [Noether 1929]. In the former, we meet the distinction between left and right modules, and, in the latter, the definitions of 'opposite ring' and 'bimodule' appear in their present-day form.

1.2.1 The definition

Let R be a ring. A *right R-module* is a set M which has addition and *scalar multiplication* on the right by elements of R; thus if $m, n \in M$ and $r \in R$, there are elements $m + n \in M$ and $mr \in M$.

Under addition, M must be an abelian group, which means that axioms (A1–5) of (1.1.1) above must be satisfied. The scalar multiplication must satisfy

(SM^{r}1) $m(rs) = (mr)s$ for all $m \in M$ and $r, s \in R$,

(SM^{r}2) $(m + n)r = mr + nr$ and $m(r + s) = mr + ms$ for all $m, n \in M$ and $r, s \in R$,

(SM^{r}3) $m1 = m$ for all m in M, where 1 is the identity element in R.

(The last axiom is sometimes stated as M is *unital*.)

A *left R-module* is defined in a similar fashion, except that we have a left scalar multiplication: if m is in M and r in R, then $rm \in M$. The axioms are

(SM$^\ell$1) $(rs)m = r(sm)$ for all $m \in M$ and $r, s \in R$,

(SM$^\ell$2) $r(m + n) = rm + rn$ and $(r + s)m = rm + sm$ for all $m, n \in M$ and $r, s \in R$,

(SM$^\ell$3) $1m = m$ for all m in M.

We encounter both left and right modules in this text; if we need to indicate which side we are working on, the notation $_RM$ is used to show that M is a left R-module and M_R to show that it is a right R-module. In either event, M may be referred to as a module *over R*.

1.2.2 *Some first examples*

Here are some basic examples of modules.

(i)   A ring R can be viewed as a right R-module, using its addition and multiplication as a ring to define its addition and scalar multiplication as a right module. More generally, any right ideal of R is a right R-module, the zero ideal being thought of as the *zero module* $0 = \{0\}$.

Likewise, left ideals of R can be considered to be left R-modules.

(ii) Any (additive) abelian group A can be regarded as a left $\mathbb{Z}$-module as follows. For any $a \in A$, first define $0 \cdot a = 0$, and then put $na = (n-1)a + a$ for $n > 0$ and $na = -(-n)a$ for $n < 0$. The axioms for scalar multiplication can be verified by an induction argument. The $\mathbb{Z}$-module structure of A is completely determined by the addition in A, and we can regard A equally as a right module with $an = na$. Thus, a (left or right) $\mathbb{Z}$-module is in essence the same thing as an abelian group.

(iii) When the ring of scalars is a field $\mathcal{K}$, a $\mathcal{K}$-module is more familiarly called a *vector space* over $\mathcal{K}$.

(iv) Some interesting examples of modules arise through polynomials acting on spaces. Let $\mathcal{K}^n$ be the space of column vectors of length n, which it is convenient to regard as a right $\mathcal{K}$-space. Given a vector x in $\mathcal{K}^n$ and an $n \times n$ matrix A over $\mathcal{K}$, the products $Ax, A^2x, \ldots$ are all elements of $\mathcal{K}^n$, which can therefore be regarded as a *right* module over the polynomial ring $\mathcal{K}[T]$ by the rule

$$xf(T) = xf_0 + Axf_1 + \cdots + A^r xf_r,$$

where

$$f(T) = f_0 + f_1T + \cdots + f_rT^r \in \mathcal{K}[T].$$

Informally, T is said to *act as A*.

Similarly, the space of row vectors $^n\mathcal{K}$ can be made into a left $\mathcal{K}[T]$-module with the corresponding action. This idea is developed in Exercises 1.2.6 and 1.2.8 below.

1.2.3 Bimodules

Given rings R and S, an *R-S-bimodule* M is a left R-, right S-module such that $r(ms) = (rm)s$ always. We indicate that M is a bimodule by writing $_RM_S$.

To illustrate this concept, we introduce some modules that play a key role in this text. Let R^n be the *free right R-module of rank* n, that is, the set of $n \times 1$ matrices over R. The usual addition and multiplication of matrices makes R^n into a left $M_n(R)$-module and an $M_n(R)$-R-bimodule. Similarly, the set ^nR of $1 \times n$ matrices gives the *free left R-module of rank* n, which is an R-$M_n(R)$-bimodule.

1.2.4 Homomorphisms of modules

Since we have two kinds of module under consideration, we introduce two notations for homomorphisms between modules. The rule is that homomorphisms of modules will generally be written *opposite the scalars*.

Thus, if $\alpha : M_R \to N_R$ is a homomorphism of right R-modules, we write αm for the image of m in N, and we require that

(Hom^{r}1) $\alpha(m + n) = \alpha m + \alpha n$ for all m and n in M,
(Hom^{r}2) $\alpha(mr) = (\alpha m)r$ for all m in M and r in R.

In the left-handed case, the image is written $m\alpha$, and we require that

(Hom$^\ell$1) $(m + n)\alpha = m\alpha + n\alpha$ for all m and n in M,
(Hom$^\ell$2) $(rm)\alpha = r(m\alpha)$ for all m in M and r in R.

It is occasionally convenient to depart from this principle when there is a strong tradition in support of the 'wrong' notation. Note that homomorphisms of other objects, such as rings, groups, bimodules, . . . , will be written on the left unless circumstances dictate otherwise.

Some basic examples of homomorphisms are given by the action of matrices on free modules. If A is an $m \times n$ matrix with entries in the ring R, the map $\alpha : R^n \to R^m$ given by $\alpha(x) = Ax$ is a homomorphism of right R-modules, while $\alpha' : {}^mR \to {}^nR$, $x\alpha' = xA$, is a homomorphism of left modules.

1.2.5 The composition of homomorphisms

The definition of the *composition product*, or *composite*, or simply *product*, of two homomorphisms α and β will depend on the side on which they operate. If they are written on the left, the rule is that

$$(\alpha\beta)m = \alpha(\beta m) \text{ for all } m \in M_R,$$

and, if they are written on the right,

$$m(\alpha\beta) = (m\alpha)\beta \text{ for all } m \in {}_R M.$$

Although it might seem at first to be unnecessarily confusing to have two interpretations of the product $\alpha\beta$ by composition, there is a considerable gain in that many formulas have a more 'natural' appearance. For example, let A and B be matrices over a ring R, of sizes $m \times n$ and $k \times m$ respectively. These act on the left of the right R-modules R^n and R^m, yielding homomorphisms

$$\alpha : R^n \to R^m, \ \alpha(x) = Ax, \text{ and } \beta : R^m \to R^k, \ \beta(x) = Bx.$$

The product $\beta\alpha$ is then the homomorphism corresponding to the matrix product BA. On the other hand, we can define homomorphisms of left R-modules by right action, giving

$$\beta' : {}^k R \to {}^m R, \ x\beta' = xB \text{ and } \alpha' : {}^m R \to {}^n R, \ x\alpha' = xA,$$

with product $\beta'\alpha'$ again corresponding to BA. This naturality is also visible in the formulas for the matrices of products of homomorphisms given in (2.2.6) and (2.2.14). (An alternative approach, based on use of the opposite ring, which allows one to work with all operators confined to one side, is mentioned in Exercise 1.2.14 below.)

As a further illustration, we exhibit some natural bimodule structures.

Let $\text{Hom}(M_R, N_R)$ be the set of R-module homomorphisms between two right R-modules M_R and N_R. This set is an abelian group under the rule

$$(\alpha + \beta)m = \alpha m + \beta m, \text{ where } m \in M \text{ and } \alpha, \beta \in \text{Hom}(M_R, N_R),$$

which has for zero element the *zero homomorphism* $0 : M \to N$ with $0(m) = 0$ for all m in M.

An R-homomorphism $\alpha : M_R \to M_R$ of a module to itself is called an *endomorphism* of M_R. We often write $\text{End}(M_R) = \text{Hom}(M_R, M_R)$, the *endomorphism ring* of M; this is indeed a ring, with multiplication defined by the rule

$$(\alpha\beta)m = \alpha(\beta m),$$

and identity element the *identity homomorphism* id_M of M:

$$id_M(m) = m \text{ for all } m \text{ in } M.$$

Notice that M is then an $\text{End}(M_R)$-R-bimodule.

Now suppose that M and N have bimodule structures, say M is an S-R-bimodule and N a T-R-bimodule for some rings S and T. Then $\text{Hom}(M_R, N_R)$ becomes a T-S-bimodule by the rule

$$(t\alpha s)m = t(\alpha(sm)), \ m \in M, \ s \in S, \ \alpha \in \text{Hom}(M_R, N_R), \ t \in T.$$

In particular, $\text{Hom}(M_R, N_R)$ is always an $\text{End}(N_R)$-$\text{End}(M_R)$-bimodule.

Corresponding results hold for left modules with the appropriate changes of notation. The set of homomorphisms from a left R-module $_RM$ to a left R-module $_RN$ is written as $\text{Hom}(_RM, _RN)$, with the endomorphism ring being written $\text{End}(_RM)$.

The multiplication in $\text{End}(_RM)$ is given by

$$m(\alpha\beta) = (m\alpha)\beta,$$

and M is then an R-$\text{End}(_RM)$-bimodule. Given bimodules $_RM_S$ and $_RN_T$, we regard $\text{Hom}(_RM, _RN)$ as an S-T-bimodule by the rule

$$m(s\alpha t) = ((ms)\alpha)t.$$

If there is no need to remind ourselves of the side of the scalar multiplication, we write $\text{Hom}_R(M, N)$ for either set of homomorphisms, and if the ring R can also be taken for granted, we write simply $\text{Hom}(M, N)$.

1.2.6 The opposite of a ring

Most of our discussion of module theory will be given only for right modules, since it will be clear that there are corresponding results for left modules. This informal translation between left and right can be made more precise by introducing the *opposite ring* R° of the ring R. The elements of R° are symbols r°, in bijective correspondence with the elements r of R, and addition and multiplication are given by

$$r^\circ + s^\circ = (r+s)^\circ \text{ and } r^\circ s^\circ = (sr)^\circ.$$

This process extends to modules. Given a right R-module M, we create a left R°-module M° whose elements m° are in bijective correspondence with the elements m of M, with addition and scalar multiplication given by

$$m^\circ + n^\circ = (m+n)^\circ \text{ and } r^\circ m^\circ = (mr)^\circ.$$

Similarly, we can turn a left module into a right module. Clearly, $R^{\circ\circ}$ and R are isomorphic as rings, and so $M^{\circ\circ}$ is a right R-module that is isomorphic to the original module M.

If $\alpha : M \to N$ is a homomorphism of right R-modules, we define a homomorphism $\alpha^\circ : M^\circ \to N^\circ$ of left R°-modules by

$$m^\circ \alpha^\circ = (\alpha m)^\circ.$$

If we identify the right R-modules $M^{\circ\circ}$ and M, then we can also identify the homomorphisms $\alpha^{\circ\circ}$ and α.

Thus, any general definition or result concerning right R-modules is equivalent to a corresponding definition or result for left R°-modules. It follows that it is usually enough to give statements and arguments for right modules only.

However, we do sometimes give a separate treatment of a topic for left modules, particularly if we make substantial use of the left-handed version.

It can be surprising to discover that, for a particular ring R, the properties of left R-modules may differ considerably from those of right R-modules; we meet examples of such rings in (3.2.12).

An abuse of language. We sometimes identify the additive groups M and M° through the evident correspondence $m \leftrightarrow m^\circ$, and say that '$M$ becomes a left R°-module by the rule $r^\circ \cdot m = m \cdot r$'. This identification is never made for the ring R itself, save in the circumstance we meet in the next subsection.

1.2.7 Balanced bimodules

Suppose that R is a commutative ring. Then R° and R are isomorphic rings, and we can make the identification $R^\circ = R$. Then any right R-module M is equally a left R-module, with $rm = mr$, for all r in R and m in M. Thus M can be regarded as an R-R-bimodule, or R-bimodule for short.

An R-bimodule over a commutative ring R which has the property that $rm = mr$ for all m in M and r in R is called a *balanced* or *symmetric* bimodule.

Although it is usually the case that a bimodule over a commutative ring is balanced, there are circumstances where this is not so. For example, let R be the polynomial ring $\mathcal{K}[T]$ over a field $\mathcal{K}$, and take M to be R itself. Regard M as a left module using the usual multiplication in R, but as a right module by letting T act as 0: $f(T) \cdot g(T) = f(T)g_0$, where $g(T) = g_0 + g_1 T + \cdots + g_r T^r$. Then M is a bimodule, but not balanced.

1.2.8 Submodules and generators

Taking advantage of the freedom to switch between right and left modules granted by the methods of (1.2.6), especially when dealing in generalities, we now confine the discussion to right modules.

Given a module M over some fixed but arbitrary ring R, a *submodule* of M is a subset M' of M for which we have

(SM1) $0 \in M'$,

(SM2) if $m, n \in M'$, then $m + n \in M'$ also,

(SM3) if $m \in M'$ and $r \in R$, then $mr \in M'$ also.

Two extreme examples are the *zero submodule* $0 = \{0\}$ and M itself; if $M' \neq M$, we say that M' is a *proper* submodule of M.

It is clear that a submodule is itself an R-module. If we think of R as a right module over itself, then it is also clear that a right ideal $\mathfrak{a}$ in R is the same thing as an R-submodule of R.

Given a finite set $\{M_1, \ldots, M_k\}$ of submodules of M, their *sum* is defined to be

$$M_1 + \cdots + M_k = \{m_1 + \cdots + m_k \mid m_i \in M_i \text{ for } 1 \le i \le k\}.$$

It is easily verified that the sum is also a submodule of M.

This definition can be extended to an infinite set $\{M_i \mid i \in I\}$ of submodules:

$$\sum_{i \in I} M_i = \{\sum_i m_i \mid m_i \in M_i, \text{ and } m_i = 0 \text{ for almost all } i\}.$$

Again, it is straightforward to show that the sum is a submodule of M.

Note that the set-theoretic intersection $\bigcap_{i \in I} M_i$ is always a submodule of M, regardless of whether or not the index set I is finite.

It is often convenient to specify a module or one of its submodules in terms of *generators*.

Given an element x of a module M,

$$xR = \{xr \mid r \in R\}$$

is the *cyclic submodule* generated by x, and M itself is cyclic if $M = xR$ for some x.

Given a subset $X = \{x_i \mid i \in I\}$ of the R-module M, where I is some index set (not necessarily finite), the submodule *generated* or *spanned* by X is

$$\mathrm{Sp}_R(X) = \sum_{i \in I} x_i R = \{\sum_{i \in I} x_i r_i \mid r_i \in R, \text{ and } x_i r_i = 0 \text{ for almost all } i\}.$$

If $\mathrm{Sp}_R(X) = M$, then X is said to be a *set of generators* for M. If we have $M = x_1 R + \cdots + x_k R$ for some finite set $X = \{x_1, \ldots, x_k\}$, then M is said to be *finitely generated*.

These definitions of course generalize those given in (1.1.5) for ideals, a cyclic ideal being the same thing as a principal ideal.

1.2.9 Kernel and image

Let $\alpha : M \to N$ be a homomorphism of (right) modules. The *kernel* of α is defined as

$$\mathrm{Ker}\,\alpha = \{m \mid \alpha m = 0\}$$

and the *image* $\mathrm{Im}\,\alpha$ is

$$\mathrm{Im}\,\alpha = \{n \mid n = \alpha m \text{ for some } m \in M\}.$$

A routine verification confirms that $\mathrm{Ker}\,\alpha$ is a submodule of M and that $\mathrm{Im}\,\alpha$ is a submodule of N.

It is easy to see that α is injective if and only if $\mathrm{Ker}\,\alpha = 0$, and it is a tautology that α is surjective if and only if $\mathrm{Im}\,\alpha = N$.

As in the case of a ring homomorphism, a module homomorphism $\alpha : M \to N$ is said to be an *isomorphism* if there is an inverse homomorphism $\alpha^{-1} : N \to M$, that is,

$$\alpha^{-1}\alpha = id_M \text{ and } \alpha\alpha^{-1} = id_N.$$

This happens precisely when α is both injective and surjective.

An isomorphism $\alpha : M \to M$ of a module with itself, that is, an invertible endomorphism, is termed an *automorphism* of M.

1.2.10 Quotient modules

Given a submodule M' of a right R-module M, we construct the *quotient module* (sometimes called the *factor* or *residue module* M/M' by analogy with the construction of the residue ring $R/\mathfrak{a}$ (1.1.8)). Define an equivalence relation on M by the rule that $m \equiv n$ if and only if $m - n \in M'$. The equivalence class of m is then

$$m + M' = \{m + m' \mid m' \in M'\},$$

which is often denoted by $\overline{m}$. Then M/M' is the set of all such classes, with

$$\overline{m} + \overline{n} = \overline{m+n} \text{ and } \overline{m}r = \overline{mr}.$$

An easy exercise shows that M/M' is an R-module.

The *canonical homomorphism* $\pi : M \to M/M'$ is defined by $\pi m = \overline{m}$. It is readily verified that π is a surjective R-module homomorphism and that $M' = \operatorname{Ker} \pi$.

As with rings, the following result is very useful.

1.2.11 The Induced Mapping Theorem for Modules

Let $\alpha : M \to N$ be a homomorphism of right R-modules. Then there is a unique injective module homomorphism

$$\overline{\alpha} : M/\operatorname{Ker}\alpha \to N$$

so that $\overline{\alpha}\pi = \alpha$; $\overline{\alpha}$ is given by $\overline{\alpha}(\overline{m}) = \alpha(m)$. □

Note. $\overline{\alpha}$ is often called the homomorphism *induced* by α.

1.2.12 Images and inverse images

It will be useful to know the relationship between the sets of submodules of two modules M and N which arises when there is a homomorphism $\alpha : M \to N$ between them.

If M' is a submodule of M, then the image of M' is

$$\alpha M' = \{\alpha m' \mid m' \in M'\},$$

which is clearly a submodule of N.

Given a submodule N' of N, the *inverse image* of N' is

$$\alpha^{-1} N' = \{m \in M \mid \alpha m \in N'\},$$

a submodule of M. (Of course, α itself need not have an inverse.)

The proof of the following result is a matter of routine checking.

1.2.13 Proposition

Let $\alpha : M \to N$ be a homomorphism of R-modules.

(i) *If M' is a submodule of M, then $\alpha^{-1}\alpha M' = M' + \operatorname{Ker}\alpha$.*

(ii) *If N' is a submodule of N, then $\alpha\alpha^{-1}N' = N' \cap \operatorname{Im}\alpha$.*

(iii) *Suppose that α is surjective. Then α induces a bijection between the set of submodules M' of M with $\operatorname{Ker}\alpha \subseteq M'$ and the set of submodules of N. Further, this bijection preserves inclusion of submodules:*

$$\operatorname{Ker}\alpha \subseteq M' \subseteq M'' \iff \alpha M' \subseteq \alpha M''.$$ □

1.2.14 *Change of rings*

Change of rings is a useful technical device for transferring structure from one ring to another. Suppose that $f : R \to S$ is a ring homomorphism and that N is a (right) S-module. Then N can be made into an R-module by the rule

$$nr = n(fr), \text{ for } n \in N \text{ and } r \in R.$$

We sometimes say that N is an R-module by *restriction of scalars*, especially if the map f is an injection.

Write $\mathfrak{a} = \operatorname{Ker} f$. Then $\mathfrak{a}$ *annihilates* N, that is,

$$n\mathfrak{a} = 0 \text{ for each } n \in N \text{ and } a \in \mathfrak{a}.$$

In the other direction, given an R-module M we define

$$M\mathfrak{a} = \{m_1 a_1 + \cdots + m_k a_k \mid k \geq 1,\, m_1, \ldots, m_k \in M,\, a_1, \ldots, a_k \in \mathfrak{a}\};$$

$M\mathfrak{a}$ is a submodule of M and it is 0 precisely when $\mathfrak{a}$ annihilates M. It is easy to establish the following lemma.

1.2.15 Lemma

Let $\mathfrak{a}$ be a twosided ideal of a ring R and let M be a right R-module.

(i) *If* $M\mathfrak{a} = 0$, *then* M *can be regarded as a right* $(R/\mathfrak{a})$-*module by the rule*

$$m\bar{r} = mr \text{ for } m \text{ in } M \text{ and } \bar{r} \text{ in } R/\mathfrak{a},$$

(ii) $M/M\mathfrak{a}$ *can be regarded as a right* $(R/\mathfrak{a})$-*module by the rule*

$$\overline{m} \cdot \bar{r} = \overline{mr} \text{ for } \overline{m} \text{ in } M/M\mathfrak{a} \text{ and } \bar{r} \text{ in } R/\mathfrak{a}. \qquad \square$$

1.2.16 *Irreducible modules*

A right R-module N is said to be *irreducible* (or *simple*) if N is nonzero, and N has no submodules other than 0 and N itself. The detection of all the irreducible R-modules for a given ring R is an important basic step in the description of all the R-modules.

It is evident that an irreducible module must be cyclic, since any nonzero element x in N will be a generator. We therefore have an R-module homomorphism $\pi : R \to N$, defined by $\pi(r) = xr$. The kernel of π is the *annihilator* $\operatorname{Ann}(x)$ of x:

$$\operatorname{Ann}(x) = \{r \in R \mid xr = 0\},$$

which is a right ideal of R.

By (1.2.13), any right ideal $\mathfrak{a}$ of R with $\operatorname{Ann}(x) \subset \mathfrak{a} \subset R$ would give a nontrivial submodule of the irreducible module N. Thus $\text{Ann}(x)$ must be a *maximal right ideal* of R, that is, a proper right ideal $\mathfrak{m}$ of R which is not contained properly in any other proper right ideal of R.

Conversely, if $\mathfrak{m}$ is a maximal right ideal of R, then $R/\mathfrak{m}$ is an irreducible right R-module.

It is clear that irreducible left R-modules correspond likewise to maximal left ideals of R.

In Exercise 1.1.3, we introduced simple rings and maximal twosided ideals of a ring: if $\mathfrak{m}$ is such an ideal, then the residue ring $R/\mathfrak{m}$ is a simple ring. We can view $R/\mathfrak{m}$ also as a right R-module, by change of rings, but it is unlikely that $R/\mathfrak{m}$ will be irreducible as a module. This can be seen from Exercise 1.1.4.

However, when R is commutative, any ideal of R is simultaneously left, right and twosided, and a simple residue ring $R/\mathfrak{m}$ will be irreducible as a module also. In fact, it is easy to show that $R/\mathfrak{m}$ must be a field.

Sometimes, we need to recognize an irreducible module as a quotient module of a noncyclic module. Let M be an arbitrary nonzero right R-module. A submodule M' of M is *maximal* if $M' \neq M$ and there is no submodule L with $M' \subset L \subset M$. The previous discussion gives the following useful lemma.

1.2.17 Lemma

Let M be a nonzero right R-module and let M' be a submodule of M. Then

(i) *M is irreducible if and only if 0 is a maximal submodule of M,*

(ii) *M/M' is irreducible if and only if M' is maximal.* $\square$

1.2.18 Maximal elements in ordered sets

The discussion in (1.2.16) above suggests that it will be valuable to have a method for obtaining the maximal submodules of a given module. In general, there is no 'constructive' way to do this, but an argument from logic, Zorn's Lemma, shows that, under reasonable conditions, maximal submodules exist.

Let X be a set and let S be a nonempty set of subsets of X. A *maximal element* of S is a member $X_{\max}$ of S such that $X_{\max}$ is not properly contained in any other member of S. Thus, if we take S to be the set of proper submodules of a right module M, a maximal element of S is a maximal submodule of M. Maximal left, right or twosided ideals of a ring (as defined in Exercise 1.1.3) can be interpreted in the same way.

In general, such a set S need not have maximal elements (see Exercise 1.2.5). Zorn's Lemma gives a convenient sufficient condition for their existence.

We need a preliminary definition. A set Λ is *totally ordered*, or simply *ordered*, if there is a relation $\leq$ between its members such that

(TO1) for any $\lambda, \mu \in \Lambda$, either $\lambda \leq \mu$ or $\mu \leq \lambda$,

(TO2) if both $\lambda \leq \mu$ and $\mu \leq \lambda$, then $\lambda = \mu$,

(TO3) if $\lambda \leq \mu$ and $\mu \leq \nu$, then $\lambda \leq \nu$.

We allow the empty set $\emptyset$ to be regarded as an ordered set and we may write $\lambda \geq \mu$ in place of $\mu \leq \lambda$.

The term 'totally ordered' is used when we wish to emphasize the difference between ordered sets and *partially ordered* sets, which need not satisfy (TO1).
. The ordered sets that we use most often are sets of natural numbers with their natural order. We write $\mathbb{N} = \{1, 2, \ldots\}$ for the set of all natural numbers if the order is unimportant, but $\omega = \{1, 2, \ldots\}$ for $\mathbb{N}$ considered as an ordered set, with the expected ordering $1 \leq 2 \leq 3 \leq \cdots$.

Let S be a nonempty set of subsets of a set X, as above. A *chain* in S is a sequence $\{X_\lambda\}$ of members of S, indexed by an ordered set Λ, such that $X_\lambda \subseteq X_\mu$ whenever $\lambda \leq \mu$. The set S is *inductive* if $\bigcup_{\lambda \in \Lambda} X_\lambda \in S$ for any such chain.

1.2.19 Zorn's Lemma
An inductive set S has a maximal element.   ○

The name of this statement, although widely used (allegedly first by Lefschetz), has attracted the attention of historians [Campbell 1978]. As a 'maximum principle', it was first brought to prominence, and used for algebraic purposes, in [Zorn 1935], apparently in ignorance of its previous usage in topology, most notably in [Kuratowski 1922]. Zorn attributed to Artin the realization that the 'lemma' is in fact equivalent to the Axiom of Choice (see [Jech 1973]). Zorn's contribution was to observe that it is more suited to algebraic applications like ours.

Here is a first application, in which the familiar fact that a finite-dimensional vector space has a basis is extended to arbitrary vector spaces.

Recall that a subset $X = \{x_i \mid i \in I\}$ of a $\mathcal{K}$-space V spans a subspace $\mathrm{Sp}(X)$ of V which consists of all the vectors of the form $v = \sum_{i \in I} k_i x_i$, where each k_i is in $\mathcal{K}$, with almost all $k_i = 0$. The set X is *linearly independent* if the only expression for the zero vector as a member of $\mathrm{Sp}(X)$ is the trivial one in which all coefficients are 0, that is, if $0 = \sum_I k_i x_i$, then $k_i = 0$ for all i.

(A linearly independent subset of V which spans V is often called a *basis* of V, particularly in texts on linear algebra. However, we find it essential to restrict the usage of the term 'basis' to *ordered* linearly independent spanning sets, which explains the wording of the next theorem. The theory of bases over general rings is considered in section 2.2.)

1.2.20 Theorem

Let V be a vector space over a field $\mathcal{K}$, and let Y be any linearly independent subset of V. Then V has a linearly independent spanning set which contains Y.

Proof Take S to be the set of linearly independent subsets of V which contain Y. Except in the trivial case when $V = 0 = \mathrm{Sp}(\emptyset)$, S has nonempty members. To see that S is inductive, we need to verify that $X = \bigcup_{\lambda \in \Lambda} X_\lambda \in S$ for a chain $\{X_\lambda\}$ in S. But this is clear, since any relation $0 = \sum_I k_i x_i$ among the elements of X can involve only a finite number of nonzero coefficients, and so must already hold in some X_λ.

By Zorn's Lemma, S has a maximal element, B say. If $\mathrm{Sp}(B) \neq V$, take any $v \in V$ with $v \notin \mathrm{Sp}(B)$. Then $B \cup \{v\} \in S$, a contradiction. $\quad\square$

The next result confirms the existence of maximal submodules in favourable circumstances.

1.2.21 Lemma

Let L be a proper submodule of a finitely generated right R-module M. Then L is contained in a maximal submodule of M.

Proof Since M is finitely generated, there is a minimal subset $\{x_0, \ldots, x_s\}$ of M such that $x_0 R + \cdots + x_s R + L = M$. Let S be the set of submodules X of M that contain $x_1 R + \cdots + x_s R + L$ but do not contain x_0. It is obvious that S is inductive, and so S has a maximal member, X' say. Any submodule properly containing X' must contain x_0 also, so X' is maximal among all the proper submodules of M. $\quad\square$

The case $M = R$ deserves special mention. [Zorn 1935] gives it as the first application of his 'maximum principle'.

1.2.22 Corollary

Let $\mathfrak{a}$ be a proper left, right or twosided ideal of a ring R. Then $\mathfrak{a}$ is contained in a maximal ideal of R which is correspondingly left, right or twosided. $\quad\square$

1.2.23 Torsion-free modules and spaces over the field of fractions

In (1.1.12), we indicated the construction of the field of fractions $\mathcal{K}$ of a commutative domain $\mathcal{O}$. We now generalize this construction to show that, under a natural condition, an $\mathcal{O}$-module M can be embedded in a vector space V over K, and that M spans V.

The requirement is that the (right) $\mathcal{O}$-module M is *torsion-free*, which means that the following condition is met.

(TF) If there is an equation $mr = 0$ with m in M and r in $\mathcal{O}$, then either $m = 0$ or $r = 0$.

Let M be a torsion-free module and let $\Sigma = \mathcal{O} \setminus \{0\}$ be the set of nonzero elements in $\mathcal{O}$. We define a relation $\sim$ on the set of pairs $(m, s) \in M \times \Sigma$ by $(m, s) \sim (m', s')$ if and only if there are elements u and u' in Σ with $mu = m'u'$ and $su = s'u'$ $(\in \Sigma)$. It is easy to see that $\sim$ is an equivalence relation. Write m/s for the equivalence class of (m, s) under this relation and V for the set of equivalence classes.

Define addition in V by

$$m/s + n/t = (mt + ns)/st, \quad m, n \in M, \quad s, t \in \mathcal{O},$$

and scalar multiplication by an element a/t of $\mathcal{K}$ by

$$(m/s)(a/t) = ma/st, \quad m \in M, \quad a, s, t \in \mathcal{O}.$$

These operations are well-defined and make V into a vector space over $\mathcal{K}$, with zero element $0/1$.

Define $\nu : M \to V$ by $\nu(m) = m/1$. Then ν is a homomorphism of $\mathcal{O}$-modules, V being an $\mathcal{O}$-module by restriction of scalars.

If $\nu m = 0/1$, then there are nonzero elements u and u' in $\mathcal{O}$ with $mu = 0u'$ and $1u = 1u'$, that is, $mu = 0$. It follows that $m = 0$ and that ν is an injection. We can therefore identify an element m of M with its image $m/1$ in V. Making this identification, we see that M spans V as a $\mathcal{K}$-space. Further, if M is a finitely generated $\mathcal{O}$-module, then V is a finite-dimensional space.

Exercises

1.2.1 Show that the irreducible $\mathbb{Z}$-modules are those of the form $\mathbb{Z}/p\mathbb{Z}$ where p is a prime number.

1.2.2 **The Isomorphism Theorems**

These are three very useful results, each of which is a consequence of the Induced Mapping Theorem (1.2.11). In these statements, R is an arbitrary ring.

(a) **The First Isomorphism Theorem**

Let $\alpha : M \to N$ be a homomorphism of right R-modules. Show that α induces an isomorphism $\bar{\alpha} : M/\operatorname{Ker}\alpha \to \operatorname{Im}\alpha$ with $\bar{\alpha}(m) = \alpha(m)$.

(b) **The Second Isomorphism Theorem**

Let K and L both be submodules of the right R-module M, and recall that $K + L = \{k + \ell \mid k \in K, \ell \in L\}$.

Define $\beta : K \to (K + L)/L$ by $\beta k = \overline{k}$. Verify that β is a surjective homomorphism of R-modules with $\operatorname{Ker} \beta = K \cap L$, and deduce that $K/(K \cap L) \cong (K + L)/L$.

(c) **The Third Isomorphism Theorem**

Suppose now that $K \subseteq L \subseteq M$. Show that the canonical map $\iota : L/K \to M/K$ is an injective homomorphism. Regard ι as the inclusion map (that is, think of L/K as a submodule of M/K).

Define $\gamma : M/K \to M/L$ by $\gamma(m + K) = m + L$. Prove that $\operatorname{Ker} \gamma = L/K$, and deduce that

$$(M/K)/(L/K) \cong M/L.$$

1.2.3 Let L be a submodule of a module M, and suppose that M/L is finitely generated. Show that L is contained in a maximal submodule of M.

1.2.4 Let $\mathfrak{a}$ be a proper ideal of a ring R. Using Exercise 1.1.3, show that there are a simple ring S and a surjective ring homomorphism $\pi : R \to S$ with $\pi(\mathfrak{a}) = 0$.

1.2.5 Let $\mathcal{K}$ be a field and let V be an infinite-dimensional vector space over $\mathcal{K}$. Show that the set S of finite-dimensional subspaces of V contains no maximal elements.

Let T be the set of proper subspaces W of V such that the quotient space V/W is finite-dimensional. Show that every proper subspace of V is contained in a maximal element of $S \cup T$, but $S \cup T$ is not inductive.

1.2.6 **Endomorphisms and polynomials**

The following definition generalizes that given in part (iv) of (1.2.2).

Let M be a right R-module and let α be an endomorphism of M. Show that M becomes a right $R[T]$-module by the rule

$$m \cdot (f_0 + f_1 T + \cdots + f_k T^k) = m f_0 + \alpha(m f_1) + \cdots + \alpha^k(m f_k).$$

(*Note.* The appearance of $\alpha, \alpha^2, \ldots$ on the left is intentional!)

Show also that if M is a right $R[T]$-module, then M is a right R-module and the map α given by $\alpha m = mT$ is an R-module endomorphism of M.

Informally, T is said to *act as* α.

Suppose that M and N are right $R[T]$-modules, with T acting as α and β respectively. Prove that $\pi : M \to N$ is an $R[T]$-module homomorphism if and only if it is an R-module homomorphism with $\pi\alpha = \beta\pi$.

1.2.7   This exercise reviews some facts from elementary linear algebra concerning the matrix representation of a linear map and the way in which it is affected by a change of basis. We give a detailed discussion of these topics in section 2.2 (see (2.2.5) and (2.2.7)) for arbitrary rings, but it will be useful to invoke them for illustrative purposes beforehand.

Let V and W be finite-dimensional vector spaces over a field $\mathcal{K}$, with bases $\{f_1, \ldots, f_n\}$ and $\{g_1, \ldots, g_m\}$ respectively. Given a vector v in V, write $v = f_1 x_1 + \cdots + f_n x_n$ (we think of V as a right module) and $x = \begin{pmatrix} x_1 \\ \vdots \\ x_n \end{pmatrix}$, the *coordinate vector* of v.

Let $\pi : V \to W$ be a linear map and write

$$\pi f_j = \sum_{i=1}^{m} g_i p_{ij} \text{ for } j = 1, \ldots, n;$$

put $P = (p_{ij})$, the $m \times n$ *matrix* of π with respect to the given bases.

Verify that the coordinate vector of πv is Px.

Let $\{f_1', \ldots, f_n'\}$ and $\{g_1', \ldots, g_m'\}$ also be bases of V and W respectively, and write

$$f_j' = \sum_{i=1}^{n} f_i \gamma_{ij}, \ g_j = \sum_{i=1}^{m} g_i' \theta_{ij} \text{ and } \Gamma = (\gamma_{ij}), \ \Theta = (\theta_{ij}).$$

Show that Γ and Θ are invertible.

For v in V, let x' be the coordinate vector of v with respect to the new basis. Verify that $x = \Gamma x'$.

Show that the matrix of π with respect to the new bases is $\Theta P \Gamma$.

1.2.8   **Some explicit modules**

This and the exercises immediately following give some concrete examples of modules. They use the facts from elementary linear algebra whose proofs were sketched in the preceding exercise.

(a) Let A be an $n \times n$ matrix over a field $\mathcal{K}$, and let M be the $\mathcal{K}[T]$-module obtained from the vector space (of column vectors) $\mathcal{K}^n$, on which T acts as A.

Show that a subspace U of $\mathcal{K}^n$ is a submodule L of M if and only if $AU \subseteq U$, that is, U is an *invariant subspace* for A. (It is often useful to distinguish between the submodule L and its *underlying subspace* U, since a subspace may underlie several different submodules.)

(b) Let $A = \begin{pmatrix} 1 & 1 \\ 0 & 1 \end{pmatrix}$, $B = \begin{pmatrix} 1 & 0 \\ 0 & 1 \end{pmatrix}$, $C = \begin{pmatrix} 0 & 1 \\ 1 & 0 \end{pmatrix}$, $D = \begin{pmatrix} -1 & 0 \\ 0 & -1 \end{pmatrix}$, and $E = \begin{pmatrix} 0 & -1 \\ 1 & 0 \end{pmatrix}$. Using these matrices in turn, $\mathbb{R}^2$ is made into an $\mathbb{R}[T]$-module in five different ways; write L, M, N, P and Q for the corresponding $\mathbb{R}[T]$-modules.

Find all the one-dimensional subspaces of $\mathbb{R}^2$ which are submodules of L (if any). Repeat for M, N, P and Q.

(c) Let M be $\mathbb{C}^3$ made into a $\mathbb{C}[T]$-module using the matrix $A = \begin{pmatrix} 0 & 1 & 1 \\ 0 & 0 & 1 \\ 0 & 0 & 0 \end{pmatrix}$, and for any vector $v \in \mathbb{C}^3$ let $L(v)$ be the $\mathbb{C}[T]$-submodule of M generated by v. Write L_0 for the special submodule given by $\begin{pmatrix} 1 \\ 0 \\ 0 \end{pmatrix}$.

Show that for $v \neq 0$, $L_0 \subseteq L(v)$ always.

Find all v with $\dim(L(v)) = 2$.

(d) Let N be $\mathbb{C}^3$ made into a $\mathbb{C}[T]$-module using the matrix
$$B = \begin{pmatrix} 0 & 1 & 0 \\ 0 & 0 & 1 \\ 1 & 0 & 0 \end{pmatrix}.$$

Find three submodules of N that are one-dimensional as $\mathbb{C}$-spaces, give a single vector which generates N as $\mathbb{C}[T]$-module and show that the submodule generated by $\begin{pmatrix} 2 \\ -1 \\ -1 \end{pmatrix}$ is two-dimensional.

(e) An extension of part (c) above. Let
$$A = \begin{pmatrix} 1 & 1 & 1 & \cdots & 1 \\ 0 & 1 & 1 & \cdots & 1 \\ 0 & 0 & 1 & \cdots & 1 \\ \vdots & \vdots & \vdots & \ddots & \vdots \\ 0 & 0 & 0 & \cdots & 1 \end{pmatrix}$$

be the $n \times n$ matrix with all entries on or above the diagonal equal to 1, the rest being 0. Make $\mathcal{K}^n$ into a $\mathcal{K}[T]$-module M

by the action of A. For $i = 1, \ldots, n$, let M_i be the subspace $\mathrm{Sp}(e_1, \ldots, e_i)$, where

$$e_1 = \begin{pmatrix} 1 \\ 0 \\ \vdots \\ 0 \\ 0 \end{pmatrix}, \ldots, e_n = \begin{pmatrix} 0 \\ 0 \\ \vdots \\ 0 \\ 1 \end{pmatrix}$$

are the standard basis vectors for $\mathcal{K}^n$ (and $M_0 = 0$).

Verify that each M_i is a submodule of M.

Give the matrices that represent the action of T on M_i and on M/M_i for each i.

Fill in the details of the following inductive proof that if L is a $\mathcal{K}[T]$-submodule of M, then $L = M_i$ for some i.

(1) By induction hypothesis, the result is true for submodules of M_{n-1} and of M/M_1.
(2) By considering M/M_1, we have $M_1 + L = M_i$ for some i.
(3) If L is not some M_i, then we have both $M_1 + L = M$ and $M_1 \cap L = 0$.
(4) This cannot happen for any L: write $e_2 = ke_1 + \ell$, with $k \in \mathcal{K}$ and $\ell \in L$, and show that k and ℓ are unique. Then express Ae_2 in two ways.

1.2.9 Block forms of matrices

(a) Let A be an $n \times n$ matrix over a field $\mathcal{K}$, let M be the $\mathcal{K}[T]$-module obtained from $\mathcal{K}^n$ with T acting as A, and suppose that L is a submodule of M, with subspace U.

Using a change of basis argument, show that there is an invertible $n \times n$ matrix Γ so that $\Gamma^{-1}A\Gamma = \begin{pmatrix} B & C \\ 0 & D \end{pmatrix}$, where B gives the action of T on U with respect to a basis of U, and D gives the action of T on V/U. (Choose a basis of U first and extend it to a basis of V; the additional basis elements correspond to a basis for V/U.)

(b) More generally, show that if $0 = L_0 \subset L_1 \subset \cdots \subset L_k = M$ is a chain of submodules of M, then there is an an invertible $n \times n$ matrix Γ such that $\Gamma^{-1}A\Gamma = (B_{ij})$, a $k \times k$ block matrix with $B_{ij} = 0$ for $i > j$. (The matrix (B_{ij}) is known as an *upper triangular block matrix*, and $\Gamma^{-1}A\Gamma$ is said to be *conjugate* to A.)

(c) Determine whether or not the matrices $C = \begin{pmatrix} 0 & 1 \\ 1 & 0 \end{pmatrix}$ and $E = \begin{pmatrix} 0 & -1 \\ 1 & 0 \end{pmatrix}$ of Exercise 1.2.8, part (b), are conjugate to upper triangular matrices over $\mathbb{R}$.

Show that the matrix $\begin{pmatrix} 0 & 1 & 0 \\ 0 & 0 & 1 \\ 1 & 0 & 0 \end{pmatrix}$ is conjugate to a diagonal matrix over $\mathbb{C}$ but not over $\mathbb{R}$.

1.2.10 Some explicit linear maps

(a) Let A be an $n \times n$ matrix over a field $\mathcal{K}$ and let M be the $\mathcal{K}[T]$-module obtained from $\mathcal{K}^n$ with T acting as A. Further, let B be an $m \times m$ matrix and let N be $\mathcal{K}^m$ viewed as a $\mathcal{K}[T]$-module with T acting as B.

Verify that $\mathcal{K}[T]$-module homomorphisms from M to N are given by the $m \times n$ matrices P such that $PA = BP$ (see Exercise 1.2.6 above).

(b) Let L, M, N, P and Q be the $\mathbb{R}[T]$-modules defined in part (b) of Exercise 1.2.8. For all pairs $X, Y \in \{L, M, N, P, Q\}$, find $\mathrm{Hom}_{\mathbb{R}[T]}(X, Y)$.

(c) Let $A = \begin{pmatrix} 0 & 1 \\ 1 & 0 \end{pmatrix}$ and $B = \begin{pmatrix} 0 & 1 & 0 \\ 0 & 0 & 0 \\ 0 & 0 & 1 \end{pmatrix}$, and let M and N be the corresponding $\mathbb{R}[X]$-modules. Show that $\mathrm{Hom}_{\mathbb{R}[T]}(M, N)$ is a vector space over $\mathbb{R}$ of dimension 1, and that it contains no injective maps and no surjective maps.

1.2.11 Bimodules and endomorphisms

Prove that the following statements about a module M_R and a ring S are equivalent.

(a) M_R is an S-R-bimodule.

(b) There is a ring homomorphism $S \to \mathrm{End}(M_R)$.

Let $E = \mathrm{End}(M_R)$. Show that there is a ring homomorphism from R to $\mathrm{End}(_E M)$ whose kernel is the annihilator $\mathrm{Ann}(M)$ of M.

What is the annihilator of M as a left E-module?

1.2.12 Let $\{M_i \mid i \in I\}$ be a set of submodules of the module M. Show that $\sum_{i \in I} M_i$ is the smallest submodule N of M with $M_i \subseteq N$ for all i, and that $\bigcap_{i \in I} M_i$ is the largest submodule L with $L \subseteq M_i$ for all i.

1.2.13 Opposites

(a) Let M be a right R-module and regard M as a left R°-module in the standard way. Is M an R°-R-bimodule?

(b) An *anti-automorphism* of a ring R is a bijective map ϵ from R to R such that, for all r and s in R,

$$\epsilon(r+s) = \epsilon r + \epsilon s \quad \text{and} \quad \epsilon(rs) = \epsilon(s)\epsilon(r).$$

Show that if there is such an anti-automorphism of R, then there is a ring isomorphism $f : R \to R^\circ$ given by $f(r) = (\epsilon r)^\circ$. Conversely, given a ring isomorphism from R to R°, it must arise through some anti-automorphism.

(c) Let $R = M_n(\mathcal{O})$, where $\mathcal{O}$ is commutative. Show that taking the transpose of a matrix defines an isomorphism $R \cong R^\circ$.

1.2.14 Onesided working

Instead of the 'operators opposite scalars' convention that we employ in this text, it is possible to write all scalars and operators on, say, the left, by using the opposite ring. This system of notation has the advantage that the meaning of a product $\alpha\beta$ of maps is unambiguous, but the disadvantage that opposite rings keep turning up for artificial reasons.

In the onesided notation, an (R, S)-bimodule M is simultaneously a left R- and a left S-module, and the actions of R and S commute with each other: $(r(sm)) = s(rm)$. Thus an (R, S)-bimodule is an R-S°-bimodule.

A homomorphism $\alpha : M \to N$ of R-modules now has

$$\alpha(rm) = r(\alpha m)$$

always; thus α is a right R°-module homomorphism.

Let M be the free left R-module ^nR of (finite) rank n. As right operators, the endomorphism ring $E = \text{End}(M)$ is the matrix ring $M_n(R)$ acting in the natural way. (For details, see (2.2.9).) In onesided notation, we instead let E° act on the left. This all works, as we can see from the correspondence $(a_{ij})^\circ \leftrightarrow (a_{ji}^\circ)$.

More precisely, we define the *transpose* $A = (a_{ij})$ over R to be the matrix $A^t = (a_{ji}^\circ)$, which has entries in R°. Then, generalizing part (b) of the previous exercise, taking the transpose of a matrix gives an isomorphism between $M_n(R)^\circ$ and $M_n(R^\circ)$. This isomorphism is used again in Exercise 2.2.7.

1.2.15 Modules over nonunital rings

Let R be a nonunital ring. A right module M over R is defined as in (1.2.1), except that the requirement that M be unital, that is, $m1 = m$ always, must of course be dropped.

The axioms for a submodule are unchanged, but the appearance of the submodule generated by a subset of M will change.

Let m be in M. Show that $mr + ma \in M$ for any element $r \in R$ and any integer $a \in \mathbb{Z}$ (where the expression ma is defined inductively as in (1.2.2)).

Deduce that the cyclic module generated by m is

$$\{mr + ma \mid r \in R, a \in \mathbb{Z}\}.$$

Describe the submodule generated by an arbitrary subset of M.

Let $\overline{R}$ be the enveloping ring of R (Exercise 1.1.5). By defining

$$m \cdot (r, a) = mr + ma, \ (r, a) \in \overline{R}, \ m \in M,$$

show that M becomes an $\overline{R}$-module. Prove that an R-submodule of M is also an $\overline{R}$-submodule, and viceversa.

A homomorphism of modules over the nonunital ring R is defined as in the unital case (1.2.4). Show that $\alpha : M \to N$ is a homomorphism of right R-modules if and only if α is a homomorphism of right $\overline{R}$-modules.

2

DIRECT SUMS AND SHORT EXACT SEQUENCES

Now that we have seen the basic definitions and examples of rings and modules, we can start to enquire about their deeper properties. A fundamental problem is to find, for a specified ring R, a satisfactory way of listing all the possible R-modules. Some modules are easy to find; for example, R itself can be viewed as a right R-module, as can the quotient module $R/\mathfrak{a}$ for any right ideal $\mathfrak{a}$ of R. Thus a possible approach to the solution of the problem is to ask whether or not an arbitrary module can be expressed in terms of some basic modules which are themselves readily described, and then to ask if such an expression is unique.

Obviously this programme requires some information about the ring, and, even then, it may be possible to obtain only partial results. Some classes of rings whose modules can be classified satisfactorily are discussed in subsequent chapters.

In this chapter, we set the foundations for the classification of modules by discussing two fundamental ways in which modules can be constructed from, or related to, more basic modules. The first of these is the formation of a direct sum, which, as the name suggests, is the most elementary way of assembling two modules to make a third. The second, more general, method is to use a short exact sequence. This construction relates triples of modules, one of which can often be regarded as being defined by the other two.

The direct sum construction gives us, in particular, the free modules R^n, $n > 1$, which are the analogues for an arbitrary ring of the vector spaces $\mathcal{K}^n$ over a field $\mathcal{K}$. We look at the relations between the various bases of a free module and the matrix description of the homomorphisms between two free modules. A general module can be presented in terms of such a homomorphism, whose matrix may, after manipulation, reveal the structure of the original module.

Since we work with arbitrary rings, we devote some attention to the invariant basis problem: can we have $R^m \cong R^n$ without $m = n$?

We also discuss the projective modules, which share many of the properties of free modules, and in some way are the proper generalization of a vector space for an arbitrary ring.

For the final topic in this chapter, we show how to construct rings by forming direct products – this is the analogue for rings of the direct sum construction for modules.

2.1 DIRECT SUMS AND FREE MODULES

The formation of the direct sum of a set of modules is the most straightforward method of constructing a module from component modules which are, in some way, more elementary than the new module. With this tool in our hands, we can then contemplate a three-part programme for the analysis of the modules over any particular ring. First, find those modules which cannot be expressed as a direct sum except trivially – these are the 'indecomposable' modules. Next, give a procedure to decompose any module into indecomposable parts. Finally, determine if such a decomposition is uniquely determined by the original module. This programme is (of course) impossible to implement over an arbitrary ring. Later in this text, we exhibit some rings over which it can be carried through, in particular Artinian semisimple rings and Dedekind domains, and we discuss some examples of unexpected behaviour.

This section is devoted to a discussion of the formalism of the construction and recognition of direct sums. We also describe free modules as special cases of direct sums. The ring of scalars R is an arbitrary ring, except where specified otherwise. We work entirely with right R-modules, there being an obvious set of parallel definitions and results for left modules. Some concrete examples of direct sums over the integers and over polynomial rings are given in the exercises.

2.1.1 Internal direct sums

Let R be a ring. A right R-module M is the *internal direct sum* of a finite set of right R-submodules $M_1, \ldots, M_k$, written as

$$M = M_1 \oplus \cdots \oplus M_k,$$

if the following conditions hold:

(DS1) $M = M_1 + \cdots + M_k$,

(DS2) $(M_1 + \cdots + M_{i-1} + M_{i+1} + \cdots + M_k) \cap M_i = 0$ for each $i = 1, \ldots, k$.

The expression $M = M_1 \oplus \cdots \oplus M_k$ is often called an *internal direct decomposition* of M, and the submodules $M_1, \ldots, M_k$ are referred to as the *summands* or *direct factors* of M.

Suppose that $M = M_1 \oplus \cdots \oplus M_k$, and take an element m of M. By (DS1), we can write

$$m = m_1 + \cdots + m_k$$

with $m_i \in M_i$ for each i. If also

$$m = m'_1 + \cdots + m'_k$$

with $m'_i \in M_i$, then

$$
\begin{aligned}
m_i - m'_i &= (m'_1 - m_1) + \cdots + (m'_{i-1} - m_{i-1}) \\
&\quad + (m'_{i+1} - m_{i+1}) + \cdots + (m'_k - m_k) \\
&= 0
\end{aligned}
$$

by (DS2), so that the expression for m is unique.

This observation gives one implication in the following restatement of the conditions for an internal direct sum, the other being equally obvious.

(DSU) A right R-module M is the internal direct sum of its submodules $M_1, \ldots, M_k$ if and only if each element m of M can be written uniquely in the form $m = m_1 + \cdots + m_k$ with $m_i \in M_i$ for each i.

We allow 'degenerate' direct sums in which one or more of the submodules happens to be the zero submodule; in particular, we are permitted to write $M = M \oplus 0$ if need be. We also allow the trivial cases $k = 0$ (so $M = 0$), and $k = 1$ (then $M = M_1$).

A module M is said to be *decomposable* if it can be written as an internal direct sum $M = L \oplus N$ of *nonzero* submodules L and N. In this case, N is said to be a *complement* of L in M, or to *complement* L. The first example below shows that a complement need not be unique.

If a nonzero module M is not decomposable, it is said to be *indecomposable*.

2.1.2 Examples: vector spaces

(i) Let $\mathcal{K}$ be a field, and $V = \mathcal{K}^2$ the two-dimensional vector space over $\mathcal{K}$. Clearly, $\mathcal{K}$ is indecomposable as a $\mathcal{K}$-module, but V has many expressions as an internal direct sum. For example, take $L = \begin{pmatrix} 1 \\ 0 \end{pmatrix} \mathcal{K}$ and $N(x) = \begin{pmatrix} x \\ 1 \end{pmatrix} \mathcal{K}$ for any x in $\mathcal{K}$; then $V = L \oplus N(x)$ for each x.

(ii) Let $\mathcal{K}$ be a field and take $V = \mathcal{K}^n$ for some integer $n > 1$. Choose an integer r with $1 < r < n$, let U be the subspace of V spanned by the standard basis vectors $e_1, \ldots, e_r$ and let W be spanned by $e_{r+1}, \ldots, e_n$, so that $V = U \oplus W$ as a $\mathcal{K}$-space. Now let A be an $n \times n$ matrix over $\mathcal{K}$, and regard V as a module over the polynomial ring $\mathcal{K}[T]$ with T acting as A, as in (iv) of (1.2.2).

Then it is easy to verify that the decomposition $V = U \oplus W$ expresses V as a direct sum of $\mathcal{K}[T]$-submodules precisely when $A = \begin{pmatrix} B & 0 \\ 0 & D \end{pmatrix}$, with B an $r \times r$ matrix and D an $(n - r) \times (n - r)$ matrix, B and D giving the action of T on U and W respectively. This observation will be refined further in the exercises, where some explicit decompositions can be found.

2.1.3 Examples: abelian groups

As noted in part (ii) of (1.2.2), an abelian group is essentially the same thing as a $\mathbb{Z}$-module. Here, we give some first results on the decomposability of abelian groups which will be developed in the exercises to this section, and in much greater generality in our discussion of Dedekind domains in Chapter 6.

(i) View the ring of integers $\mathbb{Z}$ as a right $\mathbb{Z}$-module. A (right) submodule of $\mathbb{Z}$ is the same as an ideal, and so has the form $a\mathbb{Z}$ for some integer a. Since $a\mathbb{Z} \cap b\mathbb{Z} \neq 0$ for any two nonzero ideals, it follows that $\mathbb{Z}$ is indecomposable.

(ii) Let p be a prime number and r be a natural number. If L is a submodule of the quotient module $\mathbb{Z}/p^r\mathbb{Z}$, the inverse image of L (1.2.12) under the canonical homomorphism from $\mathbb{Z}$ to $\mathbb{Z}/p^r\mathbb{Z}$ must be an ideal $a\mathbb{Z}$ of $\mathbb{Z}$ containing $p^r\mathbb{Z}$. By the unique factorization of integers, $a = p^i$ for some i with $0 \leq i \leq r$. Therefore any submodule must have the form $p^i\mathbb{Z}/p^r\mathbb{Z}$, $0 \leq i \leq r$, and so no two submodules can complement one another. Thus $\mathbb{Z}/p^r\mathbb{Z}$ is indecomposable.

(iii) Let a and b be coprime integers. By elementary number theory, there are integers r, s with $1 = ar + bs$. Suppose that M is an abelian group with $Mab = 0$. For m in M, we have $m = m \cdot 1 \in Ma + Mb$, so $M = Ma + Mb$.

An equally devious argument gives $Ma \cap Mb = 0$, hence $M = Ma \oplus Mb$. Also, $M/Ma \cong Mb$ and viceversa.

2.1.4 The uniqueness of summands

Example (i) of (2.1.2) shows that the indecomposable summands of a vector space V are not unique. However, the theory of bases of vector spaces tells us that any indecomposable summand of any $\mathcal{K}$-space must be isomorphic to $\mathcal{K}$ itself. Thus the direct summands of a finite-dimensional space are unique up to isomorphism, even though they are not absolutely unique.

Uniqueness results of this kind hold only for very special types of ring and module. In (3.3.8), we have an example of a ring R for which $R^2 = M_1 \oplus M_2 = N_1 \oplus N_2$, $M_1 \cong N_1$ but $M_2 \not\cong N_2$, while the theory of Dedekind domains provides examples in which $M_i \not\cong N_j$ for $i, j \in \{1, 2\}$ (6.1.9).

2.1.5 External direct sums

It is unlikely that the members of an arbitrary finite set $\{L_1, \ldots, L_k\}$ of right R-modules all happen to be submodules of some R-module M, and even less likely that they give an internal direct decomposition of M. However, it is possible to manufacture a module which is the internal direct sum of submodules $M_1, \ldots, M_k$ with $L_i \cong M_i$ as R-modules for each i.

We define the *external direct sum* $L_1 \odot \cdots \odot L_k$ to be the set of all k-tuples

$$(\ell_1, \ldots, \ell_k), \text{ with } \ell_i \in L_i \text{ for each } i,$$

with componentwise addition and scalar multiplication:

$$(\ell_1, \ldots, \ell_k) + (n_1, \ldots, n_k) = (\ell_1 + n_1, \ldots, \ell_k + n_k), \ \ell_i, n_i \in L_i \text{ for all } i,$$

and

$$(\ell_1, \ldots, \ell_k)r = (\ell_1 r, \ldots, \ell_k r) \text{ for } r \in R.$$

It is easy to verify that $L_1 \odot \cdots \odot L_k$ is a right R-module with this addition and scalar multiplication.

The modules $L_1, \ldots, L_k$ are not themselves submodules of $L_1 \odot \cdots \odot L_k$ since an element ℓ_1 of L_1 is logically distinct from the k-tuple $(\ell_1, 0, \ldots, 0)$, even when $k = 1$.

However, if we put

$$M_i = \{(0, \ldots, 0, \ell_i, 0, \ldots, 0) \mid \ell_i \in L_i\} \text{ for } i = 1, \ldots, k,$$

then it is readily verified that there is an R-module isomorphism $L_i \cong M_i$ for each i, and that the external direct sum $L_1 \odot \cdots \odot L_k$ is the internal direct sum $M_1 \oplus \cdots \oplus M_k$.

In the other direction, given an internal direct decomposition

$$M = M_1 \oplus \cdots \oplus M_k$$

of a right R-module, it is clear from condition (DSU) of (2.1.1) that there is an isomorphism of right R-modules

$$M \cong M_1 \odot \cdots \odot M_k,$$

given by the map $m_1 + \cdots + m_k \mapsto (m_1, \ldots, m_k)$.

2.1.6 Standard inclusions and projections

It is very useful to reformulate the relationship between internal and external direct sums in terms of certain inclusion and projection homomorphisms.

Suppose that $M = L_1 \odot \cdots \odot L_k$ is an external direct sum of right R-modules. For $i = 1, \ldots, k$, the *standard inclusions* $\sigma_i : L_i \to M$ are given by $\sigma_i(\ell_i) = (0, \ldots, 0, \ell_i, 0, \ldots, 0)$, $\ell_i \in L_i$, and the *standard projections* $\pi_i : M \to L_i$ are given by $\pi_i(\ell_1, \ldots, \ell_k) = \ell_i$.

It is easy to verify that all the maps σ_i and π_i are homomorphisms of right R-modules, and that the following identities hold.

(SIP1) $\pi_i \sigma_j = \begin{cases} id_{L_i} & i = j, \\ 0 & i \neq j, \end{cases}$

(SIP2) $id_M = \sigma_1 \pi_1 + \cdots + \sigma_k \pi_k$.

In general, given a module M, a collection

$$\{L_1, \ldots, L_k; \sigma_1, \ldots, \sigma_k; \pi_1, \ldots, \pi_k\}$$

of modules and homomorphisms satisfying the above conditions is termed a *full set of inclusions and projections* for M. The modules L_i are usually omitted from the notation since they are recovered as $L_i = \pi_i M$ for each i.

Notice that we again allow the trivial cases $k = 0$ ($M = 0$) and $k = 1$, when both the inclusion and projection are the identity map on M. We also permit one or more of the summands L_i to be zero, when the corresponding inclusion and projection will be zero homomorphisms. In particular, there is no claim that the integer k is 'best possible'.

The proof of the next result is straightforward.

2.1.7 Proposition

Let M and $L_1, \ldots, L_k$ be right R-modules. Then the following statements are equivalent.

(i) *There are R-module homomorphisms $\sigma_1, \ldots, \sigma_k, \pi_1, \ldots, \pi_k$ such that $\{L_1, \ldots, L_k; \sigma_1, \ldots, \sigma_k; \pi_1, \ldots, \pi_k\}$ is a full set of inclusions and projections for M.*

(ii) *$M \cong L_1 \odot \cdots \odot L_k$ as a right R-module, the isomorphism being given by $m \mapsto (\pi_1 m, \ldots, \pi_k m)$.*

(iii) *$M = M_1 \oplus \cdots \oplus M_k$ where $M_i = \sigma_i L_i$ for $i = 1, \ldots, k$.* $\qquad \Box$

2.1.8 Notation

As we have seen, there is a very close connection between the internal and the external direct sum decompositions of a module. We shall therefore use the notation $M = M_1 \oplus \cdots \oplus M_k$ for both the internal and the external direct sum of modules for the remainder of this text.

The context should make it clear which type of direct sum is intended. For example, an expression such as $L \oplus L$ must be interpreted as an external sum, since it makes little sense otherwise, while an expression such as $M = mR \oplus nR$, with m and n in M, will be an internal direct sum.

Suppose that, for $i = 1, \ldots, k$, $\alpha_i : L_i \to M_i$ is an R-module homomorphism. Then we denote by

$$\alpha_1 \oplus \cdots \oplus \alpha_k : L_1 \oplus \cdots \oplus L_k \to M_1 \oplus \cdots \oplus M_k$$

the R-module homomorphism obtained by letting α_i act on the ith component, that is,

$$(\alpha_1 \oplus \cdots \oplus \alpha_k)(\ell_1 \oplus \cdots \oplus \ell_k) = \alpha_1(\ell_1) \oplus \cdots \oplus \alpha_k(\ell_k).$$

2.1.9 Idempotents

Another useful characterization of direct sum decompositions of a module M is provided by idempotent elements of the endomorphism ring $\mathrm{End}(M)$ (1.2.5).

Let $\{\sigma_1, \ldots, \sigma_k; \pi_1, \ldots, \pi_k\}$ be a full set of inclusions and projections for M, and let $e_i = \sigma_i \pi_i$ for $i = 1, \ldots, k$. Then $e_i \in \mathrm{End}(M)$, and, by (SIP2) above, we have

(Idp1) $1 = e_1 + \cdots + e_k$, where $1 = id_M$ is the identity element of the ring $\mathrm{End}(M)$.

Using (SIP1), we also obtain

(Idp2) $e_i e_j = \begin{cases} e_i & i = j, \\ 0 & i \neq j. \end{cases}$

In general, an element e of an arbitrary ring S is *idempotent* if $e^2 = e$, and two idempotents e, f of S are *orthogonal* if $ef = 0 = fe$. A set of idempotents $\{e_1, \ldots, e_k\}$ satisfying both (Idp1) and (Idp2) is called a *full set of orthogonal idempotents* of S. An explicit example is given in Exercise 2.1.5 below.

Note that, conversely, a full set of orthogonal idempotents of $\text{End}(M)$ gives rise to a full set of inclusions and projections for M: for each i, take $L_i = e_i M$, π_i to be $\pi_i : m \mapsto e_i m$ and σ_i to be the evident inclusion map.

These remarks make the next result plain.

2.1.10 Proposition

Let M be a right R-module. Then there is a bijective correspondence between

(i) *direct sum decompositions $M = M_1 \oplus \cdots \oplus M_k$*
    and

(ii) *full sets of orthogonal idempotents*

$$\{e_1, \ldots, e_k\} \quad in \quad \text{End}(M),$$

under which $M_i = e_i M$ for all i. □

2.1.11 Infinite direct sums

From time to time we need to consider direct sum decompositions with an infinite number of terms.

Let M be a right R-module, and let $\{M_i \mid i \in I\}$ be a collection of submodules of M, indexed by some set I which need not be finite. By definition (1.2.8), the sum of this set of submodules is

$$\sum_{i \in I} M_i = \{\sum m_i \mid m_i = 0 \text{ for almost all } i\}.$$

We write $I \setminus \{i\}$ to denote the set obtained by omitting the element i from I. Then M is the *internal direct sum* of the set $\{M_i \mid i \in I\}$, written as $M = \bigoplus_{i \in I} M_i$, if the following conditions are satisfied.

(DS$^\infty$1) $M = \sum_I M_i$,
(DS$^\infty$2) $M_i \cap (\sum_{h \in I \setminus \{i\}} M_h) = 0$ for every $i \in I$.

These conditions reduce to (DS1) and (DS2) when $I = \{1, \ldots, k\}$ is finite. As in (2.1.1), the definition can be restated in terms of a single condition.

(DS$^\infty$U) A right R-module M is the internal direct sum of its submodules M_i ($i \in I$) if and only if each element m of M can be written uniquely in the form $m = \sum_{i \in I} m_i$ with $m_i \in M_i$ for each i and almost all $m_i = 0$.

Given an infinite collection $\{L_i \mid i \in I\}$ of right R-modules, we can perform two 'external' constructions.

The *direct product* of $\{L_i \mid i \in I\}$ is the *cartesian product*

$$\prod_I L_i = \{(\ell_i) \mid \ell_i \in L_i\},$$

in which two expressions (ℓ_i) and (m_i) are equal if and only if $\ell_i = m_i$ for all $i \in I$. Addition and scalar multiplication are defined componentwise:

$$(\ell_i) + (n_i) = (\ell_i + n_i), \quad \ell_i, n_i \in L_i$$

and

$$(\ell_i)r = (\ell_i r), \quad r \in R.$$

It is easily verified that $\prod_I L_i$ is a right R-module.

The *external direct sum* of $\{L_i \mid i \in I\}$ is

$$\bigodot_I L_i = \{(\ell_i) \mid \ell_i \in L_i, \ell_i = 0 \text{ for almost all } i\}.$$

Clearly, the external direct sum is a submodule of the direct product, and they are equal whenever I is finite.

As in the finite case, the notions of an internal and an external direct sum are interchangeable. We record this fact in the following result, which holds regardless of the size of the index set.

2.1.12 Proposition
Let M be right R-module, and let I be an index set.

(i) *If M has an internal direct sum decomposition $M = \bigoplus_I M_i$ in terms of its submodules $\{M_i \mid i \in I\}$, then M is isomorphic to their external direct sum: $M \cong \bigodot_I M_i$.*

(ii) *If there is an isomorphism $\theta : \bigodot_I L_i \to M$ of right R-modules, then $M = \bigoplus_I M_i$ where $M_i = \theta(L_i)$ for all i in I.* □

2.1.13 Remarks
(i) Following our convention in the case that I is finite (2.1.8), we write

$M = \bigoplus_I M_i$ to indicate either an internal or an external direct sum, relying on the context to make it clear which is meant.

(ii) The characterizations of finite direct sums in terms of standard inclusions and projections (2.1.7) and idempotents (2.1.10) cannot be extended to the infinite case, since it is impossible to attach a meaning to an infinite sum of maps or idempotents in our purely algebraic context. This point is also considered in Exercise 2.3.1.

(iii) The 'obvious' fact that an infinite direct sum $\bigoplus_I M_i$ is smaller than the direct product $\prod_I M_i$ is most easily confirmed by counting arguments. For instance, take $I = \mathbb{N}$, the natural numbers, and the ring R and all the modules M_i to be the integers $\mathbb{Z}$. Then it is wellknown that the direct product $\prod_{\mathbb{N}} \mathbb{Z}$ has the same cardinality as the set of real numbers $\mathbb{R}$, while $\bigoplus_{\mathbb{N}} \mathbb{Z}$ has the same cardinality as $\mathbb{Z}$. The latter assertion is often seen in the equivalent form 'the set of polynomials over $\mathbb{Z}$ is countable', there being an evident correspondence between the sequences belonging to $\bigoplus_{\mathbb{N}} \mathbb{Z}$ and polynomials – see (3.2.1) for more details.

(iv) The definition of internal and external direct sums extends in the expected way to other algebraically defined objects, such as bimodules, rings and groups, but with some variations in terminology according to the type of structure under consideration. For rings, we speak of direct products rather than direct sums; these are considered in detail in section 2.6.

For groups, we also prefer to speak of direct products. If $\{G_i \mid i \in I\}$ is a set of multiplicative groups, the *direct product* $\prod_{i \in I} G_i$ is again a multiplicative group with componentwise multiplication.

A much more general analysis of direct sums and products, which reveals the reasons for these variations in terminology, can be made in the language of category theory. We discuss these constructions in [BK: CM], §1.4.

2.1.14 Ordered index sets

In our discussion of direct sums over an arbitrary index set I, we have not so far assumed that I has any ordering. As there are circumstances in which it is advantageous to take an infinite direct sum or product over an ordered set, or over a finite ordered set other than $\{1, \ldots, k\}$, we very quickly review the definitions and notation.

Let Λ be an ordered set (1.2.18), infinite or finite. Without the order, Λ can be viewed simply as a set I. We take as granted the converse, that any unordered set I can be regarded as an ordered set Λ. This seemingly modest assertion follows from the Axiom of Choice, and the interested reader might like to give a proof based on Zorn's Lemma (1.2.19), using ordered subsets and order-preserving inclusions.

A module M is the internal direct sum of an ordered set $\{M_\lambda \mid \lambda \in \Lambda\}$ of submodules of M if it is already the internal direct sum of the corresponding unordered set $\{M_i \mid i \in I\}$. We write $M = \bigoplus_{\lambda \in \Lambda} M_\lambda$ if we wish to draw attention to the ordering.

Given an infinite ordered collection $\{L_\lambda \mid \lambda \in \Lambda\}$ of right R-modules, their direct product is again the cartesian product

$$\prod_\Lambda L_\lambda = \{(\ell_\lambda) \mid \ell_\lambda \in L_\lambda\},$$

where the sequences (ℓ_λ) must now be intepreted as ordered sequences. The external direct sum of $\{L_\lambda \mid \lambda \in \Lambda\}$ is

$$\bigodot_\Lambda L_\lambda = \{(\ell_\lambda) \mid \ell_\lambda \in L_\lambda,\ \ell_\lambda = 0 \text{ for almost all } \lambda\}.$$

The reader will have no difficulty in extending the exchangeability of internal and external direct sums promised in (2.1.12) to ordered index sets. As in the unordered case, we prefer to use the internal direct sum notation $M = \bigoplus_\Lambda M_\lambda$ unless there is a strong reason to do otherwise.

2.1.15 The module L^Λ

An important case where it is more convenient to work with an ordered index set arises when we wish to construct the direct sum or product of copies of a single module.

Let R be any ring, L any right R-module, and let Λ be an ordered set. The *direct product of Λ copies of L* is

$$L^{(\Lambda)} = \{(\ell_\lambda) \mid \ell_\lambda \in L\};$$

here, it is essential to view the sequences (ℓ_λ) as ordered sequences.

The *direct sum of Λ copies of L* is

$$L^\Lambda = \{(\ell_\lambda) \mid \ell_\lambda \in L,\ \ell_\lambda = 0 \text{ for almost all } \lambda\};$$

clearly, both the direct product and direct sum are themselves right R-modules under the obvious componentwise rules for addition and scalar multiplication.

It will usually be convenient to interpret elements of L^λ (or even, occasionally, of $L^{(\Lambda)}$) as $\Lambda \times 1$ matrices, that is, as 'column vectors' indexed by Λ. This is particularly advantageous when we wish to view the endomorphisms of L^Λ as matrices – see Exercise 2.1.6 below. In the case that L is the ring R itself, viewed as a right R-module, we obtain the *standard free R-module R^Λ* on Λ (this terminology will be justified shortly).

In the case that $\Lambda = \{1, \dots, k\}$ is finite, we use the notations L^k and R^k (as in (1.2.3)), and if $k = 0$, we put $L^0 = 0$, the zero module.

2.1.16 The module $\mathrm{Fr}_R(X)$

Suppose now that we wish to take a direct sum of copies of R indexed by an unordered set X. The simplest construction is to form, for each x in X, a formal cyclic module $xR = \{xr \mid r \in R\}$, where $xr = xs$ in xR if and only if $r = s$ in R. We then define the direct sum indexed by X to be $\mathrm{Fr}_R(X) = \bigoplus_X xR$ (external direct sum). It is clear that if Λ is any convenient ordering of X, we have $\mathrm{Fr}_R(X) = R^\Lambda$ as a right R-module.

The module $\mathrm{Fr}_R(X)$ is called the *standard free right R-module* on X. Note that elements of $\mathrm{Fr}_R(X)$ are formal sums $m = \sum_{x \in X} xr_x(m)$ with $r_x(m) \in R$, almost all $r_x(m)$ being 0, and that $m = n$ in $\mathrm{Fr}_R(X)$ if and only $r_x(m) = r_x(n)$ for all x in X. In particular, if $X = \{x\}$ has one member, $\mathrm{Fr}_R(\{x\}) = xR$ is a cyclic right module isomorphic to R. (By convention, $\mathrm{Fr}_R(\emptyset) = 0$.)

Properly speaking, the set X is not actually a subset of $\mathrm{Fr}_R(X)$. However, there is a *canonical embedding* $\iota_X : X \hookrightarrow \mathrm{Fr}_R(X)$, the element $\iota_X(x)$ being defined by the requirements that

$$r_y(\iota_X(x)) = \begin{cases} 1 & \text{if } y = x, \\ 0 & \text{if } y \neq x. \end{cases}$$

Sometimes, X is regarded as a subset of $\mathrm{Fr}_R(X)$ using this embedding.

2.1.17 Left-handed notation

It is clear that all the preceding definitions and results have counterparts for left modules, which we do not state separately. However, we do need some special notation to enable the reader to distinguish between the cases in which the ring R is to be viewed as a right R-module and those in which it is a left R-module.

We therefore write $^{(\Lambda)}R$ and $^\Lambda R$ for the direct product and direct sum respectively of Λ copies of R viewed as left R-module. Both are again left R-modules, $^\Lambda R$ being the *standard free left R-module* on Λ. If $\Lambda = \{1, \dots, k\}$, we write $^k R$. We usually view $^{(\Lambda)}R$ as the set of $1 \times \Lambda$ matrices over R, that is, the set of 'row vectors' of 'length Λ'.

2.1.18 Free generating sets

Although we have introduced the 'standard free modules', we have not as yet given any characterizations or properties of free modules in general. This is our next task.

Let R be a ring and let M be a right R-module. A subset X of M is said to be *free*, or *independent*, if there are no relations of the form

$$x_1 r_1 + \cdots + x_k r_k = 0$$

for any $k \geq 1$ and distinct $x_1, \ldots, x_k \in X$, and $r_1, \ldots, r_k \in R$, except the trivial relations in which $r_1 = \cdots = r_k = 0$.

If there is a danger of ambiguity about the coefficient ring, the terms *R-free* and *R-independent* may be used. By convention, the empty set $\emptyset$ is regarded as a free subset of any module; we can then claim that any subset of a free set is again free.

When the coefficient ring R is a field $\mathcal{K}$, a $\mathcal{K}$-module is a vector space V and a free subset of V is otherwise known as a $(\mathcal{K}\text{-})$*linearly independent* subset of V.

Unlike the situation with vector spaces, there is no reason why a particular module need contain any nonempty free subsets. This can be seen by considering the $\mathbb{Z}$-modules $\mathbb{Z}/a\mathbb{Z}$, $a \neq 0$. However, $\mathbb{Z}/a\mathbb{Z}$ is also a ring, and, when viewed as a module over itself, certainly does contain a free subset, $\{\overline{1}\}$.

2.1.19 Free modules

A right R-module M is said to be *free* if it possesses a free generating set X; informally, we say that M *is free on* X. It is easy to see that M can then be expressed as an internal direct sum $M = \bigoplus_{x \in X} xR$, where, for each x, $xR \cong R$ as a right R-module. Thus each element m of M can be written in the form $m = \sum_{x \in X} x r_x(m)$ where the coefficients $r_x(m)$ are uniquely determined by m, almost all being 0.

Given a set X, the standard free module $\mathrm{Fr}_R(X)$ is evidently free on the set X, provided that we identify each element x of X with its image in $\mathrm{Fr}_R(X)$ under the canonical embedding ι_X.

We collect together some alternative characterizations of free modules for future use. The arguments are all immediate from the preceding discussion.

2.1.20 Proposition

Let M be a right R-module and suppose that X is a subset of M. Then the following assertions are equivalent.

(i) M *is free on* X.

(ii) $M = \bigoplus_{x \in X} xR$, *where, for each* $x, xR \cong R$ *as a right* R-*module by the map* $xr \mapsto r$.

(iii) *Each element* m *of* M *can be written in the form*

$$m = \sum_{x \in X} xr_x(m)$$

where the coefficients $r_x(m)$ *are uniquely determined by* m, *almost all being* 0.

(iv) *There is a unique isomorphism*

$$\theta : M \xrightarrow{\cong} \mathrm{Fr}_R(X)$$

of right R-*modules in which* $\theta(x) = \iota_X(x)$ *for all* x *in* X, *where* $\iota_X : X \hookrightarrow \mathrm{Fr}_R(X)$ *is the canonical embedding.* $\square$

2.1.21 Extending maps

Another very useful property of free generating sets is that homomorphisms can be defined arbitrarily on them. To state the result precisely, we need a definition.

Given a subset X of a right R-module M, another right R-module N, and a map (of sets) $\zeta : X \to N$, we say that an R-module homomorphism $\alpha : M \to N$ *extends* ζ if $\alpha(x) = \zeta(x)$ for all x in X. We also use the terms ζ *extends to* M or α *is an extension of* ζ.

The result is as follows.

2.1.22 Theorem

Let X *be a subset of a right* R-*module* M. *Then the following statements are equivalent.*

(i) *For every right* R-*module* N, *every mapping* ζ *from* X *to* N *extends uniquely to an* R-*module homomorphism* $\alpha : M \to N$.

(ii) X *is a free generating set for* M.

Proof

(ii) $\Rightarrow$ (i).   By the implication (i) $\Rightarrow$ (iii) of the preceding proposition, an element m of M can be written as $m = \sum_{x \in X} xr_x(m)$ where the coefficients $r_x(m)$ are uniquely determined by m, almost all being 0. Define $\alpha : M \to N$ by $\alpha(m) = \sum_{x \in X} \zeta(x)r_x(m)$; then α is the unique extension of ζ.

(i) $\Rightarrow$ (ii).   We argue by contradiction. Suppose first that X does not generate M, and let L be the submodule of M which is generated by X. Then the canonical homomorphism $\pi : M \to M/L$ is nonzero, and both π and the zero homomorphism are extensions of the map $0 : x \mapsto 0$ from X to M/L, a contradiction. Thus X does generate M.

Next suppose that X is not independent, and consider a nontrivial relation $x_1 r_1 + \cdots + x_k r_k = 0$, with $x_1, \ldots, x_k \in X$ and $r_1, \ldots, r_k \in R$ not all 0. Define $\zeta : X \to R^k$ as follows:

$$x_1 \mapsto \begin{pmatrix} 1 \\ 0 \\ \vdots \\ 0 \end{pmatrix}, \ldots, x_k \mapsto \begin{pmatrix} 0 \\ \vdots \\ 0 \\ 1 \end{pmatrix}, x \mapsto 0 = \begin{pmatrix} 0 \\ 0 \\ \vdots \\ 0 \end{pmatrix} \text{ if } x \notin \{x_1, \ldots, x_k\}.$$

Evidently, any R-homomorphism α extending ζ would send the zero element of M to the nonzero element $\begin{pmatrix} r_1 \\ \vdots \\ r_k \end{pmatrix}$ of R^k, which is impossible. □

The above result has a consequence which is important in the investigation of the functorial properties of the free module construction ([BK: CM], (1.2.3), (1.4.3)).

2.1.23 Corollary

Let X and Y be sets, $\chi : X \to Y$ a mapping. Then there is a unique R-homomorphism $\mathrm{Fr}_R(\chi) : \mathrm{Fr}_R(X) \to \mathrm{Fr}_R(Y)$ with the following properties.

(i) *The diagram*

$$
\begin{array}{ccc}
X & \hookrightarrow & \mathrm{Fr}_R(X) \\
\chi \downarrow & & \downarrow \mathrm{Fr}_R(\chi) \\
Y & \hookrightarrow & \mathrm{Fr}_R(Y)
\end{array}
$$

commutes, where the horizontal maps are the canonical inclusions ι_X and ι_Y,

(ii) $\mathrm{Fr}_R(id_X) = id_{\mathrm{Fr}_R(X)}$,

(iii) $\mathrm{Fr}_R(\theta \chi) = \mathrm{Fr}_R(\theta) \, \mathrm{Fr}_R(\chi)$ *for any map $\theta : Y \to Z$ of sets.*

Proof

Composition of χ with the canonical inclusion ι_Y provides a set mapping from X to the R-module $\mathrm{Fr}_R(Y)$. Appealing to (i) of the preceding theorem, we obtain the R-homomorphism $\mathrm{Fr}_R(\chi)$ as the unique R-module homomorphism extension of this map. By construction, the diagram in (i) commutes.

Properties (ii) and (iii) now follow easily from the uniqueness of the extension of a map.   □

Exercises

2.1.1 **Block forms of matrices**

This exercise, together with Exercises 2 and 3 below, follows on from Exercise 1.2.9 and expands the illustration given in (ii) of (2.1.2).

Let A be an $n \times n$ matrix over a field K and let M be the $K[T]$-module obtained from K^n with T acting as A.

(a) Show that $M = L \oplus N$ as a $K[T]$-module if and only if the submodules L and N have underlying subspaces U and W with $K^n = U \oplus W$ as a K-space and $AU \subseteq U$, $AW \subseteq W$.

(b) Using a change of basis argument, show that $M = L \oplus N$ as a $K[T]$-module if and only if there is an invertible $n \times n$ matrix Γ so that $\Gamma^{-1} A \Gamma = \begin{pmatrix} B & 0 \\ 0 & D \end{pmatrix}$, where B and D give the action of T on U and W respectively relative to some bases.

(c) More generally, show that, if $M = L_1 \oplus \cdots \oplus L_k$, then there is an invertible $n \times n$ matrix Γ such that $\Gamma^{-1} A \Gamma = (B_{ij})$, a $k \times k$ *block diagonal matrix*, that is, $B_{ij} = 0$ for $i \neq j$. Prove the converse also.

(d) Deduce that $M = L_1 \oplus \cdots \oplus L_n$, where each submodule L_i is one-dimensional as a K-space, if and only if there is an invertible $n \times n$ matrix Γ such that $\Gamma^{-1} A \Gamma = \operatorname{diag}(\lambda_1, \ldots, \lambda_n)$, an $n \times n$ diagonal matrix with entries the eigenvalues $\lambda_1, \ldots, \lambda_n$ of A.

(e) Take $A = \begin{pmatrix} 0 & 1 & 0 \\ 0 & 0 & 1 \\ 1 & 0 & 0 \end{pmatrix}$. Determine the direct sum decomposition of M when the field of coefficients is taken in turn to be $\mathbb{C}$, $\mathbb{R}$, $\mathbb{Z}/2\mathbb{Z}$ or $\mathbb{Z}/3\mathbb{Z}$.

2.1.2 Let $A = \begin{pmatrix} 1 & 1 \\ 0 & 1 \end{pmatrix}$, $C = \begin{pmatrix} 0 & 1 \\ 1 & 0 \end{pmatrix}$ and $E = \begin{pmatrix} 0 & -1 \\ 1 & 0 \end{pmatrix}$. Using these matrices in turn, $\mathbb{R}^2$ is made into three $\mathbb{R}[T]$-modules: L, N and Q as in Exercise 1.2.8.

Determine which of these modules is decomposable, giving an explicit decomposition when possible.

Repeat the exercise with the field of coefficients taken in turn to be $\mathbb{C}, \mathbb{Z}/2\mathbb{Z}$ or $\mathbb{Z}/3\mathbb{Z}$.

2.1.3   Using part (e) of Exercise 1.2.8, show that the $\mathcal{K}[T]$-module given by

the matrix $A = \begin{pmatrix} 1 & 1 & 1 & \cdots & 1 & 1 \\ 0 & 1 & 1 & \cdots & 1 & 1 \\ 0 & 0 & 1 & \cdots & 1 & 1 \\ \vdots & \vdots & \vdots & \ddots & \vdots & \vdots \\ 0 & 0 & 0 & \cdots & 0 & 1 \end{pmatrix}$ is always indecomposable,

whatever the field $\mathcal{K}$.

2.1.4   **Abelian groups**

This exercise expands the example given in (2.1.3). We assume some elementary properties of the ring $\mathbb{Z}$ of integers.

(a) Let M be an abelian group and suppose that $Mc = 0$ for some positive integer c. Write the prime factorization of c in the form

$$c = q_1 \cdots q_k$$

where $p_1, \ldots, p_k$ are distinct primes and $q_i = p_i^{r(i)}$ for each i. Put

$$\widehat{q_i} = q_1 \cdots q_{i-1} q_{i+1} \cdots q_k \quad \text{for all } i.$$

Show that $M = M_1 \oplus \cdots \oplus M_k$ with $M_i = M\widehat{q_i}$ for each i.

(b) Suppose M is a cyclic abelian group (that is, a cyclic $\mathbb{Z}$-module). Show that either $M \cong \mathbb{Z}$, or there exist prime numbers $p_1, \ldots, p_k$ such that

$$M \cong \mathbb{Z}/\mathbb{Z}p_1^{r(1)} \oplus \cdots \oplus \mathbb{Z}/\mathbb{Z}p_k^{r(k)}$$

for some exponents $r(1), \ldots, r(k)$.

2.1.5   **Explicit idempotents**

Following from the preceding exercise, we see some examples of the idempotents promised by the theory in (2.1.9).

Let M be an abelian group with $Mc = 0$ for some positive integer c, and put $c = ab$ for coprime integers a, b. Write $1 = ar + bs$, and define endomorphisms α and β of M by $\alpha(m) = arm$ and $\beta(m) = bsm$. Verify that $\{\alpha, \beta\}$ is a set of projections for the direct sum decomposition $M = Ma \oplus Mb$ of M.

Hence find the full set of orthogonal idempotents of $\mathrm{End}(M)$ corresponding to the decomposition $M = M_1 \oplus \cdots \oplus M_k$ as in the previous exercise.

2.1.6   **Endomorphisms of direct sums**

It is sometimes useful to interpret endomorphisms of a direct sum as matrices. This will be done in the next section for free modules, using bases; here, we give an approach for arbitrary finite direct sums which (of necessity) avoids the use of bases.

(i) Let $M = M_1 \oplus M_2$, an internal direct sum of right R-modules, and let $\{\sigma_1, \sigma_2, \pi_1, \pi_2\}$ be the corresponding set of inclusions and projections. Given an endomomorphism μ of M, define $\mu_{ij} = \pi_i \mu \sigma_j$, an R-homomorphism from M_j to M_i, $i, j = 1, 2$. Show that for $m = m_1 + m_2 \in M$, with $m_1 \in M_1, m_2 \in M_2$, we have

$$\mu(m) = (\mu_{11}m_1 + \mu_{12}m_2) + (\mu_{21}m_1 + \mu_{22}m_2),$$

where $\mu_{11}m_1 + \mu_{12}m_2$ is in M_1 and $\mu_{21}m_1 + \mu_{22}m_2$ is in M_2. Viewing M as a 'column space' $\begin{pmatrix} M_1 \\ M_2 \end{pmatrix}$, show that μ can be represented as a matrix $\begin{pmatrix} \mu_{11} & \mu_{12} \\ \mu_{21} & \mu_{22} \end{pmatrix}$. Deduce that the ring of endomorphisms $\mathrm{End}(M)$ of M can be written as a ring of 2×2 matrices

$$\mathrm{End}(M) = \begin{pmatrix} \mathrm{End}(M_1) & \mathrm{Hom}(M_2, M_1) \\ \mathrm{Hom}(M_1, M_2) & \mathrm{End}(M_2) \end{pmatrix}$$

where $\mathrm{Hom}(M_1, M_2)$ is the set of all R-module maps from M_1 to M_2, etc. (The apparent transposition of indices arises because we write homomorphisms of right modules on the left – see (1.2.5).)

(ii) Evaluate the ring $\mathrm{End}(M)$ when M is in turn the $\mathbb{Z}$-modules $\mathbb{Z} \oplus \mathbb{Z}/q\mathbb{Z}$, $\mathbb{Z}/p\mathbb{Z} \oplus \mathbb{Z}/q\mathbb{Z}$ and $\mathbb{Z}/p^h\mathbb{Z} \oplus \mathbb{Z}/p^k\mathbb{Z}$, where p and q are distinct prime numbers and $h, k \geq 1$.

(iii) Let $M = L^k$, the direct sum of k copies of a right R-module L. Show that $\mathrm{End}(M) \cong M_k(\mathrm{End}(L))$.

Find the matrices corresponding to the idempotents

$$e_i = \sigma_i \pi_i, \ i = 1, \ldots, k,$$

arising from a full set of inclusions and projections

$$\{\sigma_1, \ldots, \sigma_k, \pi_1, \ldots, \pi_k\}$$

for M (2.1.9).

(iv) Let $M = M_1 \oplus \cdots \oplus M_k$. Show that $\mathrm{End}(M)$ can be described as a 'block matrix ring' (E_{ij}) with $E_{ij} = \mathrm{Hom}(M_j, M_i)$ for all i, j.

(v) Compute $\mathrm{End}(M)$ for $M = \mathbb{Z}/\mathbb{Z}p_1^{r(1)} \oplus \cdots \oplus \mathbb{Z}/\mathbb{Z}p_k^{r(k)}$, with $p_1, \ldots, p_k$ distinct prime numbers.

2.1.7   Let $L = L_1 \oplus \cdots \oplus L_h$, $M = M_1 \oplus \cdots \oplus M_k$ and $N = N_1 \oplus \cdots \oplus N_\ell$ be direct sum decompositions of right R-modules. Extend part (iv) of the previous exercise to describe the R-module homomorphisms from L to M in a matrix form. Verify that for homomorphisms $\lambda : L \to M$ and $\mu : M \to N$, the matrix of the product $\mu\lambda$ is the product of the corresponding matrices.

2.1.8   Let R be a ring and let X and Y be disjoint sets (that is, $X \cap Y = \emptyset$). Show that

$$\mathrm{Fr}_R(X \cup Y) \cong \mathrm{Fr}_R(X) \oplus \mathrm{Fr}_R(Y).$$

More generally, suppose that $X = \bigcup_{\lambda \in \Lambda} X_\lambda$ is a union of pairwise disjoint sets (that is, $X_\lambda \cap X_\nu = \emptyset$ if $\lambda \neq \nu$). Prove that

$$\mathrm{Fr}_R(X) = \bigoplus_\Lambda \mathrm{Fr}_R(X_\lambda).$$

2.2 MATRICES, BASES, HOMOMORPHISMS OF FREE MODULES

Homomorphisms between free modules can be helpful in describing the structure of a module. This is always the case for modules over a Euclidean domain, as we see in section 3.3, and, for arbitrary rings, for those modules of main interest to us. In turn, a homomorphism between free modules has associated with it a set of matrices which describe how the homomorphism relates the various choices of bases for the free modules. If it is possible to find one such associated matrix of suitably simplified form, then we gain a good description of the original module. The best results are obtained when there is an associated matrix that is diagonal.

In general, such a simplified associated matrix is called a matrix in *normal form*; the nature of the normal form can vary according to the problem under consideration. A more precise definition is given in 2.2.10 below.

With this motivation, we now discuss in detail the relationship between matrices and homomorphisms of free modules. Most of what follows is in essence the same as the discussion of bases and linear transformations of vector spaces that is met in a first course on linear algebra. However, there are two aspects where it differs.

One is that the coefficient ring R need not be commutative. This means that some care must be exercised when defining the matrix of a homomorphism, or else some horribly unnatural formulas can arise. (Our convention that homomorphisms be written opposite scalars is chosen to help us avoid such formulas.)

The other difference is that we allow the possibility that $R^s \cong R^t$ even if $s \neq t$; that is, a free module need not have unique rank. A discussion of the circumstances in which this possibility can happen, together with some examples, is given in the next section.

Although our main concern is with free modules having finite bases, we also make some remarks about infinite bases and the related infinite matrices.

As usual, we concentrate our discussion on right R-modules for some fixed but arbitrary ring R. However, we do need to use the corresponding results for left R-modules, which we therefore outline briefly to help the reader make the transition.

2.2.1 Bases

In the previous section, we defined a free R-module to be a module M which possesses a free generating set X (2.1.19). However, there are many circumstances where it is important to take the elements of a free generating set in a fixed order. This is particularly the case when we give the description of homomorphisms between free modules in terms of matrices later in this chapter.

We therefore define a *basis* of a free module M to be an ordered free generating set of M, that is, a free generating set $F = \{f_\lambda \mid \lambda \in \Lambda\}$ indexed by some ordered set Λ (1.2.18).

This distinction between a free generating set and a basis is often obscured, especially in an undergraduate course concerned with finite-dimensional vector spaces. The reason is that a finite set is almost invariably indexed by a set of integers $\{1, \ldots, k\}$ and thus acquires an ordering by default.

The point of the distinction can be illustrated geometrically by considering the subsets $\left\{ \begin{pmatrix} 1 \\ 0 \end{pmatrix}, \begin{pmatrix} 0 \\ 1 \end{pmatrix} \right\}$ and $\left\{ \begin{pmatrix} 0 \\ 1 \end{pmatrix}, \begin{pmatrix} 1 \\ 0 \end{pmatrix} \right\}$ of $\mathbb{R}^2$. These are the same when viewed as unordered free generating sets of $\mathbb{R}^2$, but must be different bases of $\mathbb{R}^2$, or else it would be impossible to distinguish, for example, between clockwise and anticlockwise rotations of the Euclidean plane.

Note that a free generating set X of a module M can be made arbitrarily into a basis of M by choosing any ordering $\{x_\lambda \mid \lambda \in \Lambda\}$ for X.

2.2.2 Standard bases

Let Λ be an ordered set and let R be an arbitrary ring. In (2.1.15) we defined the standard free right R-module R^Λ on Λ to be the direct sum of Λ copies of R:

$$R^\Lambda = \{(r_\lambda) \mid r_\lambda \in R,\ r_\lambda = 0 \text{ for almost all } \lambda\},$$

with the expected componentwise rules for addition and scalar multiplication.

We may interpret the elements of R^Λ as $\Lambda \times 1$ 'column-finite' matrices, that is, as column vectors indexed by Λ and having only a finite number of nonzero entries.

The *standard basis* $\{e_\lambda \mid \lambda \in \Lambda\}$ of R^Λ comprises the 'standard unit vectors' $e_\lambda = (e_{\kappa,\lambda})_{\kappa \in \Lambda}$ which are defined given by

$$e_{\kappa,\lambda} = \begin{cases} 1 & \text{if } \kappa = \lambda, \\ 0 & \text{if } \kappa \neq \lambda. \end{cases}$$

It is clear that $\{e_\lambda\}$ is indeed a basis of R^Λ. (This confirms that R^Λ is actually a free module according to our definitions!)

When $\Lambda = \{1, \ldots, s\}$, we regain the familiar standard basis $\{e_1, \ldots, e_s\}$ of R^s with

$$e_1 = \begin{pmatrix} 1 \\ \vdots \\ 0 \end{pmatrix}, \ldots, e_s = \begin{pmatrix} 0 \\ \vdots \\ 1 \end{pmatrix}.$$

(For $s = 0$, we have $R^0 = 0$, the zero module, and the 'standard basis' is taken to be the empty set).

The exponent s is called the *rank* of R^s; more generally, if the free module M has a basis with s elements, then s is the rank of M. We are committing a minor abuse of language here, since the rank of a module is not necessarily unique. However, it is uniquely defined for most rings occurring in this text. We do attribute a rank to a module only when it has a finite basis.

2.2.3 Coordinates

Let $F = \{f_\lambda \mid \lambda \in \Lambda\}$ be a basis of a free right R-module M. By (iii) of (2.1.20), an element m of M can be written in the form $m = \sum_{x \in X} f_\lambda r_\lambda(m)$ where the coefficients $r_\lambda(m) \in R$ are uniquely determined by m, almost all being 0.

The corresponding element $_F(m) = (r_\lambda(m))$ of R^Λ is called the *coordinate vector* of m with respect to the given basis F. When $\Lambda = \{1, \ldots, s\}$, the coordinate vector is written as $\begin{pmatrix} r_1(m) \\ \vdots \\ r_s(m) \end{pmatrix}$.

Let $\theta : M \to R^\Lambda$ be the map which assigns to each element m its coordinate vector relative to F. An easy verification shows that θ is an R-module homomorphism, and the uniqueness of the coordinates guarantees that θ is an injection. Since any vector in R^Λ arises as the coordinate vector of an element of M, it follows that θ is an isomorphism of R-modules. By construction, $\theta(f_\lambda) = e_\lambda$ for all $\lambda \in \Lambda$.

It is clear that this procedure is reversible: given an R-module isomorphism $\theta : M \to R^\Lambda$, define elements f_λ in M by $\theta(f_\lambda) = e_\lambda$ for all $\lambda \in \Lambda$. Then $\{f_\lambda\}$ is a basis for M and $\theta(m)$ is the coordinate vector of an element m of M.

We summarize these observations (and a little more) in the following result, which parallels the unordered version given in (2.1.20).

2.2.4 Proposition
Let M be a right R-module and let $F = \{f_\lambda \mid \lambda \in \Lambda\}$ be a subset of M, where Λ is an ordered set. Then the following assertions are equivalent.

(i) *F is a basis for M.*

(ii) *$M = \bigoplus_{\lambda \in \Lambda} f_\lambda R$, where, for each λ, the map $f_\lambda r \mapsto r$ defines a right R-module isomorphism from $f_\lambda R$ to R.*

(iii) *Each element m of M can be written in the form*

$$m = \sum_{\lambda \in \Lambda} f_\lambda r_\lambda(m),$$

where the coefficients $r_\lambda(m)$ are uniquely determined by m, almost all being 0.

(iv) *There is an isomorphism $\beta : M \xrightarrow{\cong} R^\Lambda$ of right R-modules in which $\beta(f_\lambda) = e_\lambda$, where $\{e_\lambda \mid \lambda \in \Lambda\}$ is the standard basis of R^Λ.*   □

Part (iv) above illustrates the role of the ordering of a basis. If we take F to be the standard basis of R^2, then β is the identity map, but if F is $\{e_2, e_1\}$, then β is the 'switching' map $\beta \begin{pmatrix} x \\ y \end{pmatrix} = \begin{pmatrix} y \\ x \end{pmatrix}$.

2.2.5 Matrices for homomorphisms

Let M and N be free right R-modules with finite bases $F = \{f_1, \ldots, f_s\}$ and $G = \{g_1, \ldots, g_t\}$ respectively. Given an R-module homomorphism $\alpha : M \to N$, we can write

$$\alpha f_k = g_1 a_{1k} + \cdots + g_t a_{tk} \text{ for } k = 1, \ldots, s,$$

with unique coefficients a_{jk}, $j = 1, \ldots, t$, $k = 1, \ldots, s$, and so associate with α a $t \times s$ matrix $A_{GF}(\alpha) = (a_{jk})$.

Conversely, given such a matrix (a_{jk}), we can define a homomorphism from M to N by the above formula. Thus we have a bijection

$$A_{GF} : \mathrm{Hom}_R(M, N) \longrightarrow M_{t,s}(R)$$

between the set of R-module homomorphisms from M to N and the set $M_{t,s}(R)$ of $t \times s$ matrices with coefficients in R.

In the case that M and N are the standard free modules R^s and R^t respectively, a matrix A in $M_{t,s}(R)$ gives an R-module homomorphism from R^s to R^t by left multiplication: $m \mapsto Am$. If F and G are taken to be the respective standard bases, then A is evidently its own matrix, that is, A_{GF} is the identity map on $M_{t,s}(R)$.

The basic properties of the map A_{GF} are given in the next result.

2.2.6 Theorem

Let M, N and P be finitely generated free right R-modules and take bases F of M, G of N and H of P.

(i) *For any R-module homomorphisms $\alpha, \alpha' : M \to N$,*

$$A_{GF}(\alpha + \alpha') = A_{GF}(\alpha) + A_{GF}(\alpha').$$

(ii) *If $\beta : N \to P$ is also an R-module homomorphism,*

$$A_{HF}(\beta\alpha) = A_{HG}(\beta)A_{GF}(\alpha).$$

(iii) *Let s be the number of elements in F. Then $A_{FF}(id_M) = I_s$, the $s \times s$ identity matrix.*

(iv) *Let $\sigma : M \to N$ be an isomorphism. Then the image σF of F is a basis of N, and*

$$A_{G,\sigma F}(id_N) = A_{GF}(\sigma) = A_{\sigma^{-1}G,F}(id_M).$$

(v) *For any $m \in M$,*

$$A_{GF}(\alpha)_F(m) = {}_G(\alpha m).$$

Proof

The verifications are routine checking. Part (i) is immediate from the definition of the sum of homomorphisms: $(\alpha + \alpha')(m) = \alpha(m) + \alpha'(m)$ for m in M. We give the argument for part (ii) in some detail as it illustrates the purpose behind our fundamental convention of writing homomorphisms opposite scalars (1.2.4). Write $F = \{f_1, \ldots, f_s\}$, $G = \{g_1, \ldots, g_t\}$ and $H = \{h_1, \ldots, h_u\}$.

Put $A_{GF}(\alpha) = A = (a_{jk})$ and $A_{HG}(\beta) = B = (b_{ij})$, so that $\alpha f_k = \sum_{j=1}^{t} g_j a_{jk}$ for $k = 1, \ldots, s$ and $\beta g_j = \sum_{i=1}^{u} h_i b_{ij}$ for $j = 1, \ldots, t$.

Then, for any k,

$$(\beta \alpha) f_k = \beta(\alpha f_k) = \beta(\sum_{j=1}^{t} g_j a_{jk}) = \sum_{j=1}^{t}(\beta g_j) a_{jk} = \sum_{i=1}^{u} h_i (\sum_{j=1}^{t} b_{ij} a_{jk}),$$

which proves the assertion.

Next, (iii) and (iv) follow readily from the definition in (2.2.5).

Finally, (v) may be proved, like (ii), by brute force. A more sophisticated method is to observe that an element m of M defines an R-module homomorphism $\gamma_m : R^1 \to M$ by $\gamma_m(1) = m$. Then $_F(m) = A_{FE}(\gamma_m)$, where $E = \{1\}$ is the standard basis of R^1, and $_G(\alpha m) = A_{GE}(\gamma_{\alpha m})$, so the assertion follows from (ii) and the identity $\alpha \gamma_m = \gamma_{\alpha m}$. $\qquad \qquad \square$

The above theory has two important special cases in which we take $M = N$. In the first, we consider $A_{GF}(\alpha)$ when α is the identity map id_M and F and G can be arbitrary, while, in the second, we allow α to be arbitrary but set $G = F$.

2.2.7 Change of basis

We wish to consider the transitions between the various bases of a fixed free (nonzero) right R-module M. Since we need to work with invertible matrices which are not necessarily square, we remind the reader that a $t \times s$ matrix $\Gamma = (\gamma_{ij})$ is a *twosided invertible matrix* if there is an $s \times t$ matrix Θ with $\Theta\Gamma = I_s$ and $\Gamma\Theta = I_t$, where I_s and I_t are the identity matrices of the appropriate sizes. The matrix Θ is then unique and we write $\Theta = \Gamma^{-1}$ as expected.

Two bases $F = \{f_1, \ldots, f_s\}$ and $G = \{g_1, \ldots, g_t\}$ of M determine a $t \times s$ matrix $A_{GF}(id_M)$ and an $s \times t$ matrix $A_{FG}(id_M)$. From (ii) and (iii) of the theorem above, we have

$$A_{GF}(id_M) A_{FG}(id_M) = A_{GG}(id_M) = I_t$$

and

$$A_{FG}(id_M)A_{GF}(id_M) = A_{FF}(id_M) = I_s.$$

These equations show $\Gamma = A_{GF}(id_M)$ to be a twosided invertible matrix, with inverse $A_{FG}(id_M)$. Each is known as a *transition matrix*.

On the other hand, suppose that we have just one basis $G = \{g_1, \ldots, g_t\}$ of M and a twosided invertible $t \times s$ matrix $\Gamma = (\gamma_{ij})$. Then we can construct a basis $F = \{f_1, \ldots, f_s\}$ of M by writing

$$f_j = \sum_{i=1}^{t} g_i \gamma_{ij} \text{ for } j = 1, \ldots, s.$$

To see that F is in fact a basis, we note that the coordinate vector of f_j with respect to G is $_G(f_j) = \Gamma e_j$, the jth column of Γ. Then the homomorphism $\beta : M \to R^s$ given by $\beta(m) = \Gamma^{-1}(_G(m))$ clearly sends each f_j to the standard basis element e_j, so F is a basis by (iv) of (2.2.4).

We summarize as follows.

2.2.8 Theorem

Let M be a free right R-module with basis $G = \{g_1, \ldots, g_t\}$. Then there is a bijective correspondence between

(i) *bases of M of the form $F = \{f_1, \ldots, f_s\}$*
     and
(ii) *twosided invertible $t \times s$ matrices $\Gamma = (\gamma_{ij})$,*

given by the formulas $\Gamma = A_{GF}(id_M)$ and

$$f_j = \sum_{i=1}^{t} g_i \gamma_{ij} \text{ for } j = 1, \ldots, s. \qquad \square$$

2.2.9 Matrices of endomorphisms

The other important special case of the preceding discussion concerns the endomorphisms of a free module M. We know that the set $\text{End}(M_R)$ of right R-module endomorphisms of M is a ring, and that M is an $\text{End}(M_R)$-R-bimodule (1.2.5). Choose a basis $F = \{f_1, \ldots, f_s\}$ of M, and define

$$\mu_F = A_{FF} : \text{End}(M) \to M_s(R),$$

so that $\mu_F(\alpha) = (a_{jk})$ with

$$\alpha f_k = f_1 a_{1k} + \cdots + f_s a_{sk} \text{ for } k = 1, \ldots, s.$$

By (2.2.5), μ_F is a bijection, and Theorem 2.2.6 shows that it is a ring homomorphism, hence an isomorphism of rings.

If G is another basis of M, with say t elements, then $\mu_G(\alpha) = \Gamma \mu_F(\alpha) \Gamma^{-1}$ for the invertible $t \times s$ matrix $\Gamma = A_{GF}(id_M)$; that is, the matrices representing α are *conjugate*. We sometimes say that the maps μ_G and μ_F are conjugate, as Γ does not depend on the choice of α.

An instructive elementary example is given by a division ring $\mathcal{D}$, considered as a rank one module over itself. Take g to be be some element not in the centre of $\mathcal{D}$ and put $G = \{g\}$, and take $F = \{1\}$, the identity element of $\mathcal{D}$. Both G and F are one-element bases of $\mathcal{D}$. The matrix $\Gamma = A_{GF}(id_{\mathcal{D}})$ is simply the 1×1 matrix (g^{-1}).

Each element d of $\mathcal{D}$ can be viewed as a right $\mathcal{D}$-module endomorphism of $\mathcal{D}$ by left multiplication: $x \mapsto dx$ for $x \in \mathcal{D}$. Since $d1 = 1d$ always, we have $\mu_F(d) = d$. However, $dg = g \cdot g^{-1} dg$, so that $\mu_G(d) = g^{-1} dg$.

2.2.10 Normal forms of matrices

Two matrices A, B are said to be *equivalent* or *associated* if there are invertible matrices Γ, Θ so that $\Gamma A \Theta = B$. Evidently, equivalence is an equivalence relation on the set of matrices over R. If we can single out a distinguished representative of the equivalence class of A, then we have a *normal* or *canonical* form for A. This canonical form should have useful properties – for example, it should be a diagonal matrix if possible. The reader will probably be familiar with some canonical forms for a matrix over a field (see Exercise 2.2.1 below), and later in this text we obtain canonical forms for matrices over Euclidean domains (3.3.2).

Equivalence of matrices has a nice interpretation in terms of homomorphisms. We say that a matrix A represents a homomorphism $\alpha : M \to N$ of free R-modules if there are bases F of M and G of N such that $A = A_{GF}(\alpha)$. A fixed homomorphism may be represented by many matrices, according to the choice of bases, and likewise, a given matrix may represent many homomorphisms. The connection between homomorphisms and their representing matrices is as follows.

2.2.11 Proposition

Let A and B be matrices over a ring R, and let M and N be free right R-modules. Suppose also that A represents the homomorphism $\alpha : M \to N$.

Then A is equivalent to B if and only if B also represents α.

Proof

We call on the correspondence between invertible matrices and bases given in (2.2.8). Suppose that B represents α. Then there are bases F, $\overline{F}$ of M and G, $\overline{G}$ of N with $A = A_{GF}(\alpha)$ and $B = A_{\overline{GF}}(\alpha)$. Applying part (ii) of Theorem 2.2.6 to the product $id_N \cdot \alpha \cdot id_M = \alpha$ gives

$$A_{\overline{GF}}(\alpha) = A_{\overline{G}G}(id_N)A_{GF}(\alpha)A_{F\overline{F}}(id_M),$$

and the matrices $A_{\overline{G}G}(id_N)$ and $A_{F\overline{F}}(id_M)$ are invertible.

Conversely, if $\Gamma A \Theta = B$ for invertible Γ, Θ, then there are bases $\overline{G}$ and $\overline{F}$ with $\Gamma = A_{\overline{G}G}(id_N)$ and $\Theta = A_{F\overline{F}}(id_M)$, from which $B = A_{\overline{GF}}(\alpha)$. $\square$

Thus a canonical form for any matrix which represents a homomorphism α also serves as a canonical form for α itself. Under favourable circumstances it affords a good description of Ker α, Im α, and related modules. This approach is exploited in section 3.3 to derive the structure theory for modules over a Euclidean domain.

Another canonical form problem arises through conjugacy: recall that two square matrices A, B are conjugate (or *similar*) if there is an invertible matrix Γ with $B = \Gamma A \Gamma^{-1}$. By an obvious modification of the preceding result, A and B are conjugate precisely when there are an endomorphism α of a free module M and bases F, G of M with $A = \mu_F(\alpha)$ and $B = \mu_G(\alpha)$. The problem of determining a canonical form under conjugacy is much harder than that for equivalence. We indicate how a solution for matrices over fields follows from the structure theory of modules over polynomial rings in a series of exercises in section 3.3.

2.2.12 Scalar matrices and endomorphisms

There are some points about the realization of an element of the coefficient ring R as a scalar matrix in $M_s(R)$ or as an endomorphism of a free module M which are obscured when R is commutative, but need some care when R is not commutative.

A *standard scalar matrix* in $M_s(R)$ is any diagonal matrix

$$\iota(r) = \begin{pmatrix} r & 0 & 0 & \cdots & 0 \\ 0 & r & 0 & \cdots & 0 \\ 0 & 0 & r & \cdots & 0 \\ \vdots & \vdots & \vdots & \ddots & \vdots \\ 0 & 0 & 0 & \cdots & r \end{pmatrix}$$

with constant diagonal term. The map ι is called the *standard embedding*, and it is clear that ι is an injective ring homomorphism whose image is the subring $\iota(R)$ of $M_s(R)$ consisting of all the standard scalar matrices.

Using ι, we define an action of R on $M_s(R)$ by the rules

$$rA = \iota(r)A \text{ and } Ar = A\iota(r) \text{ for } A \in M_s(R),$$

so that $M_s(R)$ becomes an R-R-bimodule. (In effect, we are using 'restriction of scalars along ι' as in (1.2.14).) With this action, we have the familiar form of a scalar matrix:

$$\iota(r) = rI_s = I_s r,$$

where I_s is the $s \times s$ identity matrix.

Next, recall that the set of standard matrix units for $M_s(R)$ (Exercise 1.1.4) is the set $\{e_{hi} \mid 1 \le h, i \le s\}$ where for each h and i, e_{hi} is the matrix with (h, i) th entry 1 and all other entries 0, and that these matrices satisfy the relations

$$e_{hi}e_{jk} = \begin{cases} e_{hk} & i = j \\ 0 & i \ne j \end{cases}$$

and

$$I_s = e_{11} + \cdots + e_{ss}.$$

Thus we obtain the usual form $A = \sum_{i,j} e_{ij}a_{ij}$ of a matrix $A = (a_{ij}), a_{ij} \in R$, where $e_{ij}a_{ij}$ is the matrix with entry a_{ij} in the (i, j) th place and all other entries 0.

In the case that R is commutative, direct calculation shows that $\iota(R) = Z(M_s(R))$, the centre of the matrix ring, and so the set of scalar matrices is an invariant of the matrix ring. On the other hand, when R is not commutative, we find only that $\iota(R)$ is the *centralizer* of the set $\{e_{ij}\}$, that is,

$$\iota(R) = \{A \in M_s(R) \mid Ae_{ij} = e_{ij}A \text{ for all } i, j\}.$$

(The centre of $M_s(R)$ is $\iota(Z(R))$.)

For an arbitrary ring T, a *set of matrix units* for T is any collection of elements $\{\phi_{hi} \mid h, i = 1, \ldots, u\}$ of T satisfying relations which are analogous to those holding for the standard matrix units, that is,

$$\phi_{hi}\phi_{jk} = \begin{cases} \phi_{hk} & i = j \\ 0 & i \ne j \end{cases}$$

and

$$1_T = \phi_{11} + \cdots + \phi_{uu}.$$

As we indicate in Exercise 2.2.4 below, given a set of matrix units for T, it is then possible to describe T as a matrix ring $M_u(S)$, but even if $T = M_s(R)$ originally, it is possible that $u \neq s$ or $S \neq R$. A trivial example is provided by the set $\{1\}$, which is a set of matrix units. Less trivial examples are given in Exercises 2.2.5 and 2.3.3. An example in which R and S are nonisomorphic rings with $M_2(R) \cong M_2(S)$ is given in [Smith 1981].

We must also be careful when we consider the embedding of R in the endomorphism ring $\mathrm{End}(M)$ of a free right R-module M. Suppose that R is commutative and take any element r of R. The formula

$$\lambda(r)m = mr \text{ for } m \in M$$

defines an R-module homomorphism of M and the map λ is an injective ring homomorphism $\lambda : R \to \mathrm{End}(M)$. However, if R is not commutative, the formula for $\lambda(r)$ no longer defines an R-module homomorphism (unless r is in the centre of R) and we must proceed differently, as follows.

Choose a basis $F = \{f_1, \ldots, f_s\}$ of M, and for each element $r \in R$ define

$$\lambda_F(r)f_i = f_i r \text{ for } i = 1, \ldots, s.$$

Then $\lambda_F(r)$ extends to an R-module homomorphism and $\lambda_F : R \to \mathrm{End}(M)$ is the desired injective ring homomorphism.

Notice that the operators $\lambda_F(r)$ correspond to the standard scalar matrices under the isomorphism $\mu_F : \mathrm{End}(M) \to M_s(R)$ of (2.2.9), since $\mu_F \lambda_F(r) = \iota(r)$ always.

2.2.13 Infinite bases

The preceding discussion has a straightforward generalization when we allow free modules with infinite bases, which we review briefly.

Suppose that M and N are free right R-modules with bases $F = \{f_\lambda \mid \lambda \in \Lambda\}$ and $G = \{g_\sigma \mid \sigma \in \Sigma\}$ respectively, where the index sets Λ and Σ are ordered sets which may be infinite. Suppose also that $\alpha : M \to N$ is an R-module homomorphism. We write

$$\alpha f_\lambda = \sum_{\sigma \in \Sigma} g_\sigma a_{\sigma\lambda}, \qquad \lambda \in \Lambda,$$

with unique coefficients $a_{\sigma\lambda}$, $\sigma \in \Sigma$, $\lambda \in \Lambda$, where for each λ only finitely many coefficients are nonzero. We thus associate with α a $\Sigma \times \Lambda$ matrix $A_{GF}(\alpha) = (a_{\sigma\lambda})$ which is *column-finite*, that is, there are only finitely many nonzero entries in each of its columns.

Conversely, given such a matrix $(a_{\sigma\lambda})$, we can define a homomorphism from

M to N by the above formula. Thus we have a bijection

$$A_{GF} : \operatorname{Hom}_R(M, N) \to M_{\Sigma, \Lambda}^{cf}(R)$$

between the set of R-module homomorphisms from M to N and the set $M_{\Sigma, \Lambda}^{cf}(R)$ of $\Sigma \times \Lambda$ column-finite matrices with coefficients in R. The formal properties of the map A_{GF} are exactly the same as those listed for finite bases in (2.2.6).

Again, there are two cases of special interest when $M = N$. First, taking $\alpha = id_M$, we obtain the transition matrices $A_{GF}(id_M) \in M_{\Sigma, \Lambda}^{cf}(R)$ and $A_{FG}(id_N) \in M_{\Lambda, \Sigma}^{cf}(R)$. These two matrices are mutually inverse. Further, given a fixed basis G of M (indexed by Σ), there is a bijective correspondence between invertible matrices in $M_{\Sigma, \Lambda}^{cf}(R)$ and bases of M indexed by Λ, as in (2.2.8). Second, taking $G = F$, we note that, for each basis F of M, there is a ring isomorphism $\mu_F = A_{FF} : \operatorname{End}(M_R) \to M_\Lambda^{cf}(R)$ between the endomorphism ring of M and the ring $M_\Lambda^{cf}(R)$ of $\Lambda \times \Lambda$ column-finite matrices over R.

2.2.14 Free left modules

Since we use free left modules and their endomorphisms extensively when we discuss the Morita theory in [BK: CM], we give a fairly detailed summary of the way in which our definitions and results for right modules are transcribed for left modules. We concentrate on modules of finite rank.

We view the standard free left R-module ${}^s R$ of finite rank s as the 'row-space' of $1 \times s$ matrices over R. The standard basis of ${}^s R$ is $\{e_i\}$ where $e_i = (0, \ldots, 0, 1, 0, \ldots, 0)$, the nonzero term being in the ith place.

If $F = \{f_1, \ldots, f_s\}$ is a basis of a free left R-module M, then an element m in M can be written as

$$m = r_1 f_1 + \cdots + r_s f_s$$

with unique coefficients. The vector

$$(m)^F = (r_1, \ldots, r_s) \in {}^s R$$

is the coordinate vector of m, and by taking coordinates with respect to the basis F, we obtain an isomorphism $M \cong {}^s R$.

Clearly, there is a bijection between bases for the free module M and isomorphisms $\theta : M \cong {}^s R$ (s not necessarily constant) with $\theta(f_i) = e_i$ for each i.

Now let M and N be free left R-modules, with bases $F = \{f_1, \ldots, f_s\}$ and $G = \{g_1, \ldots, g_t\}$ respectively, and let $\alpha : M \to N$ be an R-module homomorphism. We can represent α by an $s \times t$ matrix $A^{FG}(\alpha) = (a_{ij})$ whose entries are defined by the equations

$$f_i\alpha = \sum_{j=1}^{t} a_{ij}g_j \text{ for } i = 1, \ldots, s.$$

The properties of the maps A^{FG} are parallel to those of the maps A_{GF} for right modules listed in (2.2.6), (2.2.8) and (2.2.11), but with some notational changes because homomorphisms of left modules are written on the right, and compose accordingly. For an important example, we highlight the left-handed version of the multiplication formula given in (ii) of (2.2.6).

Let M, N and P be free left R-modules with bases F, G, and H respectively, and let $\alpha : M \to N$ and $\beta : N \to P$ be R-module homomorphisms. Then

$$A^{FH}(\alpha\beta) = A^{FG}(\alpha)A^{GH}(\beta).$$

(Recall that the the composite $\alpha\beta$ means 'first α, then β' when we are dealing with left modules.)

Each basis F of M gives rise to a ring isomorphism

$$\mu^F = A^{FF} : \text{End}(_RM) \to M_s(R).$$

It is straightforward to extend the above remarks to free left modules with infinite bases. The only significant point of difference is that, for a basis F of M indexed by Λ and a basis G of N indexed by Σ, the map A^{FG} on $\text{Hom}(M,N)$ takes values in the set $M_{\Lambda,\Sigma}^{rf}(R)$ of *row-finite* $\Lambda \times \Sigma$ matrices over R, that is, the matrices with only a finite number of nonzero entries in each row. In particular, the ring isomorphism $\mu^F : \text{End}(_RM) \cong M_{\Lambda}^{rf}(R)$ has values in the ring of row-finite $\Lambda \times \Lambda$ matrices over R.

Exercises

2.2.1   **A canonical form**

Let V and W be vector spaces over a field $\mathcal{K}$, of finite dimensions s and t respectively, and let $\alpha : V \to W$ be a linear transformation.

Choose a basis $\{g_1, \ldots, g_k\}$ of $\text{Im}\,\alpha$ and extend it to a basis $G = \{g_1, \ldots, g_t\}$ of W. Choose $\{f_1, \ldots, f_k\}$ in V with each $\alpha(f_i) = g_i$. Verify that $\{f_1, \ldots, f_k\}$ is linearly independent, and extend it to a basis $F = \{f_1, \ldots, f_s\}$ of V.

Show that $A_{GF}(\alpha) = \begin{pmatrix} I_k & 0 \\ 0 & 0 \end{pmatrix}$. Deduce that a $t \times s$ matrix A over

$\mathcal{K}$ is equivalent to exactly one matrix in the canonical form $\begin{pmatrix} I_k & 0 \\ 0 & 0 \end{pmatrix}$

(2.2.10).

Show also that there are direct decompositions of vector spaces $V = U \oplus \operatorname{Ker} \alpha$ and $W = \operatorname{Im} \alpha \oplus Y$ with $\dim U = k$ and $\dim Y = t - k$.

2.2.2   Let R be a ring and let M be a free right R-module, with finite basis $F = \{f_1, \ldots, f_m\}$. Let π be a permutation of $\{1, \ldots, m\}$ and write $\pi F = \{f_{\pi 1}, \ldots, f_{\pi m}\}$.

Show that the matrix $A_{\pi F, F}(id_M)$ is a *permutation matrix*, that is, it is obtained from the identity $m \times m$ matrix by a permutation of its rows (or, equally, columns).

2.2.3   **The quaternion algebra: a division ring**

Let $\mathcal{K}$ be a subfield of the real numbers $\mathbb{R}$, or more generally, a field in which there is no nonzero solution of the equation

$$x_0^2 + x_1^2 + x_2^2 + x_3^2 = 0.$$

The *quaternion algebra*, denoted $\mathbb{H}$ for *Hamiltonians* (see Exercise 3.2.5) is defined as a vector space of dimension 4 over $\mathcal{K}$, with basis elements $1, i, j, k$. Multiplication is given by the rules that 1 acts as the identity and i, j, k multiply as follows:

$$i^2 = j^2 = k^2 = -1, ij = -ji = k,$$

these rules being extended to all of $\mathbb{H}$ by distributivity.

Then $\mathbb{H}$ is a ring. If $x = 1x_0 + ix_1 + j_2 j + kx_3$ is in $\mathbb{H}$, its *conjugate* is $\bar{x} = 1x_0 - ix_1 - jx_2 - kx_3$.

Verify that, for all $x, y \in \mathbb{H}$,

$$\overline{x + y} = \bar{x} + \bar{y}$$

and

$$\overline{xy} = \bar{y} \cdot \bar{x}.$$

(Thus conjugation is an anti-automorphism of $\mathbb{H}$.)

Put $N(x) = x \cdot \bar{x}$. Show further that

$$N(xy) = N(x)N(y)$$

and that

$$N(x) = x_0^2 + x_1^2 + x_2^2 + x_3^2 \neq 0 \text{ if } x \neq 0.$$

Deduce that $\mathbb{H}$ is a division ring.

Show that the centre of $\mathbb{H}$ is $\mathcal{K}$.

2.2.4   **Characterizing matrix rings**

Given a ring R, the matrix ring $M_s(R)$ contains a set of standard matrix units, and the ring of scalar matrices in $M_s(R)$ can be obtained as the centralizer of this set (2.2.12). The point of this exercise is to show that if a ring contains a set of matrix units, then it must be isomorphic to some $M_s(R)$.

Let T be a ring and suppose that T contains a set of matrix units $\{\phi_{hi} \mid h, i = 1, \ldots, s\}$ as in (2.2.12). Put

$$R = \{\sum_{i=1}^{s} \phi_{i1} t \phi_{1i} \mid t \in T\}.$$

Verify that R is a subring of T and that R centralizes the given matrix units.

Given $a \in T$, put

$$a_{jk} = \sum_{i=1}^{s} \phi_{ij} a \phi_{ki} \quad \text{for } j, k = 1, \ldots, s.$$

Show that each a_{jk} is in R, and that $a_{jk}\phi_{jk} = \phi_{jj}a\phi_{kk}$ always.

As in (2.2.9), define a map $\mu : T \to M_s(R)$ by $\mu(a) = (a_{ij})$. Show that μ is a ring isomorphism which maps the matrix unit ϕ_{hi} to the corresponding standard matrix unit e_{hi} and elements of R to standard scalar matrices.

2.2.5   **Scalars spoilt**

Let $\mathcal{D}$ be a (noncommutative) division ring, and let $\Gamma = \begin{pmatrix} a & b \\ 0 & c \end{pmatrix}$ with a, c both nonzero. Show that

$$\Gamma^{-1} = \begin{pmatrix} a^{-1} & -a^{-1}bc^{-1} \\ 0 & c^{-1} \end{pmatrix}.$$

Find a standard scalar matrix rI such that $\Gamma(rI)\Gamma^{-1}$ is not standard scalar. Thus, a matrix which is scalar with respect to the standard matrix units $\{e_{ij} \mid i, j = 1, 2\}$ need not be scalar with respect to the matrix units $\{\Gamma e_{ij}\Gamma^{-1}\}$.

2.2.6   Let M be a free right R-module with basis $F = \{f_1, \ldots, f_s\}$, and let $\theta : M \to R^s$ be the isomorphism such that $\theta f_i = e_i$ for all i. Show that for all $\alpha \in \text{End}(M)$, $m \in M$ and $r \in R$, we have $\mu_F(\alpha)\theta(m)r = \theta(\alpha m r)$. (Thus, '$\theta$ respects the bimodule structures'.)

2.2.7 Transposes and opposites

The use of the opposite ring to switch between left and right modules, as in (1.2.6) and Exercise 1.2.14, gives a method of deriving the formulas for left modules in (2.2.14) from those already obtained for right modules.

Recall that a right R-module M becomes a left module M° over the opposite ring R° by $r^\circ m^\circ = (mr)^\circ$.

Suppose that M is a free right R-module with basis $F = \{f_1, \ldots, f_s\}$. Show that M° is a free left R°-module with basis $F^\circ = \{f_1^\circ, \ldots, f_s^\circ\}$. Verify that $(R^s)^\circ \cong {}^s(R^\circ)$.

Recall from Exercise 1.2.14 that if $A = (a_{ij})$ is a $u \times v$ matrix over R, then the transpose A^t of A is the $v \times u$ matrix over the opposite ring R° with (j, i) th entry a_{ij}°.

Let M and N be free right R-modules with bases F and G respectively, and recall that for an R-module homomorphism $\alpha : M \to N$, the left R°-module homomorphism $\alpha^\circ : M^\circ \to N^\circ$ is given by $m^\circ \alpha^\circ = (\alpha m)^\circ$. Verify that $A^{F^\circ G^\circ}(\alpha^\circ) = (A_{GF}(\alpha))^t$.

Show also that, for an infinite index set, taking the transpose of a matrix defines an isomorphism $M_\Lambda^{cf}(R) \cong M_\Lambda^{rf}(R^\circ)$.

2.3 INVARIANT BASIS NUMBER

In our discussion of bases of free modules in the preceding section, we allowed the possibility that a free module may have finite bases of different sizes. We now give some examples of rings for which this phenomenon occurs, and some results which show when it cannot happen.

A ring R is said to have *invariant basis number* if, whenever there is an isomorphism $R^m \cong R^n$ between (standard) free modules of finite rank, then $m = n$; this means that the rank of a free module is invariant under change of basis.

A ring is called an *IBN* ring if it has invariant basis number, and a *non-IBN* ring otherwise.

Two basic results, Theorems 2.3.7 and 3.1.9, show that very many of the rings one encounters are IBN rings. However, some important examples of rings with trivial K-theory are non-IBN rings.

2.3.1 Some non-IBN rings

Let $\omega = \{1, 2, \ldots\}$ be the set of natural numbers with their familiar ordering, and let $M_\omega^{cf}(R)$ be the ring of all column-finite $\omega \times \omega$ matrices over some ring R. The discussion in (2.2.13) shows that we can identify $M_\omega^{cf}(R)$ as the endomorphism ring of the free right R-module R^ω on the set ω.

We define the *cone* of R to be the subring CR of $M_\omega^{cf}(R)$ consisting of all the matrices which are both row- and column-finite, that is, each row or column has only a finite number of nonzero entries. (Such matrices are sometimes called *locally finite*.)

Neither $M_\omega^{cf}(R)$ nor CR has invariant basis number.

2.3.2 Theorem

(ii) $M_\omega^{cf}(R) \cong (M_\omega^{cf}(R))^2$ as a right $M_\omega^{cf}(R)$-module.

(i) $CR \cong (CR)^2$ as a right CR-module.

Proof

We prove (i) only, (ii) being similar. For each index h, let P_h be the set $\{(p_{hk}) \mid k = 1, 2, \ldots\}$ of all possible hth rows of a member of $M_\omega^{cf}(R)$; as a left R-module, P_h is the direct product $\prod_\omega R$ of countably many copies of R. Right multiplication makes P_h a right $M_\omega^{cf}(R)$-module, and $M_\omega^{cf}(R)$ itself is a right $M_\omega^{cf}(R)$-submodule of the direct product $\prod_\omega P_h$. The key to the proof is the fact that $\prod_\omega R \cong \prod_\omega R \oplus \prod_\omega R$.

Write $(M_\omega^{cf}(R))^2 = (M_\omega^{cf}(R))^{(1)} \oplus (M_\omega^{cf}(R))^{(0)}$, where the bracketed superscripts are used to label copies of the ring (considered as a right $M_\omega^{cf}(R)$-submodule). We define an $M_\omega^{cf}(R)$-module homomorphism α from $M_\omega^{cf}(R)$ to $(M_\omega^{cf}(R))^2$ by sending P_{2i-j} to $(P_i)^{(j)}$ for $i \in \omega$ and $j = 0, 1$ in the obvious way: thus the row $(p_{2i-j,k})$, $k = 1, 2, \ldots$, is sent to the row $((p_{i,k}))^{(j)}$ in $(M_\omega^{cf}(R))^{(j)}$. It is clear that α is bijective. It is also an $M_\omega^{cf}(R)$-module homomorphism because $M_\omega^{cf}(R)$ acts on the right, and hence on the rows of the various copies of $M_\omega^{cf}(R)$. $\qquad\square$

2.3.3 Two non-square invertible matrices

We note from (2.2.5) that the action of the isomorphism α in the preceding theorem must be given by an invertible 2×1 matrix A over $M_\omega^{cf}(R)$, which is easily specified. Let

$$B = \begin{pmatrix} 1 & 0 & 0 & 0 & 0 & 0 & \cdots \\ 0 & 0 & 1 & 0 & 0 & 0 & \cdots \\ 0 & 0 & 0 & 0 & 1 & 0 & \cdots \\ \multicolumn{7}{c}{\cdots\cdots\cdots\cdots\cdots\cdots} \end{pmatrix} \quad \text{and} \quad C = \begin{pmatrix} 0 & 1 & 0 & 0 & 0 & 0 & \cdots \\ 0 & 0 & 0 & 1 & 0 & 0 & \cdots \\ 0 & 0 & 0 & 0 & 0 & 1 & \cdots \\ \multicolumn{7}{c}{\cdots\cdots\cdots\cdots\cdots\cdots} \end{pmatrix};$$

these are elements of $M_\omega^{cf}(R)$. Then $A = \begin{pmatrix} B \\ C \end{pmatrix}$, with inverse the transpose matrix $(B^t \; C^t)$.

2.3.4 The type

To describe some more examples of rings without invariant basis number, we introduce the type of a ring. The ring R has *type* (w, d), where w and d are in $\mathbb{N}$, if the following conditions hold:

(T1) $R^w \cong R^{w+d}$,
(T2) if $R^s \cong R^{s+c}$, then $s \geq w$, with either $s > w$ or $c \geq d$.

Thus (w, d) is the minimum pair, under the usual lexicographic ordering, such that (T1) holds. A ring with invariant basis number will be allocated the type $(\infty, 0)$. A more sophisticated view of types is given in Exercise 2.3.2 below.

Examples of rings of all possible types were given in [Leavitt 1957] and in [Leavitt 1962]; there is a nice discussion in [Cohn 1966]. We outline the construction and state the results; the proofs lie outside the domain of this text.

Let $\mathcal{K}$ be a field and let $\mathcal{K}\langle X_{ij}, Y_{uv} \rangle$ be the noncommutative polynomial ring on $2hk$ variables

$$\{X_{ij} \mid i = 1, \ldots, h, \; j = 1, \ldots, k\} \quad \text{and} \quad \{Y_{uv} \mid u = 1, \ldots, k, \; v = 1, \ldots, h\},$$

with $h < k$. Further, let $\Xi = (X_{ij})$ and $\Upsilon = (Y_{uv})$ be the matrices, $h \times k$ and $k \times h$ respectively, with these variables as entries, and let $\mathfrak{a}$ be the twosided ideal of $\mathcal{K}\langle X_{ij}, Y_{uv} \rangle$ generated by the entries of the matrices $\Xi \Upsilon - I_h$ and $\Upsilon \Xi - I_k$. We define the *Leavitt ring* of type $(h, k - h)$ to be $L_{h,k} = \mathcal{K}\langle X, Y \rangle / \mathfrak{a}$.

It is clear from (2.2.5) together with part (iv) of Proposition 2.2.4 that $(L_{h,k})^h \cong (L_{h,k})^k$.

2.3.5 Theorem
The ring $L_{h,k}$ has type $(h, k - h)$.   ◯

In the other direction, we give a criterion for the invariance of the basis number.

2.3.6 Lemma

Let $f : R \to S$ be a ring homomorphism and suppose that S has invariant basis number. Then so also does R.

Proof

Suppose that $R^m \cong R^n$. Then by Theorem 2.2.8 there is an $m \times n$ invertible matrix $\Gamma = (\gamma_{jk})$ over R that gives the corresponding change of basis. But then $f\Gamma = (f\gamma_{jk})$ is a matrix over S which gives an isomorphism between S^m and S^n.   □

2.3.7 Theorem

Suppose that R is a (nonzero) commutative ring. Then R has invariant basis number.

Proof

By Corollary 1.2.22, R has a maximal (proper) ideal $\mathfrak{m}$. The residue ring $R/\mathfrak{m}$ is a simple commutative ring, that is, a field (Exercise 1.1.3). By elementary linear algebra, fields have invariant basis number, so the theorem follows from the preceding lemma.   □

Exercises

2.3.1   **Endomorphisms of infinite direct sums**

The extension of Exercise 2.2.6 to infinite index sets is not quite straightforward, and has led some authors into a natural error.

For simplicity, we consider only the direct sum M^ω of countably many copies of a single right R-module M. The general case is given in [Fuchs 1967], Theorem 55.1, on which this account is based.

Let $A = (a_{ij})$ be an $\omega \times \omega$ matrix with entries in the ring $E = \mathrm{End}(M_R)$. Then A is said to be *column-convergent* if for each index $j = 1, 2, \ldots$ the following condition holds:

for each element m of M, $a_{ij}m = 0$ for almost all i.

This means that each column sum $\sum_i a_{ij}$ is a well-defined element of E. Let $M_\omega^{cc}(E)$ denote the set of all column-convergent $\omega \times \omega$ matrices over E.

(i)   Verify that $M_\omega^{cc}(E)$ is a ring, that $M_\omega^{cf}(E)$ is subring of $M_\omega^{cc}(E)$, and that $M_\omega^{cc}(E) \cong \mathrm{End}(M^\omega)$.

(ii) In the case that $M = R$, verify that $M_\omega^{cc}(R) = M_\omega^{cf}(R)$.

(iii) Take $M = R^\omega$, so that $E = M_\omega^{cf}(R)$. Show that $M_\omega^{cf}(E)$ is a proper subring of $M_\omega^{cc}(E)$.

Hint. Consider the matrix A (over E) with $(i,1)$th entry the standard matrix unit $e_{1,i}$ (over R) for $i = 1, 2, \ldots$, and all other entries 0.

(iv) Show that $(R^\omega)^\omega \cong R^{\omega \times \omega}$ as right R-module, for a suitable ordering of $\omega \times \omega$. Choose any (non-order-preserving) bijection $\gamma : \omega \to \omega \times \omega$. Show that γ induces a module homomorphism $R^\omega \to R^{\omega \times \omega}$ and hence a ring isomorphism

$$M_\omega^{cf}(R) \cong M_\omega^{cc}(M_\omega^{cf}(R)).$$

(v) Verify that this ring isomorphism induces a ring isomorphism $C(R) \cong C(C(R))$, where CR is the cone of R. (See [Berrick 1982], pp. 11, 84.)

Remark. The alluring fallacy is that $M_\omega^{cf}(R) \cong M_\omega^{cf}(M_\omega^{cf}(R))$. [Camillo 1984] shows that this equality cannot hold in general – we consider this point again in Exercise 4.2.5. On the other hand, [Abrams 1987] constructs examples where the equality does hold.

2.3.2 **On the type of a ring** [Berrick & Keating 1997].

(a) We first look at a particular kind of equivalence relation on the set $\mathbb{N} = \{1, 2, \ldots\}$ of natural numbers.

An equivalence relation $\sim$ on $\mathbb{N}$ is a *congruence* if $a \sim b$ and $c \sim d$ together imply that $a + c \sim b + d$.

Notice that equality is such a relation.

(b) Show that if we replace $\mathbb{N}$ by $\mathbb{Z}$ in the definition above, a congruence on $\mathbb{Z}$ is simply congruence $\mathrm{mod}\, d$ in the usual sense, where the modulus d is the smallest natural number congruent to 0 if the congruence is not equality.

(c) Verify that a congruence $\sim$ on $\mathbb{N}$ extends to a congruence $\equiv$ on $\mathbb{Z}$ by the rule that $a \equiv b$ if there is a natural number x such that

$$a + x \sim b + x.$$

(d) Suppose that $\sim$ is a congruence relation on $\mathbb{N}$ which is not equality, and let d be the modulus of the induced congruence on $\mathbb{Z}$. Let w be the least $x \in \mathbb{N}$ with $x \sim x + d$. Show that for any $e \geq w$ the equivalence class of e contains $\{e, e + d, e + 2d, \ldots\}$.

(e) Let E be any equivalence class under $\sim$ which has more than one element. Show that E is infinite, and in particular contains numbers exceeding w. So deduce from (d) that, for any $u \in E$, $u \sim u + d$. Conclude that the equivalence classes under $\sim$ are

$$\{1\}, \ldots, \{w - 1\},$$
$$\{w, w + d, w + 2d, \ldots\},$$
$$\{w + 1, w + 1 + d, w + 1 + 2d, \ldots\},$$
$$\vdots$$
$$\{w + d - 1, w + 2d - 1, w + 3d - 1, \ldots\}.$$

The pair (w, d) is the *type* of the relation. Equality is allocated the type $(\infty, 0)$.

(f) Verify that a partition of $\mathbb{N}$ as above defines a congruence on $\mathbb{N}$.

(g) Let $\sim$ and $\approx$ be congruences on $\mathbb{N}$, with types (w, d) and (u, c) respectively. Say that $\sim$ is *finer* than $\approx$ if, whenever $a \sim b$, then $a \approx b$ also.

Show that each $\approx$-equivalence class can be written as a disjoint union of $\sim$-equivalence classes.

Prove that equality is the finest congruence.

In general, deduce that $\sim$ is finer than $\approx$ if and only if both $u \le w$ and c divides d.

(h) Let R be a ring. Define a relation $\sim = \sim_R$ on $\mathbb{N}$ by $a \sim b$ if and only if there is a right R-module isomorphism $R^a \cong R^b$. Show that this relation is a congruence on $\mathbb{N}$, and that its type (w, d) is the same as the type of the ring R.

(i) Let $f : R \to S$ be a ring homomorphism. Show that $\sim_R$ is finer than $\sim_S$ and obtain the relationship between the type of S and that of R.

Remark. Although the theory of congruences on $\mathbb{N}$ is elementary, it seems to be ignored in introductory texts that treat the number system. The classification of congruences on $\mathbb{N}$ is wellknown to semi-group theorists as a special case of the description of congruences on semigroups; see [Howie 1976]. The result (e) is also given in Lemma X.3.1 of [Cohn 1981], who provides an elementary proof in the midst of more sophisticated topics.

2.3.3 Verify that the cone CR of a ring R has type $(1, 1)$, and deduce that

(i) there is an isomorphism $CR \cong (CR)^s$ of right CR-modules for any $s > 1$,

(ii) there is a ring isomorphism $CR \cong M_s(CR)$ for any $s > 1$,

(iii) CR contains an $s \times s$ system of matrix units for any $s > 1$.

2.4 SHORT EXACT SEQUENCES

2.4 SHORT EXACT SEQUENCES

Short exact sequences play a fundamental role in K-theory, and indeed in the theory of modules in general. The existence of a short exact sequence

$$0 \longrightarrow M' \longrightarrow M \longrightarrow M'' \longrightarrow 0$$

connecting M', M and M'' tells us that, roughly speaking, any two of these modules determine the third (although not always completely). Sometimes, the modules M' and M'' are regarded as belonging to some class of 'basic' modules, the module M being a more sophisticated creature. In other cases, M has the more elementary structure, for example, it might be free, and we hope to analyse M'' in terms of the relationship between M' and M. Both these points of view are taken at one time or another.

This section is devoted to an examination of the fundamental properties of short exact sequences, particularly their splittings and their construction by push-outs and pull-backs.

We work with right modules over some fixed but arbitrary ring R. It will be clear that there is a parallel discussion for left modules; since the opposite ring R° is just as arbitrary as R, results for left modules can be obtained by means of the formal switching introduced in (1.2.6).

2.4.1 The definition

A sequence

$$M' \xrightarrow{\ \alpha\ } M \xrightarrow{\ \beta\ } M''$$

of homomorphisms of right R-modules is said to be *exact* (or *exact at M*) if $\operatorname{Ker}\beta = \operatorname{Im}\alpha$.

Thus, the homomorphism β is injective precisely when the sequence

$$0 \longrightarrow M \xrightarrow{\ \beta\ } M''$$

is exact, while α is surjective if and only if the sequence

$$M' \xrightarrow{\ \alpha\ } M \longrightarrow 0$$

is exact. Combining these examples, we see that a sequence of the form

$$0 \longrightarrow M_1 \xrightarrow{\ \alpha\ } M_2 \longrightarrow 0$$

is exact at both M_1 and M_2 if and only if α is both injective and surjective, that is, α is an isomorphism from M_1 to M_2.

More generally, a sequence

$$\cdots \longrightarrow M_{i-1} \xrightarrow{\ \alpha_{i-1}\ } M_i \xrightarrow{\ \alpha_i\ } M_{i+1} \longrightarrow \cdots$$

is called an *exact sequence* if it is exact at each of its *terms* M_i.

Such a sequence may have a finite number of terms or an infinite number of terms. In the latter case, the sequence is called a *long exact sequence*, although this expression is sometimes also used for a finite exact sequence with many terms.

The term 'exact sequence' first appears in the literature in [Kelley & Pitcher 1947]. The concept itself had earlier been introduced in [Hurewicz 1941] in describing the homology groups of pairs of topological spaces.

2.4.2 Four-term sequences

Next we see what information is carried by some four-term exact sequences.

If the sequence $0 \to M' \xrightarrow{\alpha} M \xrightarrow{\beta} M''$ is exact, then α is injective and $\operatorname{Im}\alpha = \operatorname{Ker}\beta$, so there is an isomorphism $M' \cong \operatorname{Ker}\beta$.

On the other hand, when $M' \xrightarrow{\alpha} M \xrightarrow{\beta} M'' \to 0$ is exact, the homomorphism β is surjective, and so, by the First Isomorphism Theorem (Exercise 1.2.2), $M'' \cong M/\operatorname{Im}\alpha$. It is useful to define the *cokernel* of α to be $\operatorname{Cok}\alpha = M/\operatorname{Im}\alpha$; we can then write the conclusion as $M'' \cong \operatorname{Cok}\alpha$.

2.4.3 Short exact sequences

We now make the formal definition of the most useful type of exact sequence.

A *short exact sequence* is a sequence

$$\mathbf{E} \qquad 0 \longrightarrow M' \xrightarrow{\ \alpha\ } M \xrightarrow{\ \beta\ } M'' \longrightarrow 0$$

which is exact at each of the terms M', M and M''.

The existence of such a sequence is sometimes expressed by the statements 'M is an *extension* of M'' by M'', or equally, 'M is an extension of M' by M'''. There is no consensus on this terminology! (An alternative is to describe M as an extension with kernel M' and image M''.)

Note that any homomorphism $\lambda : M \to N$ gives rise to two short exact sequences, namely

$$0 \longrightarrow \operatorname{Ker}\lambda \longrightarrow M \longrightarrow \operatorname{Im}\lambda \longrightarrow 0$$

and

$$0 \longrightarrow \operatorname{Im} \lambda \longrightarrow N \longrightarrow \operatorname{Cok} \lambda \longrightarrow 0.$$

Note also that, given the exact sequence $\mathbf{E}$, the homomorphism α induces an isomorphism from M' to $\operatorname{Ker} \beta$, while β induces an isomorphism from $\operatorname{Cok} \alpha$ to M''.

An exact sequence such as $\mathbf{E}$ above gives partial information about the module M in terms of M' and M'', but, save under special circumstances, M will not be completely determined simply by M' and M''; the homomorphisms also play a crucial role. Here are two elementary illustrations.

Take the ring of scalars to be the ring of integers $\mathbb{Z}$ and let both M' and M'' be $\mathbb{Z}/2\mathbb{Z}$. We can then form two essentially distinct short exact sequences

$$0 \longrightarrow \mathbb{Z}/2\mathbb{Z} \longrightarrow \mathbb{Z}/2\mathbb{Z} \oplus \mathbb{Z}/2\mathbb{Z} \longrightarrow \mathbb{Z}/2\mathbb{Z} \longrightarrow 0$$

and

$$0 \longrightarrow \mathbb{Z}/2\mathbb{Z} \overset{\alpha}{\longrightarrow} \mathbb{Z}/4\mathbb{Z} \overset{\beta}{\longrightarrow} \mathbb{Z}/2\mathbb{Z} \longrightarrow 0.$$

The first sequence is given by the direct sum, which we discuss at length in the next subsection, while in the second, α is induced by multiplication by 2. Note that the middle terms $\mathbb{Z}/2\mathbb{Z} \oplus \mathbb{Z}/2\mathbb{Z}$ and $\mathbb{Z}/4\mathbb{Z}$ are not isomorphic.

The general problem of classifying all short exact sequences with the terms M', M'' fixed is discussed briefly in (2.4.7).

2.4.4 Direct sums and splittings

Given a pair of R-modules M' and M'', there is at least one short exact sequence $\mathbf{E}$ in which they occur. It arises from their (external) direct sum $M = M' \oplus M''$. Recall that

$$M' \oplus M'' = \{(m', m'') \mid m' \in M', m'' \in M''\}$$

with

$$(m', m'') + (n', n'') = (m' + n', m'' + m'')$$

and

$$(m', n'') \cdot r = (m'r, m''r).$$

We have then an exact sequence, the *standard split exact sequence*

$$0 \longrightarrow M' \overset{\mu}{\longrightarrow} M \overset{\pi}{\longrightarrow} M'' \longrightarrow 0$$

with

$$\mu m' = (m', 0) \text{ and } \pi(m', m'') = m''.$$

The maps μ, π are members of the standard full set of inclusions and projections $\{\mu, \sigma; \rho, \pi\}$ for M (2.1.6), the other terms being

$$\sigma : M'' \to M, \text{ with } \sigma m'' = (0, m'')$$

and

$$\rho : M \to M', \text{ with } \rho(m', m'') = m'.$$

These maps satisfy the relations

$$\rho\mu = id_{M'}, \ \pi\sigma = id_{M''} \text{ and } \mu\rho + \sigma\pi = id_M.$$

Our first task is to give a method of recognizing those short exact sequences which are essentially minor variations of the standard split exact sequence.

A short exact sequence

$$0 \longrightarrow M' \overset{\alpha}{\longrightarrow} M \overset{\beta}{\longrightarrow} M'' \longrightarrow 0$$

is said to be a *split exact sequence* if we can find homomorphisms

$$\gamma : M \longrightarrow M' \text{ and } \delta : M'' \longrightarrow M$$

such that $\{\alpha, \delta; \gamma, \beta\}$ is a full set of inclusions and projections for M. We shall soon show that it is enough to know the existence of only one of the maps γ and δ, the other being constructible. Since this fact is very useful, some special terminology is introduced.

The short exact sequence above is said to *split at* M'', or *split over* β, if there is a homomorphism $\delta : M'' \to M$ with

$$\beta\delta = id_{M''},$$

and to *split at* M', or *split over* α, if there is a homomorphism $\gamma : M \to M'$ with

$$\gamma\alpha = id_{M'}.$$

Sometimes we say instead that β or α is split, or that δ splits β, and so forth.

Here is the promised result.

2.4.5 Theorem

Let

$$0 \longrightarrow M' \overset{\alpha}{\longrightarrow} M \overset{\beta}{\longrightarrow} M'' \longrightarrow 0$$

be a short exact sequence of R-modules. Then the following statements are equivalent.

(i) *The sequence is split at M'.*

(ii) *The sequence is split at M''.*

(iii) *There are homomorphisms $\delta : M'' \to M$ and $\gamma : M \to M'$ with $\beta\delta = id_{M''}$, $\gamma\alpha = id_{M'}$ and $\alpha\gamma + \delta\beta = id_M$, that is, $\{\alpha, \delta; \gamma, \beta\}$ is a full set of inclusions and projections for M.*

If these conditions hold, then $M \cong M' \oplus M''$, and

$$0 \longrightarrow M'' \overset{\delta}{\longrightarrow} M \overset{\gamma}{\longrightarrow} M' \longrightarrow 0$$

is also a split short exact sequence of R-modules.

Proof

Suppose that (i) is true, so that there exists γ with $\gamma\alpha = id_{M'}$. Given m'' in M'', by exactness at M'' there is an element m in M with $\beta m = m''$. Define $\delta m'' = m - \alpha\gamma m$. We have to verify that this is a valid definition. If $\beta m_1 = m''$ also, then $m_1 = m + \alpha m'$ for some m' in M', making

$$\alpha\gamma m_1 = \alpha\gamma m + \alpha\gamma\alpha m' = \alpha\gamma m + \alpha m';$$

thus

$$m_1 - \alpha\gamma m_1 = m - \alpha\gamma m.$$

It is now easy to check that δ is a homomorphism which splits β, as required for (ii), and that (iii) holds.

On the other hand suppose that δ is given. If m is in M, then $m - \delta\beta m$ lies in the kernel of β, whence, by exactness, there is a unique element m' of M' with $\alpha m' = m - \delta\beta m$. Put $\gamma m = m'$. Then γ is easily seen to be a splitting homomorphism for α, and both (i) and (iii) hold.

The fact that (iii) implies both (i) and (ii) is immediate. For the remaining assertion, define a map $\chi : M \to M' \oplus M''$ by $\chi m = (\gamma m, \beta m)$; then it is straightforward to see χ is an isomorphism. $\square$

2.4.6 Dual numbers

In general, an exact sequence of R-modules is not split, as we saw from the illustration in (2.4.3). Further examples are furnished by the *ring of dual numbers* $A[\epsilon]$. Here, A is an arbitrary coefficient ring, and

$$A[\epsilon] = \{a_0 + a_1\epsilon \mid a_0, a_1 \in A\},$$

with addition given by adding coefficients, and multiplication following from the rules $\epsilon^2 = 0$ and $a\epsilon = \epsilon a$ for all a in A.

The ideal $\epsilon A[\epsilon]$ is twosided and the residue ring $A[\epsilon]/\epsilon A[\epsilon]$ is evidently isomorphic to A. We can also view A as an $A[\epsilon]$-module with ϵ acting as 0 (by restriction of scalars, using the canonical surjection from $A[\epsilon]$ to A as in (1.2.14)). Thus there is a short exact sequence of $A[\epsilon]$-modules

$$0 \xrightarrow{} A \xrightarrow{\ \alpha\ } A[\epsilon] \xrightarrow{\ \beta\ } A \xrightarrow{} 0$$

in which α is multiplication by ϵ, that is, $a \mapsto \epsilon a$, and β sends ϵ to 0. If there were a splitting homomorphism δ of β, we would have both $\delta(1) = 1 + a'\epsilon$ for some $a' \in A$ and, since δ is an $A[\epsilon]$-homomorphism, $\epsilon\delta(1) = \delta(\epsilon \cdot 1) = 0$, a contradiction.

We note in passing that, if $\lambda : A[\epsilon] \to A[\epsilon]$ is left multiplication by ϵ, then the above sequence can be identified with both the short exact sequences

$$0 \xrightarrow{} \mathrm{Ker}\,\lambda \xrightarrow{} M \xrightarrow{} \mathrm{Im}\,\lambda \xrightarrow{} 0$$

and

$$0 \xrightarrow{} \mathrm{Im}\,\lambda \xrightarrow{} N \xrightarrow{} \mathrm{Cok}\,\lambda \xrightarrow{} 0$$

arising from λ.

2.4.7 The group Ext

Given two R-modules M' and M'', the possible extensions with kernel M' and image M'' can be described by an abelian group, $\mathrm{Ext}_R^1(M'', M')$. Some preliminary steps towards its construction are indicated in Exercise 2.4.9. The basic techniques used in this construction form our next topic.

2.4.8 Pull-backs and push-outs

The construction of a pull-back or a push-out is a very useful method of manufacturing new modules and new short exact sequences.

Suppose that we are given a *pull-back diagram* of right R-modules, that is, a diagram of the form

$$
\begin{array}{ccc}
 & & L'' \\
 & & \downarrow{\scriptstyle \theta} \\
M & \xrightarrow{\ \beta\ } & M''
\end{array}
$$

where β and θ are homomorphisms of R-modules.

The *pull-back* of M and L'' over M'' is the R-module

$$M \times_{M''} L'' = \{(m, \ell'') \in M \oplus L'' \mid \beta m = \theta \ell''\}.$$

An alternative name for this construction is the *fibre product*. It is customary to omit to mention the homomorphisms β and θ, despite the fact that they have a crucial influence on the structure of the pull-back.

There are homomorphisms $\overline{\beta} : M \times_{M''} L'' \to L''$ and $\overline{\theta} : M \times_{M''} L'' \to M$, given by projection to the corresponding factors, such that there is a commutative diagram

$$
\begin{array}{ccc}
M \times_{M''} L'' & \xrightarrow{\ \overline{\beta}\ } & L'' \\
\downarrow{\overline{\theta}} & & \downarrow{\theta} \\
M & \xrightarrow{\ \beta\ } & M''
\end{array}
$$

Such a diagram is referred to as a *pull-back square, fibre square* or *cartesian square*.†

A *push-out diagram* arises from the opposite corner of the square; thus it begins with a diagram of R-modules and R-module homomorphisms as follows:

$$
\begin{array}{ccc}
L' & \xrightarrow{\ \mu\ } & L \\
\downarrow{\phi} & & \\
M' & &
\end{array}
$$

The *push-out* (of M and L over L') is defined to be

$$M' \oplus_{L'} L = (M' \oplus L)/\{(\phi\ell', -\mu\ell') \mid \ell' \in L'\}.$$

(It is easy to verify that the term in braces (curly brackets) is a submodule of the direct sum.) An alternative name is the *co-fibre-product*.

We note that there are homomorphisms $\overline{\mu} : M' \to M' \oplus_{L'} L$ and $\overline{\phi} : L \to M' \oplus_{L'} L$, induced in the obvious way by the inclusion maps to the direct

---

† The concept is of recent origin, being due to the throng of 1950s algebraic topologists. The last term may therefore be described as an attempt to put Descartes before the hordes.

sum, which give a commutative diagram

$$
\begin{array}{ccc}
L' & \xrightarrow{\ \mu\ } & L \\
\Big\downarrow{\scriptstyle \phi} & & \Big\downarrow{\scriptstyle \overline{\phi}} \\
M' & \xrightarrow{\ \overline{\mu}\ } & M' \oplus_{L'} L
\end{array}
$$

known as the *push-out, cofibre square* or *cocartesian square*.

The following result is important, despite being easy to verify. In the language of category theory, it tells us that the pull-back and push-out are 'universal constructions'. This observation provides the insight that leads to the construction of pull-backs and push-outs for structures other than modules. We consider rings in Exercise 2.6.11, and more general structures in [BK: CM].

2.4.9 Proposition

(i) *Suppose that*

$$
\begin{array}{ccc}
L_1 & \xrightarrow{\ \beta_1\ } & L'' \\
\Big\downarrow{\scriptstyle \theta_1} & & \Big\downarrow{\scriptstyle \theta} \\
M & \xrightarrow{\ \beta\ } & M''
\end{array}
$$

is a commutative square of R-modules. Then there is exactly one R-module homomorphism $\omega : L_1 \to M \times_{M''} L''$ *with* $\overline{\theta}\omega = \theta_1$ *and* $\overline{\beta}\omega = \beta_1$.

(ii) *Suppose that*

$$
\begin{array}{ccc}
L' & \xrightarrow{\ \mu\ } & L \\
\Big\downarrow{\scriptstyle \phi} & & \Big\downarrow{\scriptstyle \phi_1} \\
M' & \xrightarrow{\ \mu_1\ } & M_1
\end{array}
$$

is a commutative square of R-modules. Then there is a unique homomorphism $\xi : M' \oplus_{L'} L \to M_1$ *with* $\xi\overline{\mu} = \mu_1$ *and* $\xi\overline{\phi} = \phi_1$. $\square$

2.4.10 Base change for short exact sequences

Suppose that we are given the short exact sequence

$$\mathbf{E} \qquad 0 \longrightarrow M' \xrightarrow{\alpha} M \xrightarrow{\beta} M'' \longrightarrow 0$$

and a homomorphism $\theta : L'' \to M''$. Using the pull-back, we obtain an exact sequence

$$\theta^* \mathbf{E} \qquad 0 \longrightarrow M' \xrightarrow{\mu} M \times_{M''} L'' \xrightarrow{\overline{\beta}} L'' \longrightarrow 0,$$

where $\overline{\beta}$ is as before and $\mu m' = (\alpha m', 0)$. We call $\theta^* \mathbf{E}$ the *pull-back* of $\mathbf{E}$ along θ.

Similarly, given a short exact sequence

$$\mathbf{F} \qquad 0 \longrightarrow L' \xrightarrow{\mu} L \xrightarrow{\lambda} L'' \longrightarrow 0,$$

and a homomorphism ϕ of L' into M', we have an exact sequence

$$\phi_* \mathbf{F} \qquad 0 \longrightarrow L' \xrightarrow{\overline{\mu}} M' \oplus_{L'} L \xrightarrow{\beta} L'' \longrightarrow 0,$$

the *push-out* of $\mathbf{F}$ along ϕ.

These operations are also sometimes called *base* change and *cobase* change respectively, by analogy with topological constructions.

2.4.11 The direct sum of short exact sequences

Another method for constructing short exact sequences consists of taking term-by-term direct sums. For $i = 1, \ldots, k$, let

$$\mathbf{E}_i \qquad \longrightarrow M'_i \xrightarrow{\alpha_i} M_i \xrightarrow{\beta_i} M''_i \longrightarrow 0$$

be an exact sequence of right R-modules. Then their *direct sum* is the sequence $\mathbf{E}_1 \oplus \cdots \oplus \mathbf{E}_k$:

$$0 \to M'_1 \oplus \cdots \oplus M'_k \xrightarrow{\alpha_1 \oplus \cdots \oplus \alpha_k} M_1 \oplus \cdots \oplus M_k \xrightarrow{\beta_1 \oplus \cdots \oplus \beta_k} M''_1 \oplus \cdots \oplus M''_k \to 0,$$

which is evidently a short exact sequence.

Exercises

2.4.1   The Three Lemma

Show that any module homomorphisms $\alpha : L \to M$ and $\beta : M \to N$ induce an exact sequence

$$0 \to \operatorname{Ker} \alpha \to \operatorname{Ker} \beta\alpha \to \operatorname{Ker} \beta \to \operatorname{Cok} \alpha \to \operatorname{Cok} \beta\alpha \to \operatorname{Cok} \beta \to 0.$$

2.4.2   The Snake Lemma

This result is much used in homological algebra, where it is a fundamental tool for constructing long exact sequences. It appears as an exercise in this text since its applications are also in exercises.

Suppose that we have a commuting diagram of right R-modules and R-homomorphisms, and that both rows are exact:

$$
\begin{array}{ccccccccc}
0 & \longrightarrow & M' & \overset{\mu'}{\longrightarrow} & M & \overset{\mu}{\longrightarrow} & M'' & \longrightarrow & 0 \\
& & \downarrow{\scriptstyle \alpha'} & & \downarrow{\scriptstyle \alpha} & & \downarrow{\scriptstyle \alpha''} & & \\
0 & \longrightarrow & N' & \overset{\nu'}{\longrightarrow} & N & \overset{\nu}{\longrightarrow} & N'' & \longrightarrow & 0
\end{array}
$$

Construct a *connecting homomorphism* $\delta : \operatorname{Ker} \alpha'' \to \operatorname{Cok} \alpha'$ as follows: take $y \in \operatorname{Ker} \alpha''$, choose $m \in M$ with $\mu m = y$ and observe that $\alpha m = \nu' m'$ for some $m' \in M'$. Then put $\delta(y) = \overline{m}'$ in $\operatorname{Cok} \alpha'$.

Show that δ is a well-defined R-module homomorphism which fits into an exact sequence†

$$
\begin{array}{ccccc}
0 & \longrightarrow & \operatorname{Ker} \alpha' & \longrightarrow & \operatorname{Ker} \alpha & \longrightarrow & \operatorname{Ker} \alpha'' \\
& & & & & & \\
& & & \delta & & & \\
& & & & & & \\
& \longrightarrow & \operatorname{Cok} \alpha' & \longrightarrow & \operatorname{Cok} \alpha & \longrightarrow & \operatorname{Cok} \alpha'' & \longrightarrow & 0
\end{array}
$$

2.4.3   Show that a pull-back diagram leads to an exact sequence

$$0 \longrightarrow M \times_{M''} L'' \longrightarrow M \oplus L'' \longrightarrow M''$$

and a push-out diagram to an exact sequence

$$L' \longrightarrow M' \oplus L \longrightarrow M' \oplus_{L'} L \longrightarrow 0.$$

---

† This is presumably the kind of exercise that Mac Lane had in mind when he advised asp-iring mathematicians to chase a diagram each day before breakfast.

Give (trivial) examples to show that these sequences cannot always be extended to short exact sequences.

2.4.4   (a) Given a pull-back square

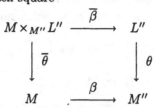

show that $\bar{\theta}$ induces an isomorphism $\operatorname{Ker}\beta \cong \operatorname{Ker}\bar{\beta}$ and that β induces an injection $\operatorname{Cok}\bar{\theta} \to \operatorname{Cok}\theta$.

(b) Given a push-out square

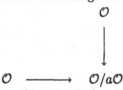

show that $\bar{\mu}$ induces an isomorphism $\operatorname{Cok}\phi \cong \operatorname{Cok}\bar{\phi}$ and that ϕ induces a surjection $\operatorname{Ker}\mu \to \operatorname{Ker}\bar{\mu}$.

2.4.5   Let $\mathcal{O}$ be a commutative domain and choose $a \in \mathcal{O}$, a nonzero.

(a)   Let L be the pull-back of the diagram

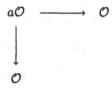

in which both maps are the canonical surjection. Show that L is generated by the pair $(1,1)$, $(0,a)$, and that the map defined by $(1,1) \mapsto (1,0)$, $(0,a) \mapsto (0,1)$ gives an isomorphism $L \cong \mathcal{O}^2$.

(b)   Let N be the push-out of the diagram

$$a\mathcal{O} \longrightarrow \mathcal{O}$$
$$\downarrow$$
$$\mathcal{O}$$

in which each map is the inclusion map, and write the image of $(x,y) \in \mathcal{O}^2$ as $[x,y] \in N$.

Show that the map $N \to \mathcal{O}$ given by $[x,y] \mapsto x+y$ is a surjective homomorphism of $\mathcal{O}$-modules, and that it is split by the map $x \mapsto [x, 0]$. Deduce that $N \cong \mathcal{O} \oplus \mathcal{O}/a\mathcal{O}$.

2.4.6   More non-split exact sequences

Verify that the sequences below are not split, except in trivial cases.

(a) Let $T = T_2(R)$ be the ring $\begin{pmatrix} R & R \\ 0 & R \end{pmatrix}$ of 2×2 *triangular matrices* over a ring R, and let $a = \begin{pmatrix} 0 & R \\ 0 & 0 \end{pmatrix}$, a (twosided) ideal of T; then there is a short exact sequence of right T-modules

$$0 \longrightarrow a \longrightarrow T \longrightarrow R \oplus R \longrightarrow 0$$

in which all the maps are the obvious ones.

(b) Let n, r and s be any positive integers. There is a short exact sequence of $\mathbb{Z}$-modules

$$0 \longrightarrow \mathbb{Z}/n^r\mathbb{Z} \longrightarrow \mathbb{Z}/n^{r+s}\mathbb{Z} \longrightarrow \mathbb{Z}/n^s\mathbb{Z} \longrightarrow 0$$

in which the first map is induced by multiplication by n^s and the second is the canonical surjection.

(c) There is an exact sequence

$$0 \longrightarrow a\mathcal{O} \longrightarrow \mathcal{O} \longrightarrow \mathcal{O}/a\mathcal{O} \longrightarrow 0$$

where $\mathcal{O}$ is a commutative domain and $a \in \mathcal{O}$, $a \neq 0$.

2.4.7   Let

$$0 \longrightarrow M' \xrightarrow{\;\alpha\;} M \xrightarrow{\;\beta\;} M'' \longrightarrow 0$$

and

$$0 \longrightarrow N' \xrightarrow{\;\gamma\;} N \xrightarrow{\;\delta\;} N'' \longrightarrow 0$$

be short exact sequences of R-modules.

Show that, when $M'' = N$, there is a short exact sequence

$$0 \longrightarrow M \times_{M''} N' \longrightarrow M \xrightarrow{\;\delta\beta\;} N'' \longrightarrow 0$$

and, when $M = N'$, there is a short exact sequence

$$0 \longrightarrow M' \xrightarrow{\;\gamma\alpha\;} M'' \longrightarrow M'' \oplus_M N \longrightarrow 0$$

2.4.8   Let $\mathcal{K}$ be a field and let L, M, N and P be $\mathcal{K}[T]$-modules which are finite-dimensional as $\mathcal{K}$-spaces. All bases mentioned in this exercise are to be $\mathcal{K}$-bases.

(a) Let

$$L \xrightarrow{\phi} N$$
$$\downarrow \mu$$
$$M$$

be a push-out diagram with both ϕ and μ injective. Take a basis $\{\ell_1, \ldots, \ell_s\}$ of L and bases

$$\{\mu\ell_1, \ldots, \mu\ell_s, m_1, \ldots, m_t\}$$

and

$$\{\phi\ell_1, \ldots, \phi\ell_s, n_1, \ldots, n_u\}$$

of M and N respectively, so that the matrices representing the actions of T on M and N have block forms $\begin{pmatrix} A & B \\ 0 & C \end{pmatrix}$ and $\begin{pmatrix} A & D \\ 0 & E \end{pmatrix}$, where A gives the action of T on L (Exercise 1.2.9) Show that, in the notation of (2.4.8), the push-out $M \oplus_L N$ has basis

$$\{\overline{\phi}\mu\ell_1, \ldots, \overline{\phi}\mu\ell_s, \overline{\phi}m_1, \ldots, \overline{\phi}m_t, \overline{\mu}n_1, \ldots, \overline{\mu}n_u\}$$

and that the matrix representing the action of T for this basis is

$$\begin{pmatrix} A & B & D \\ 0 & C & 0 \\ 0 & 0 & E \end{pmatrix}.$$

(b) Let

$$N$$
$$\downarrow \beta$$
$$M \xrightarrow{\theta} P$$

be a pull-back diagram with both θ and β surjective. Choose bases $\{m_1, \ldots, m_t, p_1, \ldots, p_w\}$ and $\{n_1, \ldots, n_u, q_1, \ldots, q_w\}$ of M and N respectively so that $\{\beta q_1, \ldots, \beta q_w\}$ is a basis of P and $\theta p_i = \beta q_i$ for all i.

Verify that the actions of T on M and N are represented by matrices of the forms $\begin{pmatrix} A & B \\ 0 & C \end{pmatrix}$ and $\begin{pmatrix} Z & Y \\ 0 & C \end{pmatrix}$ respectively, where C gives the action of T on P.

Show that (with an obvious and convenient abuse of notation)

$$\{m_1, \ldots, m_t, n_1, \ldots, n_u, q_1 + p_1, \ldots, q_w + p_w\}$$

is a basis for the pull-back $M \times_P N$ and that the action of T on the pull-back is given by the matrix $\begin{pmatrix} A & 0 & B \\ 0 & Z & Y \\ 0 & 0 & C \end{pmatrix}$.

2.4.9   This exercise gives some results on base and cobase change for exact sequences which are useful for the construction of the groups $\operatorname{Ext}_R^1(M'', M')$.

(a) Let

$$\mathbf{E} \qquad 0 \longrightarrow M' \overset{\alpha}{\longrightarrow} M \overset{\beta}{\longrightarrow} M'' \longrightarrow 0$$

be a short exact sequence. Show that $\alpha_* \mathbf{E}$ and $\beta^* \mathbf{E}$ are split exact sequences.

(b) Two short exact sequences

$$\mathbf{E}_1 \qquad 0 \longrightarrow M' \overset{\alpha_1}{\longrightarrow} M \overset{\beta_1}{\longrightarrow} M'' \longrightarrow 0$$

and

$$\mathbf{E}_2 \qquad 0 \longrightarrow M' \overset{\alpha_2}{\longrightarrow} M \overset{\beta_2}{\longrightarrow} M'' \longrightarrow 0$$

are said to be *equivalent* (or *congruent*) if there is an isomorphism $\chi : M_1 \to M_2$ with $\chi\alpha_1 = \alpha_2$ and $\beta_2\chi = \beta_1$. Verify that this equivalence is indeed an equivalence relation on the set $\operatorname{SES}(M'', M')$ of all extensions with kernel M' and image M''.

(c) Let $\rho : K'' \to L''$ and $\theta : L'' \to M''$ be homomorphisms. Show that the short exact sequences $\rho^*(\theta^* \mathbf{E})$ and $(\theta\rho)^* \mathbf{E}$ are equivalent. Show also that (for suitably defined maps and sequences) $(\phi\psi)_* \mathbf{F}$ is equivalent to $\phi_*(\psi_* \mathbf{F})$ and $\theta^*(\psi_* \mathbf{E})$ is equivalent to $\psi_*(\theta^* \mathbf{E})$.

(d) Let χ be an R-module automorphism of the direct sum $M' \oplus M''$. Using the notation of Exercise 2.1.6, show that χ induces an equivalence of the standard split exact sequence with itself if and only if χ can be represented in the form $\begin{pmatrix} id_{M'} & \psi \\ 0 & id_{M''} \end{pmatrix}$ for some homomorphism $\psi : M'' \to M'$.

2.5 PROJECTIVE MODULES

As we have seen in section 2.5, an arbitrary short exact sequence of modules may or may not be split. This observation suggests that it will be important to have a good understanding of those modules (if any) whose presence in a short exact sequence guarantees that the sequence splits. The expectation is fulfilled: modules arising in this way play a fundamental role in the analysis of modules in general.

This section is devoted to an exposition of the basic properties of one such type of module, a projective module. A module P is projective if any short exact sequence

$$0 \longrightarrow L \longrightarrow M \longrightarrow P \longrightarrow 0$$

terminating in P must be split.

If we require instead the splitting of all short exact sequences which start with a given module, we arrive at the definition of an injective module. We do not consider injective modules in detail in this text as they do not play a major role in K-theory, but their properties are indicated in some exercises. The third possible definition, namely, that, if a module M appears in the middle of a short exact sequence, then the sequence splits, gives the notion of a semisimple module, which we meet in section 4.1, particularly in the Complementation Lemma 4.1.14.

As usual, we confine the discussion to right modules; it will be obvious that all the definitions and results have left-handed counterparts. The ring of scalars R is arbitrary unless otherwise stated.

2.5.1 The definition and basic properties

Let R be any ring. A right R-module P is *projective* if every short exact sequence

$$0 \longrightarrow L \longrightarrow M \overset{\pi}{\longrightarrow} P \longrightarrow 0$$

of right R-modules splits at P, that is, there is an R-module homomorphism $\sigma : P \to M$ with $\pi\sigma = id_P$.

As we noted in Theorem 2.4.5, the existence of such a splitting leads to a direct sum decomposition $M \cong L \oplus P$.

The ring R itself is a projective right R-module: given any surjective homomorphism $\pi : M \to R$, choose m in M with $\pi m = 1$ and define a splitting $\sigma : R \to M$ by $\sigma r = mr$. This argument extends to show that free modules in general are projective. In passing, we remark that the existence of non-free projective modules is harder to demonstrate, as it depends on the nature of the coefficient ring R. A first example is given in Exercise 2.5.4 below.

2.5.2 Theorem

For any set X and ring R, the free right R-module $\mathrm{Fr}_R(X)$ on X is a projective R-module.

Proof

Let $\pi : M \to \mathrm{Fr}_R(X)$ be a surjective homomorphism of R-modules. For each member x of X, choose an element $\xi(x) \in M$ with $\pi\xi(x) = x$. By Theorem 2.1.22, ξ extends to a homomorphism $\sigma : \mathrm{Fr}_R(X) \to M$, and $\pi\sigma$ is the identity map on $\mathrm{Fr}_R(X)$ since it is already so on the free generating set X. □

The above result will often be applied in its equivalent form where the alternative, ordered, notation for free modules is preferred.

2.5.3 Corollary

For any ordered set Λ, the free module R^Λ is projective. In particular, for any natural number n, R^n is projective. □

We now give some alternative characterizations of projective modules.

2.5.4 Theorem

A right R-module P is projective if and only if, whenever the row is exact, the following diagram of R-modules and R-module homomorphisms can be completed:

$$
\begin{array}{ccc}
 & P & \\
\xi \swarrow & \downarrow \theta & \\
M \xrightarrow{\ \beta\ } M'' & \longrightarrow & 0
\end{array}
$$

(That is, there exists ξ with $\beta\xi = \theta$. In fact, ξ need not be unique.)

Proof

To see that the condition is sufficient for projectivity, take $M'' = P$ and θ the identity homomorphism.

In the reverse direction, we form an exact sequence **E** by taking M' to be Ker β and α to be the inclusion, and we let $\theta^*\mathbf{E}$ be its pull-back as in (2.4.10). We then have a commutative diagram

$$
\begin{array}{ccccccccc}
\theta^*\mathbf{E} & 0 & \longrightarrow & M' & \overset{\mu}{\longrightarrow} & M \times_{M''} P & \overset{\lambda}{\longrightarrow} & P & \longrightarrow & 0 \\
 & & & \Vert & & \downarrow{\scriptstyle\phi} & & \downarrow{\scriptstyle\theta} & & \\
\mathbf{E} & 0 & \longrightarrow & M' & \overset{\alpha}{\longrightarrow} & M & \overset{\beta}{\longrightarrow} & M'' & \longrightarrow & 0
\end{array}
$$

Now, λ is split, by ν say, and we take $\xi = \phi\nu$. $\qquad\square$

The next result incidentally provides an alternative proof that a free module is projective.

2.5.5 Theorem

Let Λ be any ordered set and let $\{P_\lambda \mid \lambda \in \Lambda\}$ be a set of right R-modules, R any ring.

Then the direct sum $P = \bigoplus_{\lambda \in \Lambda} P_\lambda$ is projective if and only if each P_λ is projective.

Proof

For this argument, it is more convenient to view P as an external direct sum. Extending the definitions given in the finite case (2.1.6), for each $\lambda \in \Lambda$ we have a standard inclusion $\sigma_\lambda : P_\lambda \to P$, where $\sigma_\lambda(p_\lambda)$ has p_λ in the λ-position and 0 elsewhere, and a standard projection $\pi_\lambda : P \to P_\lambda$, which sends the sequence (p_λ) to its λth entry p_λ.

Suppose that each P_λ is projective, and that we are given the solid arrows in the diagram

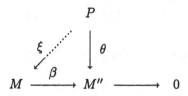

Put $\theta_\lambda = \theta\sigma_\lambda$ for each λ. Then there are homomorphisms ξ_λ which give

commutative diagrams

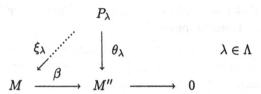

Define $\xi : P \to M$ by $\xi((p_\lambda)) = \sum_\lambda \xi_\lambda p_\lambda$. This infinite summation makes sense since only a finite number of terms p_λ are nonzero, and ξ is clearly a homomorphism which completes the original diagram.

Conversely, suppose we are given the homomorphisms θ_λ. Then we put $\theta = \sum_\lambda \theta_\lambda \pi_\lambda : P \to M''$, extend it to $\xi : P \to M$, and put $\xi_\lambda = \xi \sigma_\lambda$ for all λ. $\qquad\square$

We record a useful special case separately.

2.5.6 Corollary
Let $P = P' \oplus P''$. Then P is projective if and only if both P' and P'' are projective. $\qquad\square$

Our next characterization of projective modules requires a little preparation.

2.5.7 Lemma
An R-module M has a generating set indexed by the ordered set Λ if and and only if there is a surjective homomorphism from the free R-module R^Λ to M.

In particular, M is a finitely generated R-module with n generators if and only if there is a surjective homomorphism from R^n to M.

Proof
Suppose that $\{m_\lambda \mid m_\lambda \in M, \lambda \in \Lambda\}$ is the set of generators (1.2.8). Then the map from R^Λ to M given by $(r_\lambda) \mapsto \sum m_\lambda r_\lambda$ is a surjection of R-modules. Conversely, suppose there is such a surjection, π say. Since the standard basis $\{e_\lambda\}$ is a generating set for R^Λ, $\{\pi e_\lambda\}$ is evidently a set of generators for M. The rest is clear. $\qquad\square$

We obtain at once a very useful criterion for projectivity.

2.5.8 Theorem

(i) *A module P is projective if and only if*

$$P \oplus Q \cong R^\Lambda$$

for some module Q and set Λ.

(ii) *A module P is finitely generated and projective if and only if*

$$P \oplus Q \cong R^n$$

for some module Q and integer n. $\qquad\qquad\qquad\qquad\qquad\square$

2.5.9 Idempotents and projective modules

We can interpret the previous result in terms of idempotent elements in the endomorphism ring $\text{End}(R^n)$.

Recall that an element ϵ of $\text{End}(R^n)$ is idempotent if $\epsilon^2 = \epsilon$. Clearly, $1 - \epsilon$ is then also an idempotent, and $\{\epsilon, 1 - \epsilon\}$ constitutes a full set of idempotents for $\text{End}(R^n)$. Thus, by (2.1.10), there is a direct sum decomposition of R^n as a right R-module, given by $R^n = \epsilon R^n \oplus (1 - \epsilon)R^n$, and every direct sum decomposition of R^n arises in this way.

Combining these observations with part (ii) of the preceding theorem gives the following result.

2.5.10 Theorem

Let P be an R-module. Then P is finitely generated and projective if and only if $P \cong \epsilon R^n$ for some n and some idempotent ϵ in $\text{End}(R^n)$. $\qquad\square$

This theorem is presumably responsible for the term 'projective', since an idempotent element of $\text{End}(R^n)$ can be regarded as a *projection operator*. Thus the projective modules are those obtained from free modules by projection.

Next we consider the relation between the idempotent ϵ and the isomorphism class of the module ϵR^n. Recall that two elements η, ϵ of an arbitrary ring S are conjugate if $\eta = \alpha\epsilon\alpha^{-1}$ for some invertible element $\alpha \in S$.

2.5.11 Proposition

Let ϵ and η be idempotents in $\text{End}(R^n)$. Then the following statements are equivalent:

(i) *ϵ and η are conjugate;*

(ii) *there are R-module isomorphisms*

$$\epsilon R^n \cong \eta R^n \quad and \quad (1 - \epsilon)R^n \cong (1 - \eta)R^n.$$

Proof

(i) $\Rightarrow$ (ii):   If $\eta\alpha = \alpha\epsilon$ with α an automorphism of R^n, then the restrictions of α to ϵR^n and to $(1 - \epsilon)R^n$ induce isomorphisms as required.

(ii) $\Rightarrow$ (i):   Suppose that ξ and ζ are the isomorphisms, and put

$$\alpha = \xi\epsilon + \zeta(1 - \epsilon).$$

Then α is both injective and surjective, hence invertible, and we have $\eta\alpha = \eta\xi\epsilon = \alpha\epsilon$. $\square$

Recall that we can regard endomorphisms of R^n as members of the matrix ring $M_n(R)$ acting on 'column space' by left multiplication in the expected way. The discussion in (2.2.9) shows that two conjugate matrices ϵ, η in $M_n(R)$ can also be interpreted as matrix representations of a single endomorphism of R^n, but with respect to different n-element bases of R^n.

2.5.12 A cautionary example

The existence of an R-module isomorphism $\epsilon R^n \cong \eta R^n$ is not in itself sufficient to imply that ϵ and η are conjugate, since it is possible to have $\epsilon R^n \cong \eta R^n$ but $(1 - \epsilon)R^n \ncong (1 - \eta)R^n$. This phenomenon is known as non-cancellation.

An immediate example is provided by taking R to be the coneindexcone of a ring $C(S)$ of some nonzero ring S. Then $R^2 \cong R$ as a right R-module (2.3.2), so we can take $\epsilon = I$, the 2×2 identity matrix, and $\eta = \begin{pmatrix} 1 & 0 \\ 0 & 0 \end{pmatrix}$. Evidently, $\epsilon R^2 \cong \eta R^2$, but ϵ and η cannot be conjugate, since $(1 - \epsilon)R^2 = 0$ is not isomorphic to $(1 - \eta)R^2 = R$.

More sophisticated examples are given in (3.3.10).

2.5.13 Injective modules

Our definition of a projective module is that it is a module such that all short exact sequences

$$0 \longrightarrow L \longrightarrow M \overset{\pi}{\longrightarrow} P \longrightarrow 0$$

of right R-modules split at P.

If instead we seek splitting at the left-hand end of a short exact sequence, we arrive at the definition of an *injective* module: a right R-module I is injective if any short exact sequence

$$0 \longrightarrow I \overset{\alpha}{\longrightarrow} M \longrightarrow N \longrightarrow 0$$

of right R-modules splits at I. Despite the duality between the definitions, there was a significant delay between the introduction of injective modules in [Baer 1940] and projective modules in [Cartan & Eilenberg 1956].

We do not develop the theory of injective modules in detail in this text, since it is not important for our subsequent work. Some exercises in this section indicate their basic properties.

Detailed investigations of injective modules can be found in many books on ring theory or homological algebra, for example, [Rotman 1979], §3.

Exercises

2.5.1   Let $a \in \mathbb{Z} \setminus \{0, 1, -1\}$. Show that $\mathbb{Z}/a\mathbb{Z}$ is not a projective $\mathbb{Z}$-module.

2.5.2   Let A be a ring. Show that A is not projective when considered as a module over the ring of dual numbers $A[\epsilon]$ – see (2.4.6).

2.5.3   Following Exercise 2.4.6, show that the following modules are not projective.

   (i) $R \oplus R$, considered as a right module over the ring $\begin{pmatrix} R & R \\ 0 & R \end{pmatrix}$ of 2×2 upper triangular matrices over R.
   (ii) $\mathbb{Z}/n^s\mathbb{Z}$ considered as a $(\mathbb{Z}/n^{r+s}\mathbb{Z})$-module, where n, r and s are positive integers.
   (iii) $\mathcal{O}/a\mathcal{O}$ as an $\mathcal{O}$-module, where $\mathcal{O}$ is a commutative domain, $a \in \mathcal{O}$, and a is neither 0 nor a unit.

2.5.4   Let $\{e_{ij}\}$ be the standard set of matrix units in the matrix ring $M_n(R)$. Show that, for each $i, e_{ii}M_n(R)$ is a projective $M_n(R)$-module, the 'ith row' of $M_n(R)$, and that $e_{ii}M_n(R) \cong e_{jj}M_n(R)$ as right $M_n(R)$-modules for all pairs of indices i, j.

Now take the ring of scalars R to be a field $\mathcal{K}$. Show that

$$e_{ii}M_n(\mathcal{K}) \not\cong M_n(\mathcal{K}) \text{ for } n > 1,$$

and further that $e_{ii}M_n(\mathcal{K})$ is not (isomorphic to) a free right $M_n(\mathcal{K})$-module.

(Some hypothesis on R is needed for this 'obvious' conclusion, as can be seen from the existence of a ring S with $S \cong S^2$ as a right S-module, and hence $e_{11}M_2(S) \cong M_2(S)$ (2.3.2).)

2.5.5   Show that the rows $(R \ R)$ and $(0 \ R)$ are projective right $T_2(R)$-modules, where $T_2(R)$ is the ring of upper triangular matrices over R.

2.5.6   Let R and S be rings and let M be an R-S-bimodule. The *generalized upper triangular matrix ring* is $T = \begin{pmatrix} R & M \\ 0 & S \end{pmatrix}$, with the obvious matrix addition and multiplication. Show that the rows of T are projective right T-modules.

Let $\mathfrak{a} = \begin{pmatrix} 0 & M \\ 0 & 0 \end{pmatrix}$. Show that $\mathfrak{a}$ is an ideal of T and that $T/\mathfrak{a} \cong R \oplus S$ as T-modules. Show that $R \oplus S$ is not projective (unless $M = 0$). Give an example to show that $\mathfrak{a}$ may be projective.

The remaining exercises demonstrate some properties of injective modules.

2.5.7   Let R be an arbitrary ring.

   (i) Show that a right R-module I is injective if and only if the following diagram can be completed whenever the row is exact:

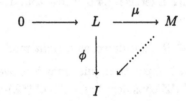

   (ii) Let $\{M_\lambda \mid \lambda \in \Lambda\}$ be a set of R-modules. Verify that the direct product $\prod_{\lambda \in \Lambda} M_\lambda$ is injective if and only if each M_λ is injective.

2.5.8   The following stronger version of (i) above is a fundamental result in the theory of injective modules. It has no counterpart for projective modules.

A right R-module I is injective if and only if, for any right ideal $\mathfrak{a}$ of R and any homomorphism $\beta : \mathfrak{a} \to I$, the diagram

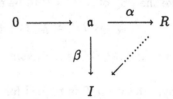

(in which α is the inclusion) can always be completed.

Here is an outline of the derivation of the nontrivial implication.

Suppose that we have a diagram as in (i) above. An extension of ϕ is defined to be a triple (M', μ', ψ') consisting of a submodule M' of M, a homomorphism $\mu' : L \to M'$ and an injective homomorphism $\psi' : M' \to I$ with $\psi'\mu' = \phi$. Let S be the set of all extensions of ϕ, and order S by stipulating that $(M', \mu', \psi') \leq (M'', \mu'', \psi'')$ if the second triple is an extension of the first. Verify that S is an inductive set and so has a maximal member (M_0, μ_0, ψ_0) by Zorn's Lemma 1.2.19.

Suppose that $M_0 \neq M$ and take $m \in M, m \notin M_0$. Check that $\mathfrak{a} = \{r \in R \mid mr \in M_0\}$ is a right ideal of R. Define $\beta : \mathfrak{a} \to I$ by $\beta(r) = \phi(mr)$ and let γ be its extension to R. Using γ and the push-out construction, find a member $(M_0 + mR, \mu, \psi)$ of S greater than (M_0, μ_0, ψ_0), a contradiction. Hence $M_0 = M$.

2.5.9   A right R-module M is *divisible* if, whenever elements m in M and $x \neq 0$ in R are given, there is an element n of M with $m = nx$.

Show that, if M is divisible, then so is any homomorphic image of M.

2.5.10 Recall that a (noncommutative) ring R is a domain if the equation $rs = 0$ holds in R only if $r = 0$ or $s = 0$.

Suppose that R is a domain. Show that for any $x \in R$ and any right R-module I, there is a bijective correspondence between right R-module homomorphisms $\beta : xR \to I$ and elements $y \in I$, given by $\beta \leftrightarrow \beta(x) = y$.

Deduce that if I is injective, then I is divisible.

2.5.11 A domain R is a *right principal ideal domain* if every right ideal $\mathfrak{a}$ is principal: $\mathfrak{a} = xR$ for some x in R. (We meet such rings in sections 3.2 and 3.3.)

Show that, for a right principal ideal domain R, an R-module I is injective if and only if it is divisible.

Deduce that the $\mathbb{Z}$-modules $\mathbb{Q}$ and $\mathbb{Q}/\mathbb{Z}$ are injective.

2.6 DIRECT PRODUCTS OF RINGS

It often happens that a ring R can be decomposed into a direct product of component rings $R_1, \ldots, R_k$. When this can be done, the module theory of R is determined by (and determines) that of its components. In this section, we give criteria that enable us to recognize when R has a direct product decomposition and we discuss the relationship between the modules over R and those over its component rings.

Again, we work with right modules only, leaving the left-handed case to the reader. The component rings $R_1, \ldots, R_k$ are arbitrary unless otherwise stated.

We remind the reader that an (additive) abelian group is the same as a $\mathbb{Z}$-module, so that direct sum and product constructions for modules apply to abelian groups. Recall also that these constructions coincide when taken over a finite indexing set. However, it is more natural to speak of a direct product rather than a direct sum of rings, for reasons which will be fully revealed when we consider sums and products from the point of view of category theory in [BK: CM], §1.4.

2.6.1 The definition

As usual with direct sum and product constructions, there are both internal and external versions, which we must distinguish while making the formal definitions, but which are often regarded as the same in practice.

Let $\{R_1, \ldots, R_k\}$ be a finite set of rings. We define the *external direct product*

$$R = R_1 \times \cdots \times R_k$$

to be the set of all k-tuples

$$r = (r_1, \ldots, r_k), \text{ where } r_i \in R_i \text{ for all } i,$$

with componentwise addition and multiplication:

$$(r_1, \ldots, r_k) + (s_1, \ldots, s_k) = (r_1 + s_1, \ldots, r_k + s_k)$$

and

$$(r_1, \ldots, r_k)(s_1, \ldots, s_k) = (r_1 s_1, \ldots, r_k s_k).$$

It is clear that R is a ring with zero element $(0, \ldots, 0)$ and identity element $(1, \ldots, 1)$. (If the indexing set is infinite, we are forced to consider sequences (r_i) in which all entries are nonzero, or else there will be no identity element. This is one reason why the name 'direct product' is more appropriate here than 'direct sum', a point considered in more detail in Exercise 2.6.9.)

The subsets

$$\mathfrak{a}_i = \{(0, \ldots, 0, r_i, 0, \ldots, 0) \mid r_i \in R_i\} \text{ for } i = 1, \ldots, k$$

are each twosided ideals of R, and each is a ring in its own right, with identity element $e_i = (0, \ldots, 0, 1, 0, \ldots, 0)$, the identity element of R_i being in the ith place. However, $\mathfrak{a}_i$ is *not* a subring of R (save in trivial exceptional cases). We also note that $\mathfrak{a}_i \mathfrak{a}_j = 0$ if $i \neq j$.

The above properties suggest the requirement to be met in order to express a given ring R as an *internal direct product*. We need a collection $\mathfrak{a}_1, \ldots, \mathfrak{a}_k$ of twosided ideals of R such that $R = \mathfrak{a}_1 \oplus \cdots \oplus \mathfrak{a}_k$, an internal direct sum of abelian groups.

Before we state the expected result relating these constructions, we introduce a new type of set of idempotents.

2.6.2 Central idempotents

Recall that an element f of a ring R is idempotent if $f^2 = f$: we say that f is *central* if $f \in Z(R)$, that is, for all $r \in R$ we have $rf = fr$.

Recall also that idempotents $f_1, \ldots, f_k$ are orthogonal if

$$f_i f_j = f_j f_i = 0$$

whenever $i \neq j$. A set $\{f_1, \ldots, f_k\}$ of orthogonal, central idempotents is called a *full set of orthogonal central idempotents* for R if further

$$f_1 + \cdots + f_k = 1_R.$$

(It will be convenient to allow trivial cases where an idempotent is 0; this corresponds to a component R_i being the zero ring.) Observe that any ring homomorphism carries a full set of orthogonal idempotents to a full set of orthogonal idempotents, and, if it is surjective, sends central idempotents to central idempotents.

It is clear that the elements $e_1, \ldots, e_k$ defined above for an external direct product $R = R_1 \times \cdots \times R_k$ constitute a full set of orthogonal central idempotents of R, which we refer to as the *standard set* for the direct product.

We now establish the equivalence of the internal and external direct product decompositions for rings. This result was first formulated in terms of idempotents in [Pierce 1881].

2.6.3 Proposition

Let R be a ring. Then the following assertions are equivalent.

(i) *There are rings $R_1, \ldots, R_k$ and a ring isomorphism $R \cong R_1 \times \cdots \times R_k$.*

(ii) *There is a full set of orthogonal central idempotents $\{f_1, \ldots, f_k\}$ for R.*

(iii) *There are (twosided) ideals $\mathfrak{a}_1, \ldots, \mathfrak{a}_k$ of R such that, as an abelian group, R is the (internal) direct sum $R = \mathfrak{a}_1 \oplus \cdots \oplus \mathfrak{a}_k$.*

When any of these three conditions holds, then the other two conditions may be interpreted so that, for each i, $\mathfrak{a}_i = f_i R$ is a ring with identity f_i and there is a ring isomorphism $f_i R \cong R_i$.

Proof

(i) $\Rightarrow$ (ii):   By hypothesis, there is a ring isomorphism

$$\theta : R_1 \times \cdots \times R_k \xrightarrow{\;\cong\;} R.$$

For each i, put $f_i = \theta(e_i)$, where $\{e_1, \ldots, e_k\}$ is the standard full set of orthogonal central idempotents for the external direct product.

(ii) $\Rightarrow$ (iii):   Given a full set of orthogonal central idempotents $\{f_1, \ldots, f_k\}$ for R, put $\mathfrak{a}_i = f_i R$ for each i, a twosided ideal by centrality of f_i. Given $r \in R$, we have $r = 1r = f_1 r + \cdots + f_k r$, which shows that $R = \mathfrak{a}_1 + \cdots + \mathfrak{a}_k$. Since $f_i^2 r = f_i r$, orthogonality gives

$$(\mathfrak{a}_1 + \cdots + \mathfrak{a}_{i-1} + \mathfrak{a}_{i+1} + \cdots + \mathfrak{a}_k) \cap \mathfrak{a}_i = 0 \text{ for all } i.$$

By (2.1.1), these two properties make R, as an abelian group, the direct sum of its subgroups $\mathfrak{a}_i$ (or, equally, the direct product of the $\mathfrak{a}_i$ since there are only finitely many terms).

(iii) $\Rightarrow$ (i):   By (iii), each element r in R can be written uniquely as $r = r_1 + \cdots + r_k$ with $r_i \in \mathfrak{a}_i$. Writing the identity element as $1 = f_1 + \cdots + f_k$ gives a set of elements satisfying

$$f_i f_j = \begin{cases} f_i & \text{if } i = j \\ 0 & \text{if } i \neq j \end{cases}$$

because $(\mathfrak{a}_1 + \cdots + \mathfrak{a}_{i-1} + \mathfrak{a}_{i+1} + \cdots + \mathfrak{a}_k) \cap \mathfrak{a}_i = 0$ for all i. Then each $\mathfrak{a}_i$ forms a ring R_i with identity element f_i, and the map $r \mapsto (r_1, \ldots, r_k)$ is readily verified to be a ring isomorphism from R to $R_1 \times \cdots \times R_k$. $\square$

2.6.4 Remarks

(a)   Because of this close relation between external direct products, internal direct products and full sets of idempotents, we often abuse notation and write

$$R = R_1 \times \cdots \times R_k$$

simply to indicate that R has a full set of orthogonal central idempotents giving the isomorphism which is implicit in this notation.

(b)   In our discussion of direct decompositions of modules, we showed that there is a bijective correspondence between full sets of orthogonal idempotents

of the endomorphism ring End(M) of a right R-module and direct decompositions of M (2.1.9). This applies to the ring R itself viewed as a right R-module:
we have End(R) $= R$ (elements of R acting on R by left multiplication), so
that a full set of orthogonal idempotents of R gives a direct decomposition of
R into right ideals (and, equally, left ideals).

Without centrality, there is no reason why the right ideal components should
be twosided ideals, so we will not have a direct decomposition of R into rings.

2.6.5 An illustration

Part (iii) of the last result suggests how direct products of rings arise in
practice. Let S be any ring and let $\mathfrak{m}$ and $\mathfrak{n}$ be twosided ideals of S. Suppose
that $\mathfrak{m}$ and $\mathfrak{n}$ are *comaximal*, that is, $S = \mathfrak{m} + \mathfrak{n}$, and consider the ring
homomorphism $\theta : S \to S/\mathfrak{m} \times S/\mathfrak{n}$ induced by the canonical surjections μ, ν
of S onto the respective factors (1.1.8).

Write $1 = m + n$ with $m \in \mathfrak{m}$ and $n \in \mathfrak{n}$. If $s \in \mathfrak{m} \cap \mathfrak{n}$, then $s = s \cdot 1 \in \mathfrak{mn}$,
so $\mathfrak{m} \cap \mathfrak{n} = \mathfrak{mn}$. Thus Ker $\theta = \mathfrak{mn}$, and there is an induced injective ring
homomorphism $\psi : S/\mathfrak{mn} \to S/\mathfrak{m} \times S/\mathfrak{n}$ (1.1.9). Next, take an element $(\mu x, \nu y)$
of the direct product, and put $s = xm + yn$ in S. Then $\mu s = \mu x$ and $\nu s = \nu y$,
which shows θ to be surjective and hence ψ to be an isomorphism of rings.
The full set of orthogonal central idempotents for $S/\mathfrak{mn}$ is $\{\psi m, \psi n\}$.

For example, if a and b are coprime integers, then we have a ring isomorphism $\mathbb{Z}/ab\mathbb{Z} \cong \mathbb{Z}/a\mathbb{Z} \times \mathbb{Z}/b\mathbb{Z}$.

The extension of the construction to three or more ideals is straightforward
– see Exercise 2.6.3 below.

2.6.6 Modules

We now look at the construction of modules over a direct product of rings
$R = R_1 \times \ldots \times R_k$. Given a set $\{M_1, \ldots, M_k\}$, where each M_i is a (right)
R_i-module, we form the R-module $M = M_1 \oplus \cdots \oplus M_k$ as follows. First take
the direct sum of $M_1, \ldots, M_k$ as $\mathbb{Z}$-modules: this is the set of all k-tuples
$m = (m_1, \ldots, m_k)$ where $m_i \in M_i$ for each i, with componentwise addition:

$$(m_1, \ldots, m_k) + (m'_1, \ldots, m'_k) = (m_1 + m'_1, \ldots, m_k + m'_k).$$

Then define the action of R by

$$(m_1, \ldots, m_k)r = (m_1 r_1, \ldots, m_k r_k) \text{ for } r = (r_1, \ldots, r_k) \in R.$$

It is straightforward to verify that M is a right R-module.

In the other direction, suppose that M is an R-module and that we are given a full set $\{f_1, \ldots, f_k\}$ of orthogonal central idempotents of R that exhibits R as a direct product of rings. For each i, put $M_i = M f_i$. Clearly, M_i is a module over $R_i = R f_i$, and there is an R-module isomorphism from M to $M_1 \oplus \cdots \oplus M_k$ given by

$$m \longmapsto (m f_1, \ldots, m f_k).$$

Again, we abuse language by treating this canonical isomorphism as an identity. This identification amounts to regarding M as the internal direct sum or product of its $\mathbb{Z}$-submodules M_i.

The above construction can also be viewed as a special case of the formation of a direct sum of modules over a fixed ring, as considered in section 2.1. For each index i, there is a canonical surjective ring homomorphism $\pi_i : R \to R_i$. An R_i-module M_i can be regarded as an R-module by restriction of scalars using π_i (1.2.14), and we can then form the R-module (external) direct sum as in (2.1.5). This procedure clearly gives the same result as before.

2.6.7 Homomorphisms

Homomorphisms of R-modules are also determined by their actions on the R_i-module components, as follows.

Let $M = M_1 \oplus \cdots \oplus M_k$ and $N = N_1 \oplus \cdots \oplus N_k$ be R-modules, with M_i and N_i R_i-modules for $i = 1, \ldots, k$. Given a collection $\{\alpha_1, \ldots, \alpha_k\}$ of R_i-module homomorphisms $\alpha_i : M_i \to N_i$, we define

$$\alpha = \alpha_1 \oplus \cdots \oplus \alpha_k : M \longrightarrow N$$

by

$$\alpha(m_1, \ldots, m_k) = (\alpha_1 m_1, \ldots, \alpha_k m_k).$$

Clearly, α is an R-module homomorphism.

In the reverse direction, given an R-module homomorphism $\alpha : M \to N$, define $\alpha_i : M_i \to N_i$ by $\alpha_i(m f_i) = (\alpha m) f_i$, where $\{f_1, \ldots, f_k\}$ is the full set of orthogonal central idempotents which expresses R as a direct product. Then each α_i is an R_i-module homomorphism and

$$\alpha m = \alpha_1(m f_1) + \cdots + \alpha_k(m f_k),$$

that is, we regain α as the direct sum

$$\alpha = \alpha_1 \oplus \cdots \oplus \alpha_k.$$

We also note that, since the idempotent f_i is central, it can be viewed as defining an R-endomorphism λ_i of a right R-module M by left multiplication:

$\lambda_i(m) = mf_i$. It is easy to see that $\{\lambda_1, \ldots, \lambda_k\}$ is a full set of orthogonal central idempotents of the endomorphism ring $\text{End}(M_R)$, thus giving a decomposition of rings

$$\text{End}_R(M) \cong \text{End}_{R_1}(M_1) \oplus \cdots \oplus \text{End}_{R_k}(M_k).$$

It may well be the case that one or more of these components is the zero ring. This happens for example if M is an R_1-module viewed as an R-module by restriction of scalars.

We summarize this discussion in the following result, the remaining details of which are left to the reader.

2.6.8 Theorem

Let $R = R_1 \times \cdots \times R_k$ be a direct product of rings and let $M = M_1 \oplus \cdots \oplus M_k$ and $N = N_1 \oplus \cdots \oplus N_k$ be decompositions of R-modules into R_i-modules. Then the following hold.

(i) *There is a natural isomorphism of abelian groups*

$$\text{Hom}_R(M, N) \cong \text{Hom}_{R_1}(M_1, N_1) \oplus \cdots \oplus \text{Hom}_{R_k}(M_k, N_k),$$

given by

$$\alpha \longleftarrow\!\!\!\longrightarrow \alpha_1 \oplus \cdots \oplus \alpha_k.$$

In particular,

$$\text{Hom}_R(M_i, N_j) = \begin{cases} \text{Hom}_{R_i}(M_i, N_i) & \text{for } i = j, \\ 0 & \text{for } i \neq j, \end{cases}$$

where, for each i, M_i and N_i are regarded as R-modules by restriction of scalars along the canonical surjection from R to R_i.

(ii) *The homomorphism α is injective or surjective if and only if each homomorphism α_i is injective or surjective respectively.*

(iii) *There is a ring isomorphism*

$$\text{End}_R(M) \cong \text{End}_{R_1}(M_1) \times \cdots \times \text{End}_{R_k}(M_k). \qquad \square$$

Note that we usually regard these isomorphisms as identities.

2.6.9 Corollary

Let R and M be as in the theorem.

(a) *M is finitely generated as an R-module if and only if each M_i is finitely generated as an R_i-module.*

(b) *M is projective as an R-module if and only if each M_i is projective as an R_i-module.*

Proof

(a) M is finitely generated if and only if there is a surjection from R^n to M for some integer n, which can happen if and only if there is a collection of surjections from $(R_i)^n$ to $M_i, 1 \le i \le k$, since $R^n \cong R_1^n \oplus \cdots \oplus R_k^n$.

(b)   This follows from the observation that a short exact sequence of R-modules

$$0 \longrightarrow K \longrightarrow L \longrightarrow M \longrightarrow 0$$

decomposes into a direct sum of short exact sequences of R_i-modules

$$0 \longrightarrow K_i \longrightarrow L_i \longrightarrow M_i \longrightarrow 0. \qquad \square$$

2.6.10 Historical note

The abstract treatment of the direct sum of modules (section 2.1) is first given in [Noether & Schmeider 1920], and the direct product of rings (section 2.6) in [Noether 1929]. Free modules, as the analogues of vector spaces for arbitrary rings, have a long history. However, the 'universal' property of the free module construction that is given in Corollary 2.1.23 was first formulated in [Samuel 1948]. The interpretation of free modules as universal objects plays an important role in [BK: CM].

The connections between bases, matrices and homomorphisms (section 2.2) are simply transcriptions of results that, presumably, have been wellknown for vector spaces since the definition of a vector space was formulated by Grassmann in 1848. Examples of rings without invariant basis number (section 2.3) are originally found in [Leavitt 1957], [Leavitt 1962], and the theory of non-IBN rings seems to have progressed little beyond the provision of examples. However, such rings occur in K-theory as rings with trivial K-groups.

Short exact sequences (section 2.4) are now an indispensable part of the vocabulary of the theory of rings and modules. They originate in homological algebra, which itself derives from the axiomatic description of the algebraic techniques that are used to compute the homology of topological spaces. The analysis of short exact sequences also lies at the heart of K-theory, which, at a foundational level, can be viewed as a method for defining invariants that are additive on short exact sequences. The prototype of such an invariant is the dimension $\dim V$ of a finite-dimensional vector space V: given a subspace U of V, we have

$$\dim U + \dim(V/U) = \dim V.$$

Exercises

2.6.1   A central idempotent f of the ring R is *primitive* if it is not the sum $f' + f''$ of two nonzero orthogonal central idempotents of R. Suppose that there is a full set $\{f_1, \ldots, f_k\}$ of orthogonal central idempotents for R which are all primitive. Show that any central idempotent in R is a sum of the members of some subset of $\{f_1, \ldots, f_k\}$.

Prove that the identity element of R is a primitive idempotent precisely when R has no nontrivial decomposition as a direct product of rings.

2.6.2   Let $\pi : R \to S$ be a homomorphism of rings. Thus S is an R-R-bimodule via π. Now suppose that π is a split R-R-bimodule surjection; that is, there is an R-R-homomorphism $\sigma : S \to R$ with $\pi\sigma = id_S$.

Show that $\{\sigma(1_S), 1_R - \sigma(1_S)\}$ is a full set of orthogonal central idempotents for R, and that $R = \operatorname{Im}\sigma \times \operatorname{Ker}\pi$ as a ring.

Conversely, verify that, if $R = S \times T$ as a ring, then the canonical homomorphism from R to S is a split R-R-bimodule surjection.

2.6.3   **Constructing direct products of rings**

This exercise embellishes the illustration given in (2.6.5). Suppose that the set $\{\mathfrak{m}_1, \ldots, \mathfrak{m}_k\}$ of twosided ideals of a ring R is *pairwise comaximal*, that is, $R = \mathfrak{m}_i + \mathfrak{m}_j$ whenever $i \neq j$.

Show that $\mathfrak{m}_i$ and $\mathfrak{m}_1 \cdots \mathfrak{m}_{i-1}\mathfrak{m}_{i+1} \cdots \mathfrak{m}_k$ are comaximal for each i, and that $\mathfrak{m}_1 \cap \cdots \cap \mathfrak{m}_k = \mathfrak{m}_1 \cdots \mathfrak{m}_k$.

Hint. Argue by induction, interpreting the identity $R = R \cdot R$ in many ways.

Hence find a ring isomorphism

$$R/\mathfrak{m}_1 \cdots \mathfrak{m}_k \cong R/\mathfrak{m}_1 \times \cdots \times R/\mathfrak{m}_k.$$

Describe the corresponding full set of orthogonal central idempotents.

Give explicit decompositions of the rings $\mathbb{Z}/6\mathbb{Z}$, $\mathbb{Z}/10\mathbb{Z}$ and $\mathbb{Z}/30\mathbb{Z}$.

2.6.4   Let $R = R_1 \times \cdots \times R_k$ be a direct product of rings. Show that

(i) $Z(R) = Z(R_1) \times \cdots \times Z(R_k)$, where $Z(R)$ is the centre of R,

(ii) $M_n(R) \cong M_n(R_1) \times \cdots \times M_n(R_k)$ for all $n > 1$.

2.6.5   Verify that a module $M = M_1 \oplus \cdots \oplus M_k$ over $R = R_1 \times \cdots \times R_k$ is an injective R-module if and only if each M_i is an injective R_i-module.

2.6.6   Let $R = S \times T$ be a direct product of nontrivial rings S and T. Show that S is a projective right R-module but that it is not isomorphic to any free right R-module.

Why does this result not conflict with the fact that there exist rings with $R \cong R^2$ as right R-modules?

2.6.7 Group rings

Let A be an arbitrary ring and let G be a group. The *group ring* AG of G with coefficients in A is defined to be the free left A-module generated by G, so that an element $x \in AG$ has the form $x = \sum_{g \in G} x_g g$ with $x_g \in A$ and almost all $x_g = 0$. We make AG into a ring by defining $ga = ag$, for a in A and g in G, and, for g, h in G, taking gh to be what it is in G. By means of the distributive and associative laws, this rule of multiplication extends uniquely to a multiplication on all of AG.

In the case that $G = \{g_1 = 1, g_2, \ldots, g_n\}$, show that, if n^{-1} exists in A, then $e = n^{-1}(g_1 + \cdots + g_n)$ is a central idempotent in AG. Identify the ring AGe.

2.6.8 Cyclic groups

Let $G = \{1, g, \ldots, g^{p-1}\}$ be the cyclic group of prime order p. Choose a primitive pth root ω of unity in the complex numbers $\mathbb{C}$, that is, $\omega^p = 1$, $\omega \neq 1$. For $h = 0, 1, \ldots, p-1$, put

$$e_h = p^{-1}(1 + \omega^h g + \omega^{2h} g + \cdots + \omega^{(p-1)h} g^{p-1}).$$

Show that $\{e_0, \ldots, e_{p-1}\}$ is a full set of primitive orthogonal central idempotents for the complex group ring $\mathbb{C}G$, and identify the rings $\mathbb{C}Ge_h$.

What happens on replacing $\mathbb{C}$ by $\mathbb{R}$ or $\mathbb{Q}$? What happens for G cyclic of order 4?

2.6.9

Let $\{R_i \mid i \in I\}$ be a set of rings indexed by an infinite set I. Show that the direct product (of abelian groups) $\prod_i R_i$ is a ring under the obvious multiplication, but that the direct sum $\bigoplus_i R_i$ is merely a nonunital ring.

These matters are considered in greater depth in [BK: CM], (5.3.4).

2.6.10

Recall from Exercise 1.1.5 that there is a standard method of embedding a nonunital ring R in its enveloping ring $\overline{R}$. Suppose that R is already a ring. Verify that there is a ring isomorphism from $\overline{R}$ to the ring direct product $R \times \mathbb{Z}$ by $(r, n) \mapsto (r + n \cdot 1_R, n)$.

2.6.11

Let $R = R_1 \times R_2$, the direct product of rings. Suppose that R_1 and R_2 have types (w_1, d_1) and (w_2, d_2) respectively. Using the methods of Exercise 2.3.2, show that R has type

$$(\max(w_1, w_2), \operatorname{lcm}(d_1, d_2)).$$

2.6.12 Pull-backs for rings

Let

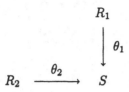

be a pull-back diagram of rings and ring homomorphisms. Show that there is a pull-back ring D which is defined by imitating the construction for modules (2.4.8).

What goes wrong with the obvious attempt to define a push-out for rings?

Suppose further that

$$
\begin{array}{ccc}
 & & M_1 \\
 & & \downarrow \mu_1 \\
M_2 & \xrightarrow{\ \mu_2\ } & N
\end{array}
$$

is a pull-back diagram of modules, each module being defined over the corresponding ring, and that the maps μ_1, μ_2 are abelian group homomorphisms which are compatible with the module structures. (Stated alternatively, μ_i is an R_i-module homomorphism, N being an R_i-module by restriction of scalars.)

Show that the (abelian group) pull-back of the diagram is a D-module and that the homomorphisms in the resulting diagram are compatible with the module structures.

2.6.13 The case $k = 2$ of Proposition 2.6.3 can be sharpened in the following way. Suppose that $\mathfrak{a}$ is a (twosided) ideal of a ring R and that $\mathfrak{a}$ is itself a ring under the multiplication in R. For nontriviality, suppose that the multiplicative identities $e = 1_{\mathfrak{a}}$ and 1_R are distinct and nonzero.

(i) Show that e is an idempotent with $\mathfrak{a} = Re = eR$.

(ii) Let $r \in R$ and write $re = ea$ and $er = be$ for $a, b \in \mathfrak{a}$. By considering eRe, deduce that e is a central idempotent.

(iii) Conclude from Proposition 2.6.3 that there are ring isomorphisms

$$R \cong eR \times (1 - e)R \cong \mathfrak{a} \times R/\mathfrak{a}.$$

(iv) Show conversely that any ring isomorphism $R \cong S_1 \times S_2$ arises from such a twosided ideal of R.

3
NOETHERIAN RINGS AND POLYNOMIAL RINGS

Now that we have the fundamental notions of ring theory and module theory at our disposal, we can discuss the properties of some particular types of ring. These rings are chosen because they play an important role in K-theory, but they are also natural objects of study for an introductory text on ring theory. Our main aim is to present the structure theory of their modules in a form which will be useful elsewhere.

The first special class that we consider comprises the Noetherian rings. This is still a very broad class, as it includes the great majority of the rings that one normally encounters.

We then investigate skew polynomial rings and their modules. A skew polynomial ring $A[T, \alpha]$ over a coefficient ring A consists of polynomials over A, but with multiplication by the variable T skewed by the action of an endomorphism α of the coefficient ring A, so that $a \cdot T = T \cdot \alpha a$. In favourable circumstances, the K-theory of the skew polynomial ring is the same as that of A. In other cases, the skew polynomial construction can be iterated to obtain examples of rings with rather complicated K-theory.

Our definitions and results are presented mainly for right modules, and it will be clear that they all have formal left-handed counterparts. However, the reader is alerted to the fact that the discussion is no longer left–right symmetric, in that, for some particular rings, the properties of left modules differ considerably from those of right modules. For instance, there are examples of rings which are right Noetherian without being left Noetherian and also of right polynomial rings which are not left polynomial rings.

3.1 NOETHERIAN RINGS

In this section, we set out the fundamental properties of Noetherian rings and modules, together with the confirmation that many interesting kinds of ring are Noetherian. We see that Noetherian rings have invariant basis number, and give a criterion for a ring to be non-Noetherian.

3.1.1 The Noetherian condition

First, the formal definitions. Let R be a ring. A right R-module M is *right Noetherian* if every R-submodule of M is finitely generated. This implies in particular that M itself is finitely generated.

The ring R is *right Noetherian* if it is itself Noetherian as a right R-module, that is, every right ideal of R is finitely generated, and left Noetherian if it is Noetherian as a left R-module.

Note that a ring may be right Noetherian but not left Noetherian; we give an example in (3.2.12) below. The term 'Noetherian ring' will mean a ring which is both left and right Noetherian. It is clear that, when R is commutative, R is left Noetherian precisely when it is right Noetherian.

Obvious examples of Noetherian rings are fields and division rings, since they have no proper ideals other than the zero ideal. The ring $\mathbb{Z}$ of integers is Noetherian, since every ideal is principal, that is, of the form $a\mathbb{Z}$ for some element a in $\mathbb{Z}$. The same argument applies to the ring of polynomials $\mathcal{K}[X]$ over a field $\mathcal{K}$. More generally, a polynomial ring in any finite set of variables over a right Noetherian ring is again right Noetherian, as we see in (3.2.4). On the other hand, an example of a non-Noetherian ring is given by $R = \mathcal{K}[X_1, X_2, ...]$, the polynomial ring in a countably infinite set of commuting variables over some field $\mathcal{K}$. The ideal $(X_1, X_2, ...)$, generated by all the variables, cannot be finitely generated, since any finite set of polynomials involves only a finite set of variables.

The Noetherian property originated, along with the notion of a module, in the work of Dedekind on factorization in rings of algebraic integers in the 1870s. A modern translation of this work is given in [Dedekind 1996], while Dedekind's collected works are published in [Dedekind 1932]. As the reader may suspect, the Noetherian property was formalized by Emmy Noether [Noether 1926] when she recast Dedekind's theory in axiomatic form. The type of ring that she introduced is now called a Dedekind domain; we consider these in depth in Chapter 5.

We first explore some formal consequences of the definitions.

3.1.2 Proposition
Let

$$0 \longrightarrow M' \overset{\alpha}{\longrightarrow} M \overset{\beta}{\longrightarrow} M'' \longrightarrow 0$$

be an exact sequence of right R-modules.
Then M is Noetherian if and only if both M' and M'' are Noetherian.

Proof

Suppose that M is Noetherian. A submodule of M' is isomorphic to a submodule of M, and so is finitely generated. A submodule N of M'' is the homomorphic image of its inverse image

$$\beta^{-1}N = \{m \in M \mid \beta m \in N\}$$

in M. Since $\beta^{-1}N$ is finitely generated, so also is N. Thus M' and M'' are Noetherian.

Conversely, consider a submodule N of M. Let $N' = N \cap \alpha M'$ and let N'' be the image of N in M'', so that there is an exact sequence

$$0 \longrightarrow N' \longrightarrow N \longrightarrow N'' \longrightarrow 0.$$

Since both N' and N'' are finitely generated, so also is N. □

3.1.3 Corollary

Let $\{M_1, \ldots, M_k\}$ be a finite set of Noetherian right R-modules. Then the direct sum $M_1 \oplus \cdots \oplus M_k$ is also Noetherian.

In particular, a free right module of finite rank over a right Noetherian ring must be Noetherian. □

In (2.5.7) we saw that a finitely generated module must be a homomorphic image of a free module of finite rank. Combining this observation with the last two results, we obtain a fundamental property of Noetherian rings.

3.1.4 Theorem

Suppose that R is a right Noetherian ring and that M is a finitely generated right R-module.

Then M is Noetherian. □

3.1.5 The ascending chain condition and the maximum condition

It is useful to have some alternative formulations of the Noetherian condition. An *ascending chain* of submodules of the right R-module M is a sequence

$$M_1 \subseteq M_2 \subseteq \cdots \subseteq M_i \subseteq M_{i+1} \subseteq \cdots,$$

each M_i being a submodule of M.

If the set of indices $\{1, \ldots, i, i+1, \ldots\}$ is finite, then the ascending chain is also said to be *finite*. If the set of indices is infinite but the set $\{M_i \mid i \in \mathbb{N}\}$ contains only finitely many distinct submodules, the ascending chain is said to *terminate* or to become *stationary*; in this case, there is an index h with $M_h = M_{h+j}$ for all $j \geq 1$. In the remaining case, that it contains infinitely many distinct terms, the chain is said to be *infinite*. A right R-module M is said to satisfy the *ascending chain condition* if every ascending chain of submodules of M either terminates or is finite.

It is sometimes convenient to restrict attention to *proper* ascending chains in M, that is, chains in which the inclusion $M_i \subset M_{i+1}$ is a strict inequality for each i. Clearly, M satisfies the ascending chain condition if and only if every proper ascending chain in M is finite.

The module M is said to satisfy the *maximum condition* if any nonempty set of submodules of M has a maximal member (with respect to inclusion).

3.1.6 Theorem
Let M be a right R-module. Then the following assertions are equivalent.

(i) *M is Noetherian.*

(ii) *M satisfies the ascending chain condition.*

(iii) *M satisfies the maximum condition.*

Proof

(i) $\Rightarrow$ (ii).   Let $M_1 \subseteq \cdots \subseteq M_i \subseteq M_{i+1} \subseteq \cdots$ be an ascending chain of submodules of M. Then $\bigcup M_i$ is a submodule of M, so it is finitely generated. But we can choose h so that each generator of $\bigcup M_i$ is in M_h, so the chain terminates there.

(ii) $\Rightarrow$ (iii).   We argue by contradiction. Suppose that X is some set of submodules of M which does not have a maximal member. Assume that for some $i \geq 1$ we have constructed a chain $M_1 \subset \cdots \subset M_i$ of members of X, with proper inclusions. Since M_i is not maximal in X, there is a further member M_{i+1} of X with $M_i \subset M_{i+1}$, and so we can construct an infinite ascending chain in M.

(iii) $\Rightarrow$ (i).   Given a submodule N of M, let X be the set of finitely generated submodules of N. Clearly, X is not empty, and so has a maximal member L, say. If $L \neq N$, there is some $n \in N$ with $n \notin L$. But then $L + nR \in X$, contrary to the maximality of L. So $L = N$ after all, making N finitely generated. $\qquad\qquad\square$

3.1.7 Corollary

A ring R is right Noetherian if and only if it has no infinite ascending chain of right ideals. ascending chain of right ideals of R terminates. □

We have the following application.

3.1.8 Proposition

Suppose that M is a Noetherian module and that ϕ is a surjective endomorphism of M. Then ϕ is an isomorphism.

Proof

Since $\operatorname{Ker}\phi \subseteq \operatorname{Ker}\phi^2 \subseteq \cdots$ is an ascending chain in M, we must have $\operatorname{Ker}\phi^i = \operatorname{Ker}\phi^{i+1}$ for some i. Suppose $m \in \operatorname{Ker}\phi$. Now $m = \phi^i n$ for some n; but then n is in $\operatorname{Ker}\phi^{i+1}$ and hence in $\operatorname{Ker}\phi^i$. Thus $m = 0$, and so $\operatorname{Ker}\phi = 0$.
□

This result has an important consequence.

3.1.9 Theorem

Let R be a nonzero right Noetherian ring. Then R has invariant basis number.

Proof

Suppose that there is an isomorphism $R^n \cong R^m$ with $m \geq n$. Clearly, we can extend the isomorphism to a surjective endomorphism ψ of R^m with kernel isomorphic to R^{m-n}. By (3.1.4), R^m is Noetherian, and so ψ is an isomorphism by the preceding result. Thus $m = n$. □

Next we look at some methods for constructing Noetherian rings. The first follows from (3.1.2) by a change of rings argument as in (1.2.14).

3.1.10 Proposition

Let $\mathfrak{a}$ be a (twosided) ideal of a right Noetherian ring R. Then the residue ring $R/\mathfrak{a}$ is also right Noetherian. □

3.1.11 Module-finite extensions

We now look at a situation in which a ring often inherits properties from a subring.

Suppose that the ring R is an extension of a ring A, that is, A is a subring of R. Then R is both a left and a right A-module, and we often view A as a 'ring of coefficients' for R. In the case that the coefficient ring is a field $\mathcal{K}$

which is contained in the centre of R, we say that R is a $\mathcal{K}$-*algebra*. Thus the matrix rings $M_n(\mathcal{K})$, with n an integer, are $\mathcal{K}$-algebras, as also are the cone $C\mathcal{K}$ and the infinite matrix rings $M_\omega^{cf}(\mathcal{K})$ and $M_\omega^{rf}(\mathcal{K})$.

Suppose that R is an extension of A. We say that R is *right module-finite* over A if R is finitely generated as a right A-module.

For example, the group ring AG (Exercise 2.6.7) of a finite group G, with coefficients in any ring A, is both left and right module-finite over A, the generators being the group elements themselves. (In this example, we identify A with the subring $A \cdot 1_G$ of AG.)

An algebra which is module-finite over its coefficient field $\mathcal{K}$ is of course referred to as a finite-dimensional $\mathcal{K}$-algebra, since it is finite-dimensional as a vector space over $\mathcal{K}$. The matrix rings $M_n(\mathcal{K})$, n an integer, are finite-dimensional, but the cone $C\mathcal{K}$ and the infinite matrix rings $M_\omega^{cf}(\mathcal{K})$ and $M_\omega^{rf}(\mathcal{K})$ are not.

Among the most frequent examples of module-finite extensions are orders. Suppose that $\mathcal{O}$ is a commutative domain (that is, $\mathcal{O}$ is a commutative ring for which, if $ab = 0$ for a, b in $\mathcal{O}$, then either $a = 0$ or $b = 0$). A ring R is an $\mathcal{O}$-*order* if the following axioms are satisfied.

(O1) $\mathcal{O}$ is a subring of the centre $Z(R)$ of R.

(O2) R is (left and right) module-finite over $\mathcal{O}$.

(O3) R is torsion-free as an $\mathcal{O}$-module, that is, whenever the equation $xr = 0$ holds with $x \in \mathcal{O}$ and $r \in R$, then either x or r is 0.

A basic example is $\mathbb{Z}G$, the integral group ring of a finite group, which is a $\mathbb{Z}$-order. Also, for any commutative domain $\mathcal{O}$ and positive integer n, the matrix ring $M_n(\mathcal{O})$ is an $\mathcal{O}$-order. When $\mathcal{O}$ is actually a field, the term 'algebra' is preferred to 'order'.

The term 'order' was first used in the second half of the 19th century to describe an arbitrary ring of algebraic integers in a number field. The usage of the term was extended in the early part of the 20th century to include noncommutative versions of rings of integers ([Deuring 1937]) – see Exercise 5.2.4. Our definition is cast in the widest (reasonable) form.

Our next result tells us that orders are often Noetherian.

3.1.12 Proposition

Suppose that R is a right module-finite extension of a right Noetherian ring A. Then R is right Noetherian.

In particular, if R is an $\mathcal{O}$-order with $\mathcal{O}$ Noetherian, then R is (left and right) Noetherian.

Proof

Note that any right ideal $\mathfrak{a}$ of R is an A-submodule of R and so by (3.1.4) a finitely generated A-module. But then $\mathfrak{a}$ is a finitely generated R-module. □

Finally, we give a simple criterion for a ring to be non-Noetherian. This is stated 'on the left' for its application in (3.2.12).

3.1.13 Proposition

Let R be a domain and suppose that there are two nonzero elements a and b of R with $Ra \cap Rb = 0$.

Then

(i) *the left ideal*

$$\mathfrak{a} = Rb + Rba + Rba^2 + Rba^3 + \cdots$$

is isomorphic to the infinite direct sum $\bigoplus_i Rba^i$,

(ii) *R is not left Noetherian.*

Proof

There is an obvious surjection from the direct sum to the given ideal. This map is also an injection, since any relation

$$r_m ba^m + \cdots + r_n ba^n = 0, \; m < n,$$

with r_m and r_n both nonzero, would give a nonzero element $r_m b$ in $Ra \cap Rb$.

Since no ideal Rba^i is zero, the direct sum is infinite, and therefore the ideal $\mathfrak{a}$ cannot be finitely generated. □

Exercises

3.1.1   Show that a ring R is right Noetherian if and only if its opposite R° is left Noetherian.

3.1.2   Let $\mathcal{K}$ be the field of fractions of a commutative domain $\mathcal{O}$. Show that $\mathcal{K}$ is not a Noetherian $\mathcal{O}$-module (except when $\mathcal{K} = \mathcal{O}$).

3.1.3   Show that a subring of a Noetherian ring need not be Noetherian.

   Heavy hint. By (1.1.12), a commutative domain is contained in a field.

3.1.4   Let $R = R_1 \times \cdots \times R_k$, a direct product of rings. Show that R is right Noetherian if and only if each R_i is right Noetherian.

3.1.5   Let $f : R \to S$ be a surjective ring homomorphism. Show that a right
         S-module M is Noetherian if and only if it is Noetherian as a right
         R-module. In particular, S is right Noetherian as a ring if and only
         if it is Noetherian as a right R-module.

         Deduce that the following are equivalent.

         (a) R is right Noetherian,
         (b) S is right Noetherian and $\operatorname{Ker} f$ is Noetherian as a right R-
              module.

3.1.6   Let R and S be rings and let M be an R-S-bimodule. When is the
         (generalized) upper triangular matrix ring $T = \begin{pmatrix} R & M \\ 0 & S \end{pmatrix}$ right
         Noetherian?

3.1.7   Let D be the pull-back of rings defined by the diagram

$$
\begin{array}{ccc}
 & & R_1 \\
 & & \downarrow {\scriptstyle \theta_1} \\
R_2 & \xrightarrow{\;\;\theta_2\;\;} & S
\end{array}
$$

         in which both θ_1 and θ_2 are surjective. Show that the ring D is right
         Noetherian if and only if both R_1 and R_2 are right Noetherian.

3.1.8   A ring may satisfy the ascending chain condition for proper twosided
         ideals without being Noetherian.

         Consider the cone CR of any ring R. By Exercise 1.1.3, CR has
         a simple residue ring S, which trivially satisfies the ascending chain
         condition on twosided ideals. By (2.3.2), CR does not have invariant
         basis number, and so S also fails to have invariant basis number
         (2.3.6). Thus S is not left or right Noetherian (3.1.9).

3.1.9   A submodule L of a right R-module M is called *intersection irre-
         ducible* (or *intersection indecomposable*) if there are no submodules
         M_1, M_2 of M with $L = M_1 \cap M_2$ and $L \neq M_1, L \neq M_2$.

         Suppose that M is Noetherian. Show that any proper submodule
         of M can be written as a finite intersection of intersection irreducible
         submodules of M.

         Hint. Show that, if not, there is a maximal counterexample.

3.2 SKEW POLYNOMIAL RINGS

We now introduce skew polynomial rings, which are both an active topic of
research and a rich source of examples in ring theory. Given an arbitrary ring

A and a ring endomorphism α of A (that is, a ring homomorphism from A to itself), the skew polynomial ring $A[T, \alpha]$ over A can be defined informally as the set of all 'right' polynomials

$$a_0 + Ta_1 + \cdots + T^k a_k,$$

where $a_0, a_1, \ldots, a_k$ are in A and $k \geq 0$, with addition given by the usual rule, but with multiplication twisted by the rule

$$a \cdot T = T \cdot \alpha a.$$

The properties of the skew polynomial ring depend both on A and on the endomorphism α. If A is right Noetherian and α is an automorphism, then Hilbert's Basis Theorem (3.2.3) tells us that $A[T, \alpha]$ is also right Noetherian. However, when α is not an automorphism, $A[T, \alpha]$ need not be right Noetherian even when A is right Noetherian.

For most of the discussion, we confine our attention to the case when the coefficient ring is a field or division ring. Then, for any endomorphism α, the skew polynomial ring is actually a right Euclidean ring, that is, it satisfies a division algorithm analogous to the familiar division algorithm on the integers $\mathbb{Z}$.

As many of our results depend only on the Euclidean property, rather than any special properties of skew polynomial rings, our results are stated and proved in the broader context of Euclidean domains. They can then be applied to $\mathbb{Z}$ itself, and also to the ring of (integral) Hurwitz quaternions, and, later, to complete discrete valuation rings.

3.2.1 The definition

The formal definition of a skew polynomial ring is as follows. Given a ring A, let $A[T, \alpha]$ be the free right A-module on the ordered set $\{0, 1, \ldots\}$ of nonnegative integers. An element $a(T)$ of $A[T, \alpha]$ is thus a sequence $a(T) = (a_0, a_1, \ldots, a_i, \ldots)$, in which all the coefficients a_i are 0 except for a finite number of indices i. The polynomial whose coefficients are all 0 is written as 0 also, and is called the *zero polynomial*.

The *degree* of a nonzero polynomial $a(T)$ is the largest index k with $a_k \neq 0$; this term a_k is known as the *leading coefficient* of $a(T)$. The zero polynomial is given the degree $-\infty$. When the leading coefficient is 1, the polynomial is called *monic*.

To write the polynomials in more familiar notation, we define formal expressions $T^0, T^1, T^2, \ldots$ by

$$T^0 = (1, 0, 0, \ldots),$$
$$T^1 = (0, 1, 0, \ldots),$$
$$\vdots$$
$$T^i = (0, \ldots 0, 1, 0, \ldots),$$

so that each T^i has degree i. We can then write the member $a(T)$ of $A[T, \alpha]$ in the form

$$a(T) = T^0 a_0 + T^1 a_1 + \cdots + T^i a_i + \cdots + T^k a_k.$$

Such expressions are not unique, since terms with coefficient 0 can always be added on. However, if

$$b(T) = T^0 b_0 + T^1 b_1 + \cdots + T^i b_i + \cdots + T^\ell b_\ell$$

is another right polynomial, then $a(T) = b(T)$ if and only if $\deg a(T) = \deg b(T)$ and $a_i = b_i$ for all $i \leq \deg a(T)$.

Now introduce multiplication in $A[T, \alpha]$ by the rule

$$T^i a_i \cdot T^j b_j = T^{i+j} \alpha^j(a_i) \cdot b_j, \quad \text{where } a_i, b_j \in A.$$

It is easy to verify that $A[T, \alpha]$ becomes a ring, with zero the zero polynomial and identity T^0, and that T^i is indeed the i th power of T. As expected, we write $T^0 = 1$ and $T^1 = T$. When α is the identity map on A, we regain the ordinary polynomial ring $A[T]$.

The ring $A[T, \alpha]$ also contains *left polynomials*, that is, elements of the form $a_0 + a_1 T + \cdots + a_k T^k$. Such an expression can be rewritten as a right polynomial

$$a_0 + T \cdot \alpha(a_1) + \cdots + T^k \cdot \alpha^k(a_k).$$

However, a right polynomial cannot in general be written as a left polynomial unless α is surjective.

Note also that the coefficients of a left polynomial need not be unique; for example, if a is in the kernel of α, the left polynomial aT is 0. Thus, in general, it is not safe to attribute a degree to a left polynomial as such. However, when α is injective, a left polynomial can be allocated the same degree as its right-handed version.

It is clear from these comments that when α is an automorphism of A, we could equally define $A[T, \alpha]$ in terms of left polynomials, using α^{-1} in place of α. When α is not an automorphism, the ring $A[T, \alpha]$ cannot be viewed as a left polynomial ring.

3.2.2 Some endomorphisms

Since the nontriviality of the theory of skew polynomial rings depends on the existence of various types of endomorphism of the coefficient rings, the reader is entitled to have a supply of rings whose endomorphisms can be described easily.

Consider an ordinary polynomial ring $Q[X]$, where Q is a field and X an indeterminate. Each polynomial $f(X)$ defines an endomorphism of $Q[X]$ by the substitution $\alpha : X \mapsto f(X)$, which is an automorphism precisely when $f(X) = aX + b$ with $a \neq 0$.

The polynomial ring $Q[X]$ is a commutative domain and so has a field of fractions, the *field of rational functions* (1.1.12). The substitution $X \mapsto f(X)/g(X)$, where $f(X)$ and $g(X)$ are polynomials and $g(X)$ is nonzero, gives an endomorphism of $Q(X)$. If $f(X) = aX + b$ and $g(X) = cX + d$ with $ad - bc \neq 0$ (so that there is no cancellation), this endomorphism is an automorphism of $Q(X)$. In fact, any (ring) automorphism of $Q(X)$ that leaves Q unchanged must arise in this way – see [Jacobson 1964], p. 158.

There is in fact an iterative process for building skew polynomial rings. Given a nontrivial automorphism α of a field $\mathcal{K}$, the noncommutative ring $\mathcal{K}[T, \alpha]$ can be shown to have a division ring of fractions $\mathcal{K}(T, \alpha)$, the ring of skew rational functions ([BK: CM], §6.2). In turn, automorphisms of $\mathcal{K}(T, \alpha)$ can be defined by substitutions, and the construction repeated.

We start with a major result, whose title is explained after Corollary 3.2.4.

3.2.3 The Hilbert Basis Theorem

Let A be a right Noetherian ring and let α be a ring automorphism of A. Then the skew polynomial ring $A[T, \alpha]$ is also right Noetherian.

Proof

Since α is an automorphism of A, we can regard an element of $A[T, \alpha]$ as a left polynomial

$$x = x_0 + x_1 T + \cdots + x_k T^k,$$

where $x_0, x_1, \ldots, x_k$ are in A and $k \geq 0$.

Let $\mathfrak{a}$ be a right ideal of $A[T, \alpha]$. We confirm that $\mathfrak{a}$ is finitely generated. For each $k = 0, 1, \ldots$, let $\mathfrak{c}_k$ be the set of all elements x_k in A that appear as the leading coefficient of an element x of $\mathfrak{a}$ when x is written as a left polynomial of degree k. For each such x_k, we choose some polynomial p_{x_k} in $\mathfrak{a}$ with degree k and x_k as leading coefficient (we take $p_0 = 0$). It is straightforward to check that $\mathfrak{c}_k$ is a right ideal in A and that $\mathfrak{c}_k \subseteq \mathfrak{c}_{k+1}$ always. Thus there is an index m with $\mathfrak{c}_m = \mathfrak{c}_{m+i}$ for $i \geq 1$.

For $k = 0, \ldots, m$, let

$$\mathfrak{c}_k = z_{k1}A + \cdots + z_{kn}A, \text{ with each } z_{kj} \in \mathfrak{c}_k.$$

(We can always add trivial terms to ensure that the number of generators is the same for each k.) Further, put

$$\mathfrak{b} = \sum_{k=1}^{m} \sum_{j=1}^{n} p_{z_{kj}} A[T, \alpha].$$

Clearly, $\mathfrak{b}$ is a finitely generated right ideal of $A[T, \alpha]$, and $\mathfrak{b} \subseteq \mathfrak{a}$.

To prove that any nonzero element x of $\mathfrak{a}$ lies in $\mathfrak{b}$, we argue by induction on its degree k. Suppose first that $k \leq m$. We can write the leading coefficient x_k of x as $x_k = \sum_j z_{kj} a_{kj}$. Let $y = \sum_j p_{z_{kj}} \alpha^k(a_{kj})$. Then y is in $\mathfrak{b}$ and y also has degree k and leading coefficient x_k. Thus $x - y$ is in $\mathfrak{a}$ and has lower degree. If instead $k > m$, we write $x_k = \sum_j z_{mj} a_{mj}$ and put $y = \sum_j p_{z_{mj}} \alpha^m(a_{mj}) T^{k-m}$. Then again y is in $\mathfrak{b}$, and y has degree k and leading coefficient x_k, so the induction goes through. $\square$

Since the map α is an automorphism, $A[T, \alpha]$ is equally the ring of left skew polynomials corresponding to α^{-1}. Thus $A[T, \alpha]$ is left Noetherian if A is left Noetherian, by a similar argument.

Recall that when α is the identity map, we regain the usual polynomial ring. Induction gives the following result.

3.2.4 Corollary
If A is right Noetherian, so also is $A[T_1, \ldots, T_n]$. $\square$

The above corollary leads to Hilbert's original form of the Basis Theorem: if $\mathfrak{a}$ is an ideal in a polynomial ring $K(T_1, \ldots, T_n)$, K a field, then $\mathfrak{a}$ is finitely generated. This result [Hilbert 1932-35] is fundamental to modern algebraic geometry in that it allows an algebraic variety to be described as the set of common zeros of a finite collection of polynomials.

We next investigate some stronger results that can be obtained by choosing the coefficient ring A to be a division ring $\mathcal{D}$. The discussion includes as a special case the ordinary polynomial ring $\mathcal{K}[T]$ over a field $\mathcal{K}$. (Our notational convention is that $\mathcal{D}$ denotes a division ring, possibly commutative, while the symbol $\mathcal{K}$ is reserved for a (commutative) field.)

Recall that the ring R is a domain if the equation $ab = 0$ implies that $a = 0$ or $b = 0$. We record

3.2.5 Lemma

(i) *Suppose that A is a domain and that α is an injective endomorphism of A. Then $A[T, \alpha]$ is also a domain.*

(ii) *If α is any endomorphism of a division ring $\mathcal{D}$, then $\mathcal{D}[T, \alpha]$ is a domain.*

Proof

The first assertion has an easy verification, and the second follows since any endomorphism of $\mathcal{D}$ must be injective.   □

3.2.6 The division algorithm

The key to computation in the ring $\mathcal{D}[T, \alpha]$ is that the *division algorithm* holds for right division: given f and g in $\mathcal{D}[T, \alpha]$, there are polynomials q and r with $f = gq + r$ and $\deg(r) < \deg(g)$, the case $r = 0$ being that where $\deg(r) = -\infty$.

To see this, write $f = T^n f_n + \cdots$ and $g = T^m g_m + \cdots$, concentrating on the terms of highest degree. Evidently, we may focus on the case that $n > m$. Then the polynomial

$$f - gT^{n-m}(\alpha^{n-m}(g_m))^{-1} f_n$$

has degree at most $n - 1$, so we proceed by induction on n.

We say that q is the *right quotient* and r the *right remainder*; it is clear that they are uniquely determined by f and g. If the right remainder is zero, then g is a *left divisor* of f (and q is a *right divisor*).

Notice that we do not claim that the division algorithm holds also for left division. Indeed, the example in (3.2.12) below shows that it cannot, in general. However, if α is an automorphism of $\mathcal{D}$, then elements of $\mathcal{D}[T, \alpha]$ can equally be regarded as left polynomials and we do have the division algorithm for left division. There is no reason why the left quotient and remainder should be the same as their right-handed counterparts.

3.2.7 Euclidean domains

Many of our results on skew polynomial rings depend on the division algorithm in a formal manner, rather than any intrinsic properties of the polynomial rings themselves. Such results can therefore be extended to domains for which the division algorithm holds, the Euclidean domains. This terminology is justified in (3.2.11) below. Examples of non-polynomial Euclidean domains are the familiar ring $\mathbb{Z}$ of ordinary integers, and the ring of Hurwitz quaternions (Exercise 3.2.5); we meet some further examples in our discussion of Dedekind domains.

The definition is as follows.

A *right Euclidean domain* is a domain R together with a function $\varphi : R \to \{0\} \cup \mathbb{N}$ that has the following properties:

(ED1) $\varphi(a) = 0 \Leftrightarrow a = 0$ for a in R,

(ED2) $\varphi(ab) \geq \varphi(a)$ for all nonzero a and b in R,

(ED3) the *division algorithm* holds for right division in R — given a and b in R, we have $a = bq + r$ for some q and r in R with $0 \leq \varphi(r) < \varphi(b)$.

A left Euclidean domain is defined by making the appropriate change to axiom (ED3). If R is both left and right Euclidean under the *same* function φ, we say that R is a (twosided) Euclidean domain.

In many examples, the function φ satisfies the following stronger version of axiom (ED2):

(ED2′) $\varphi(ab) = \varphi(a)\varphi(b)$ for all nonzero a and b in R.

3.2.8 Proposition

Let α be an endomorphism of the division ring $\mathcal{D}$. Then the skew polynomial ring $\mathcal{D}[T, \alpha]$ is a right Euclidean ring with $\varphi(f) = 2^{\deg(f)}$ for $f \neq 0$ and $\varphi(0) = 0$.

We could equally use the function $\varphi(f) = 1 + \deg f$ for $f \neq 0$, $\varphi(0) = 0$. $\square$

We next see how axiom (ED2) allows us to identify units.

3.2.9 Proposition

Let R be a right Euclidean domain.

(i) *Given two nonzero elements a, b of R, we have $\varphi(ab) = \varphi(a)$ if and only if b is a unit in R.*

(ii) *An element u of R is a unit in R if and only if*

$$\varphi(u) = \varphi(1) = \min\{\varphi(b) \mid b \neq 0\}.$$

(In practice, this minimum is normally 1.)

Proof

(i)   Suppose that $\varphi(ab) = \varphi(a)$. By applying (ED3) to the pair a, ab, we may write

$$a(1 - bq) = r,$$

where $\varphi(r) < \varphi(ab) = \varphi(a)$. So, by (ED2), $1 - bq = 0$ and b has a right inverse q. Next, by applying (ED2) to the products bq and $b = b(qb)$, we find that $\varphi(bq) = \varphi(b)$. So, by our previous argument, q also has a right inverse, b' say. Then $b = bqb' = b'$, and hence b is a unit with twosided inverse q.

Conversely, if $bc = 1$ for some c, then (ED2) yields

$$\varphi(a) = \varphi(abc) \geq \varphi(ab) \geq \varphi(a).$$

(ii)   Since $\varphi(a) = \varphi(1a) \geq \varphi(1)$ for any nonzero a, $\varphi(1)$ is the minimum as claimed. If $\varphi(u) = \varphi(1)$, then $\varphi(1u) = \varphi(1)$, and u is a unit by part (i). The rest of the proof is clear, on putting $a = 1$ in part (i).   □

Recall that a domain R is a right principal ideal domain if each right ideal $\mathfrak{a}$ is principal, that is, $\mathfrak{a} = aR$ for some element a in R.

3.2.10 Theorem

Let R be a right Euclidean domain. Then R is a principal right ideal domain and hence right Noetherian.

In particular, given a division ring $\mathcal{D}$ with endomorphism α, the skew polynomial ring $\mathcal{D}[T, \alpha]$ is a principal right ideal domain.

Proof

The argument is almost identical to that used in the familiar case where R is $\mathbb{Z}$, or the ordinary polynomial ring over a field. Given a nonzero right ideal $\mathfrak{a}$, choose any nonzero element $a \in \mathfrak{a}$ with $\varphi(a)$ minimal; then the division algorithm is used to show that a generates $\mathfrak{a}$. The fact that R is right Noetherian is immediate from the definition (3.1.1).   □

3.2.11 Euclid's algorithm

We can now explain the use of the term 'Euclidean domain'. In Book VII, Propositions 1 and 2, Euclid (in about 300BC) gave an algorithmic procedure for the computation of the greatest common divisor of two positive integers. (A modern translation can be found in [Euclid 1956].) This procedure can be carried out in any ring which satisfies the axioms for a Euclidean domain given in (3.2.7). In principle, the technique is the same as that which the reader has no doubt seen for $\mathbb{Z}$, with some complications caused by noncommutativity.

Suppose that a and b are nonzero elements of a domain R. A *common left divisor* of the pair a, b is an element x of R with $a = xy$ and $b = xz$ for elements y, z of R. A *greatest common left divisor* of a, b is a common left divisor d with the property that any common left divisor x of a, b is a left divisor of d, that is, $d = xw$ for some w in R. If we rephrase these definitions in terms of ideals, we see x is a common left divisor of a, b precisely when $aR + bR \subseteq xR$, and d is a greatest common left divisor of the pair precisely when the ideal dR is minimal among the principal right ideals containing $aR + bR$.

When R is a principal right ideal domain, in particular, a right Euclidean domain, a pair a, b will always have a greatest common left divisor, namely, any generator d of $aR + bR = dR$. In this case, d is a common left divisor which can be written in the form $d = as + bt$ for some s, t in R, and it is unique up to multiplication by a unit in R.

Euclid's algorithm is a recursive procedure for computing the greatest common left divisor of two elements a, b of a right Euclidean domain; essentially, we proceed by induction on the value of $\min(\varphi(a), \varphi(b))$, a technique which is known as 'the method of descent'. Suppose that $\varphi(a) \geq \varphi(b)$. By the division algorithm, axiom (ED3), we can write $a = bq + r$ with $0 \leq \varphi(r) < \varphi(b)$.

If $r = 0$, then b is already the desired greatest common left divisor. If not, we note first that $aR + bR = bR + rR$, so that the greatest commom left divisor of a, b is the same as that of b, r, and, second, that $\min(\varphi(b), \varphi(r)) < \min(\varphi(a), \varphi(b))$. After a finite nunber of applications of the division algorithm, we must come upon a pair c, d with d a left divisor of c and hence the greatest common left divisor of the original pair. Notice that the elements s, t in R with $d = as + bt$ can be recovered by reading the steps of the computation in reverse order.

The algorithm is often presented in tabular form in elementary texts on number theory, and the reader is invited to write it out in this way as an exercise. The method is machine-programmable for anyone who has access to, or can construct, a package for computation in integers, skew polynomial rings or integral quaternions.

[Wedderburn 1932] contains the first analysis of noncommutative Euclidean domains.

3.2.12 An example

We next exhibit a ring which is right but not left Noetherian. Even more: the ring is right but not left Euclidean. We follow [McConnell & Robson 1987], §1.2.11.

Let $\mathcal{K}$ be a field and let $A = \mathcal{K}(X)$ be the field of rational functions over $\mathcal{K}$ in a variable X (thus $\mathcal{K}(X)$ is the usual ring of fractions of the commutative polynomial ring $\mathcal{K}[X]$). The substitution $X \mapsto X^2$ defines an injective endomorphism α of A which is not surjective. Take $R = A[T, \alpha]$.

We claim that $RT \cap RTX = 0$, which follows from the observation that an element of RT must have the form

$$\left(\sum_i T^i a_i(X)\right)T = \sum_i T^{i+1} a_i(X^2),$$

where each a_i is a rational function in X, while an element of RTX can involve only odd powers of X. Proposition 3.1.13 now shows that R is not left Noetherian.

3.2.13 Inner order and the centre

As a preliminary to the determination of the twosided ideals of the skew polynomial ring $\mathcal{D}[T,\alpha]$, we must describe its centre, which depends on the nature of the endomorphism α.

Since $\mathcal{D}$ is a division ring, α must be injective. We say that α is a *proper endomorphism* if it is not an automorphism. An *inner automorphism* of $\mathcal{D}$ is an automorphism of the form

$$\alpha(d) = \zeta d \zeta^{-1}$$

for some fixed nonzero element ζ in $\mathcal{D}$. We say that an automorphism α has *finite inner order* if α^n is an inner automorphism for some positive integer n; otherwise, α has infinite inner order. The *inner order* of α is then the order of the image of α in the quotient group of the automorphism group by the group of inner automorphisms. Some nontrivial examples are indicated in Exercise 3.2.9.

We remark that a proper endomorphism can never become an automorphism by iteration; for, if α^n is an automorphism for some n, then α itself must be an automorphism. Slightly more subtle is the fact that if α is proper, then it can never induce an automorphism on any of the subrings $\alpha^s(\mathcal{D})$ of $\mathcal{D}$, $s \geq 0$, each of which is itself a division ring. (Here, we thank Vic Camillo for the next observation.) For suppose that α does act as an automorphism on $\alpha^s(\mathcal{D})$ for some s. Then, for any y in $\mathcal{D}$, there is an x in $\mathcal{D}$ with $\alpha(\alpha^s(x)) = \alpha^s(y)$. Since α is injective, $\alpha(x) = y$ already and α is surjective.

Now we can describe the centre $Z(\mathcal{D}[T,\alpha])$ of $\mathcal{D}[T,\alpha]$. We write $\mathcal{K}$ for the centre of $\mathcal{D}$, and let $\mathcal{D}^\alpha$ denote the set of elements d of $\mathcal{D}$ which are invariant under α, that is, with $\alpha(d) = d$.

3.2.14 Lemma

(i) *If α is a proper endomorphism, or if α is an automorphism with infinite inner order, then*

$$Z(\mathcal{D}[T,\alpha]) = \mathcal{K} \cap \mathcal{D}^\alpha.$$

(ii) *If α has finite inner order n, with $\alpha^n(d) = \zeta d \zeta^{-1}$, then*

$$Z(\mathcal{D}[T, \alpha]) = \mathcal{K}[T^n \zeta] \cap \mathcal{D}^\alpha [T^n \zeta].$$

Proof

If $d \in \mathcal{D}$ belongs to $Z(\mathcal{D}[T, \alpha])$, then d must be in the centre of $\mathcal{D}$, and $Td = dT = T\alpha(d)$, so d is invariant under α.

Suppose that $g = T^m g_m + \cdots + g_0$ is in the centre. Calculating Tg shows that $\alpha(g_i) = g_i$ for all i. But then $xg = gx$ for x in $\mathcal{D}$ gives

$$\alpha^i(x)g_i = g_i x \text{ for all } x,$$

so that either $g_i = 0$ or α^i is inner. This establishes assertion (i). It also shows that, in case (ii), if g_i is not zero, then $i = mn$ for some m and $g_{mn} = \zeta^m z_m$ with z_m in $Z(\mathcal{D})$. $\qquad\square$

3.2.15 Ideals

We now have enough information to determine the ideals of $\mathcal{D}[T, \alpha]$. By the right Euclidean property, a twosided ideal $\mathfrak{a}$ in $\mathcal{D}[T, \alpha]$ must be given as a right ideal $\mathfrak{a} = g(T)\mathcal{D}[T, \alpha]$ in which the generator $g(T)$ has the property that

$$\mathcal{D}[T, \alpha]g(T) \subseteq g(T)\mathcal{D}[T, \alpha].$$

Such a polynomial is called *semiinvariant*. If

$$\mathcal{D}[T, \alpha]g(T) = g(T)\mathcal{D}[T, \alpha],$$

then g is *normal* or *invariant*.

In combination with (3.2.14), the next two results characterize in turn the semiinvariant and normal polynomials of $\mathcal{D}[T, \alpha]$.

3.2.16 Theorem

The twosided ideals of $\mathcal{D}[T, \alpha]$ are precisely those principal right ideals whose generators are of the form $T^t g$, where $g \in Z(\mathcal{D}[T, \alpha])$ and $t \geq 0$.

Proof

It is easy to check that principal right ideals of this form are in fact twosided. For the converse, we may assume that the generator of the principal right ideal $\mathfrak{a}$ is of the form $g = T^t f$, where $f = 1 + \sum_{j \geq 1} T^j f_j$. Now $gT - Tg$ lies in $\mathfrak{a}$ and has degree at most $\deg(g) + 1$; so it must be of the form $g(Ta + b)$ for some $a, b \in \mathcal{D}$. However, $gT - Tg$ has no terms in degree either t or $t + 1$, forcing $a = b = 0$. Thus $gT = Tg$ and so each $f_j \in \mathcal{D}^\alpha$. It follows that $fT = Tf$. A similar comparison of lowest terms shows that, for any $d \in \mathcal{D}$, $dg - g\alpha^t(d)$ in $\mathfrak{a}$ must also be the zero polynomial. We then have

$$dfT^t = dT^t f = T^t f \alpha^t(d) = fT^t \alpha^t(d) = fdT^t,$$

whence $df = fd$ because $\mathcal{D}[T, \alpha]$ is a domain. Hence f is central, as required. $\qquad\square$

3.2.17 Corollary

(i) *When α is an automorphism, each semiinvariant polynomial of $\mathcal{D}[T, \alpha]$ is normal.*

(ii) *When α is a proper endomorphism, the normal polynomials of $\mathcal{D}[T, \alpha]$ are the constant polynomials in $\mathcal{K} \cap \mathcal{D}^\alpha$.*

Proof

(i)   This is clear from the description of semiinvariant polynomials in the theorem.

(ii)   Suppose that $T^t c$ is a semiinvariant polynomial where $c \in \mathcal{K} \cap \mathcal{D}^\alpha$, and choose $d \in \mathcal{D}$ outside the image of α. Then $T^t cd \in \mathcal{D}[T, \alpha]T^t c$ precisely when $t = 0$. $\qquad\square$

We also record the fact that a skew polynomial ring may have very few ideals in comparison with an ordinary polynomial ring.

3.2.18 Corollary

Suppose that the endomorphism α of $\mathcal{D}$ is either proper or an automorphism of infinite inner order. Then the ideals of $\mathcal{D}[T, \alpha]$ are $T^t \mathcal{D}[T, \alpha]$ for $t \geq 0$.

Proof

By (3.2.16), an ideal has a generator in $T^t Z(\mathcal{D}[T, \alpha])$ for some t; by (3.2.14), any element of $Z(\mathcal{D}[T, \alpha])$ is invertible in $\mathcal{D}[T, \alpha]$. $\qquad\square$

3.2.19 Total division

In the next section, we describe the structure of modules over $\mathcal{D}[T, \alpha]$ when α is an automorphism. Before doing this, we need a few words about division in $\mathcal{D}[T, \alpha]$. Let a and b be elements of $\mathcal{D}[T, \alpha]$. As before, a is a right divisor of b if $b = qa$ for some q, and q is then a left divisor of b.

We say that a is a *total divisor* of b if, for all x in $\mathcal{D}[T, \alpha]$, a is a left divisor of xb and a right divisor of bx. The notation is $a \| b$.

We remark in passing that there is no difficulty in extending the definitions of total divisors and of semiinvariant and normal to the elements of an arbitrary domain.

3.2.20 Lemma

Let a and b be elements of $\mathcal{D}[T, \alpha]$, where α is an automorphism of a division ring $\mathcal{D}$. Then the following are equivalent.

(i) *$a\|b$.*

(ii) *There is a normal element c of $\mathcal{D}[T, \alpha]$ such that*

$$b\mathcal{D}[T, \alpha] \subseteq c\mathcal{D}[T, \alpha] \subseteq a\mathcal{D}[T, \alpha].$$

(iii) *There is a normal element c which is divisible by a, on both the left and the right, and which divides b on both sides.*

Proof

We rely on the fact (3.2.17) that each ideal of $\mathcal{D}[T, \alpha]$ is generated by a normal polynomial.

(i) $\Rightarrow$ (ii).   From the definition,

$$\mathcal{D}[T, \alpha]b \subseteq a\mathcal{D}[T, \alpha] \text{ and } b\mathcal{D}[T, \alpha] \subseteq \mathcal{D}[T, \alpha]a;$$

thus we can take c to be the generator of the twosided ideal $\mathcal{D}[T, \alpha]b\mathcal{D}[T, \alpha]$.

(ii) $\Rightarrow$ (iii) $\Rightarrow$ (i).   Straightforward.   □

When $\mathcal{D}[T, \alpha]$ is far from commutative (as measured by the inner order of α), total division is a rare phenomenon (see (3.2.18)).

3.2.21 Proposition

Suppose that the automorphism α of a division ring $\mathcal{D}$ has infinite inner order, and let f be a monic polynomial.

Then $f\|g$ for some g if and only if $f = T^h$ for some $h \geq 0$.

Proof

From the description of normal polynomials given in (3.2.16), f must divide some T^t; but an easy argument shows that the monomial T^t has only monomial factors.   □

3.2.22 Unique factorization

The familiar result that an integer has a unique prime factorization can be extended to Euclidean domains and, indeed, to noncommutative principal ideal domains. We content ourselves with a brief sketch of the results.

For a general ring, we seek factorizations into irreducible elements: an element r of a ring R is *irreducible* (or an *atom*) if there is no factorization $r = st$ of r in R except for those trivial factorizations in which either s or t is a unit of R. (The distinction between prime and irreducible elements is discussed in section 5.1, in particular (5.1.3) and (5.1.28).)

When R is a Euclidean domain, commutative or not, we know from (3.2.9)(i) that, in a nontrivial factorization $a = bc$, we have both $\varphi(a) > \varphi(b)$ and $\varphi(a) > \varphi(c)$. Thus an induction argument on the value of $\varphi(a)$ shows that a (nonzero, nonunit) element a of R has a factorization into irreducible elements. The uniqueness of this factorization is easy to describe when R is commutative. We say that two nonzero elements b and c of R are *associates* if $b = uc$ for some unit u of R. Then, if

$$a = p_1 \cdots p_r = q_1 \cdots q_s$$

are two irreducible factorizations of a, we have $r = s$ and we can renumber the irreducible factors so that p_i and q_i are associates for $i = 1, \ldots, r$. The proof of this uniqueness result is almost exactly the same as the familiar argument over $\mathbb{Z}$ – see Exercise 3.2.1.

The situation is more complicated when R is not commutative. We say that two elements b and c of an arbitrary ring R are *right similar* if there is a right R-module isomorphism $R/bR \cong R/cR$, *left similar* if $R/Rb \cong R/Rc$, and *similar* if they are both left and right similar. Then, in a noncommutative Euclidean domain, the irreducible factors of an element are unique up to similarity. The proof of this result and some illustrations of the definitions are outlined in Exercises 4.1.3, 4.1.4 and 4.2.7.

The above results also hold for noncommutative twosided principal ideal domains, but the arguments are more technical. Details can be found in [Jacobson 1968], Chapter 3, Theorem 8, and in [Cohn 1985], Chapter 3. (We do in fact derive the unique factorization result for commutative principal ideal domains as a consequence of our discussion of Dedekind domains (5.1.25).)

3.2.23 Further developments

(1)   A skew polynomial ring as defined in this text is a special case of a more general construction, which shares the same name. In the generalization, multiplication is given by the rule

$$a \cdot T = T \cdot \alpha a + \delta a,$$

where δ is a *derivation* of A, that is, δ is an additive map from A to itself with

$$\delta(ab) = a\delta(b) + \delta(a)\alpha(b)$$

always. Such rings are also known as *Ore extensions* in tribute to the fundamental paper [Ore 1933]. Rings of this type remain an active subject of research, since they occur in the investigation of quantum groups and related concepts.

Discussions of the basic theory of the more general type of skew polynomial ring can be found in [McConnell & Robson 1987] and [Rowen 1988], while some recent developments are given in [Goodearl 1992], [Jordan 1993], and [Lam & Leroy 1992].

(2)   Very general versions of the Euclidean algorithm are considered in detail in [Cohn 1985].

Exercises

3.2.1   Let R be a commutative Euclidean domain and let p be an irreducible element of R. Prove the following assertions.

(a) The ideal pR is maximal.

(b) If $b \in R$ and p does not divide b, then

$$1 = sp + tb \text{ for some } s, t \in R.$$

(c) If $p|bc$ where $b, c \in R$, then either $p|b$ or $p|c$ (perhaps both hold).

Hence prove the claim in (3.2.22): if $a \in R$ and

$$a = p_1 \cdots p_r = q_1 \cdots q_s$$

are two irreducible factorizations of a, then $r = s$ and the irreducible factors can be re-arranged so that p_i and q_i are associates for $i = 1, \ldots, r$.

3.2.2   Let α be an inner automorphism of the division ring $\mathcal{D}$, say $\alpha(d) = \zeta d \zeta^{-1}$ for some fixed nonzero element ζ in $\mathcal{D}$. Show that $\mathcal{D}[T, \alpha] = \mathcal{D}[T\zeta]$, the ordinary polynomial ring.

3.2.3   Let α be an endomorphism of a ring A, and let

$$A^\alpha = \{a \mid \alpha(a) = a\}$$

be the set of elements which are invariant under α.

Show that α can be extended to an endomorphism, also called α, of $A[T, \alpha]$, and that $(A[T, \alpha])^\alpha = A^\alpha[T]$.

3.2.4   Show that Axiom (ED2) for a right Euclidean domain is redundant inasmuch as, if R is a domain for which $\theta : R \to \{0\} \cup \mathbb{N}$ satisfies

Axioms (ED1) and (ED3), then the function $\varphi : R \to \{0\} \cup \mathbb{N}$ defined by

$$\varphi(a) = \min\{\theta(au) \mid u \in U(R)\}$$

satisfies (ED1), (ED2) and (ED3) [Nagata 1978].

For the proof of (ED2), it is helpful to show that, if b has $\varphi(ab) \leq \varphi(ac)$ for all $c \neq 0$, then, from the division algorithm applied to the pair a, ab, we have $b \in U(R)$ (see (3.2.9)).

3.2.5   **Hurwitz quaternions.**

In the notation of Exercise 2.2.3, let $\mathbb{H} = \mathbb{Q}[i, j, k]$ be the quaternion algebra over the rational numbers and write $w = (1 + i + j + k)/2$.

The ring H of *Hurwitz* or *integral quaternions* is the subring

$$H = \mathbb{Z} \cdot 1 + \mathbb{Z}i + \mathbb{Z}j + \mathbb{Z}w.$$

Verify that H is indeed a ring.

Show that, if $x \in H$, then $\overline{x} = x_0 1 - x_1 i - x_2 j - x_3 k$ is in H and $x + \overline{x}$ and $x \cdot \overline{x} = x_0^2 + x_1^2 + x_2^2 + x_3^2$ are integers, and that x is a zero of the integer polynomial

$$c_x(T) = T^2 - (x + \overline{x})T + x\overline{x}.$$

For x in $\mathbb{H}$, put $\varphi(x) = N(x) = x \cdot \overline{x}$. Show that for any element x in $\mathbb{H}$, there is an element h of H with $\varphi(x - h) < 1$. Deduce that H is a (twosided) Euclidean domain which satisfies condition (ED2'): $\varphi(a)\varphi(b) = \varphi(ab)$.

Show that the group of units $U(H)$ of H has order 24.

Prove that the following elements of H are irreducible: $d = 1 + i$, $t = 1 + i + j$ and $f = 2 + i$.

Verify that $2 = d\overline{d}$, $3 = t\overline{t}$, $5 = f\overline{f}$ and that $d = u\overline{d}$ for some unit u of H. Show that there is no unit x of H with $t = x\overline{t}$ or $f = x\overline{f}$.

Remark. These results can be interpreted informally as meaning that 2 becomes a square in H while 3 and 5 split into distinct primes – this foreshadows phenomena that we investigate in section 5.3 for quadratic fields.

It can be shown that any odd prime p in $\mathbb{Z}$ splits into two distinct primes in H. This in essence is a consequence of Euler's Theorem that the congruence $1 + x^2 + y^2 \equiv 0 \pmod{p}$ is always solvable for odd p. A detailed discussion is given by [Dickson 1938], §91. The ring H is designated thus in honour of its first investigator, A. Hurwitz (see [Hurwitz 1896]), while the symbol $\mathbb{H}$ for the quaternion algebra commemorates Sir W. R. Hamilton, whose original account is to be found in [Hamilton 1853].

3.2.6 Opposites.

Given an endomorphism α of the ring A, define α° on the opposite ring A° by $\alpha^\circ(a^\circ) = (\alpha a)^\circ$. Verify that α° is an endomorphism of A° and that $(A[T, \alpha])^\circ$ can be identified as the skew left polynomial ring $A^\circ_\ell[U, \alpha^\circ]$ over A° corresponding to α° (in which $Ua^\circ = \alpha^\circ(a^\circ)U$).

3.2.7 Let $R = \mathcal{K}(X)[T, \alpha]$ where $\alpha(X) = X^2$, as in (3.2.12).

Find the following:

(i) the units of R;
(ii) the centre of R;
(iii) the (twosided) ideals in R;
(iv) the right ideals in R.

3.2.8 Let $\mathcal{K}$ be a field of positive characteristic p and define the endomorphism α of the field of rational functions $\mathcal{K}(X)$ by $\alpha(X) = X + 1$.

Show that α has inner order p. Repeat the previous exercise for $\mathcal{K}(X)[T, \alpha]$.

3.2.9 Inner order.

This exercise exhibits a method for constructing some nontrivial examples of automorphisms of given inner order.

Let $\mathcal{Q}$ be a field with an automorphism γ of $\mathcal{Q}$ of infinite order – for example, take $\mathcal{Q} = \mathbb{Q}(Y)$ for a variable Y and $\gamma : Y \mapsto Y + 1$. Now take $R = \mathcal{Q}[X, \gamma^n]$ for some natural number $n > 1$. Then R is a noncommutative (twosided) principal ideal domain (3.2.10), and so, by a result to be proved in [BK: CM], §6.2, R has a division ring of fractions $\mathcal{D}$.

The automorphism γ extends to R and hence to $\mathcal{D}$ by acting on the coefficients of a right polynomial in the obvious way, and, by construction, γ^n is a nontrivial inner automorphism of $\mathcal{D}$.

To see that n is the inner order of γ, consider the skew polynomial ring $\mathcal{D}[T, \gamma]$. If the inner order of γ is m, then the centre of $\mathcal{D}[T, \gamma]$ is $\mathcal{F}[T^m\zeta]$ where $\mathcal{F}$ is the centre of $\mathcal{D}$ and $\gamma^m(d) = \zeta d\zeta^{-1}$ for $d \in \mathcal{D}$, by Lemma 3.2.14. But the centre must contain $T^n X^{-1}$, which cannot happen except trivially, since the only possibility is that

$$T^n X^{-1} = (T^m \zeta)^{n/m} = T^n \zeta^{n/m}.$$

The remaining problems give counterexamples to various relaxations of the theorems on skew polynomial rings. They are taken from [McConnell & Robson 1987], §1.2.11.

3.2.10   Suppose that A is right Noetherian and that α is a surjective endomorphism of A. Show that α is also injective.

3.2.11   Define an endomorphism α of the polynomial ring $\mathcal{K}[X]$ by $\alpha(f(X)) = f(0)$. Show the following.

    (i) $R = \mathcal{K}[X][T,\alpha]$ is not a domain.

    (ii) The sums $RTX + RTX^2 + RTX^3 + \cdots$ and $TXR + T^2XR + T^3XR + \cdots$ are both direct.

    (iii) R is neither left nor right Noetherian.

3.2.12   Take instead α to be given by $X \mapsto X^2$ (so that we have a subring of the ring considered in (3.2.12)). Show that R is neither left nor right Noetherian.

3.3 MODULES OVER SKEW POLYNOMIAL RINGS

Our principal objective in this section is to describe the module theory of a skew polynomial ring $\mathcal{D}[T,\alpha]$, where $\mathcal{D}$ is a division ring and α is an automorphism of $\mathcal{D}$. However, the main tool that we use, a technique for the reduction of a matrix to diagonal form by means of row and column operations, depends only on the fact that $\mathcal{D}[T,\alpha]$ is a twosided Euclidean domain. Since we do have applications for this more general result, we state and prove our results in the wider setting. This reduction argument has the added interest that it foreshadows a topic of considerable importance in K-theory, namely, the extent to which row and column operations suffice to transform an invertible matrix into a diagonal matrix.

As we have noted at several places, a square matrix A over a field $\mathcal{K}$ gives rise to a module over the ordinary polynomial ring $\mathcal{K}[T]$, given by T acting as A on $\mathcal{K}^n$. Thus our analysis of $\mathcal{K}[T]$-modules gives parallel results about the forms into which A is transformed by conjugacy. These results are obtained in a series of exercises at the end of this section. Our analysis of modules over skew polynomial rings also allows us to exhibit some examples of the phenomenon of non-cancellation.

As usual, we state our results in terms of right modules. It will be clear that corresponding results hold for left modules, since we work with twosided Euclidean domains.

3.3.1 Elementary operations

Let R be a (twosided) Euclidean domain. Our description of R-modules comes through a matrix reduction which is achieved by a sequence of 'elementary operations' on a given matrix to diagonalize it. Specifically, we allow the following *elementary operations* to be performed on a rectangular $n \times p$ matrix A with entries in R. (These operations can of course be performed on a matrix over an arbitrary ring.)

(E1) Interchange two rows or columns of A.

(E2R) Add to any one row of A a left multiple of a different row.

(E2C) Add to any one column of A a right multiple of a different column.

The effect of any of these operations can be achieved by multiplying A by a suitable invertible matrix. For the row operations, first perform the operation on the $n \times n$ identity matrix to obtain a matrix P and then compute PA. (This works because, if P has ith row r_i, then PA has ith row $r_i A$.) For a column operation, act on the $p \times p$ identity matrix to obtain Q and then take AQ. (Again, if Q has jth column c_j, then AQ has jth column Ac_j.) Since the operations are reversible, the matrices P and Q are invertible.

The key result shows that a matrix can be transformed into a normal form (2.2.10) by elementary operations.

3.3.2 The Diagonal Reduction Theorem

Let R be a (twosided) Euclidean domain and let $A = (a_{ij})$ be an $n \times p$ matrix over R.

Then there exist invertible matrices P and Q over R such that

$$PAQ = \mathrm{diag}(e_1, \ldots, e_r, 0, \ldots, 0),$$

a diagonal matrix, with $e_1 \| e_2 \| \cdots \| e_r$. Moreover, P may be obtained by applying a sequence of row operations to the identity matrix, and Q may be obtained by applying a sequence of column operations to the identity matrix.

Proof

We may assume that $A \neq 0$. We argue by induction on the size (n, p) of A, and, for each given size, on $N(A) = \min\{\varphi(a_{ij}) \mid a_{ij} \neq 0\}$. The technique is to apply a sequence of elementary row and column operations; by the remarks above, each step can be represented by pre- or post-multiplication by an invertible matrix, and so therefore can the whole sequence.

<u>Step 1.</u>   Consider first, as the basic step in the induction on (n, p), the case where $A = (a_1, \ldots, a_p)$ is $1 \times p$. Exchange to ensure that $\varphi(a_1) = N(A)$. If

a_1 left divides a_i, then a column operation converts a_i to 0. If a_1 does not left divide some a_i, we have $a_i = a_1 q_i + r_i$ and $\varphi(r_i) < \varphi(a_1)$, by the division algorithm. Thus a column operation reduces $N(A)$. By induction on $N(A)$, we can therefore reduce to the row matrix $(e, 0, \ldots, 0)$. The case $p = 1$ is similarly established.

Now consider A in general.

<u>Step 2.</u>   Exchange rows and exchange columns so that $\varphi(a_{11}) = N(A)$. As in Step 1, we can reduce $N(A)$ unless a_{11} left divides every entry of the first row, and similarly we may assume a_{11} right divides every entry in the first column. Thus we may assume that

$$A = \left(\begin{array}{c|ccc} a_{11} & 0 & \cdots & 0 \\ \hline 0 & & & \\ \vdots & & B & \\ 0 & & & \end{array} \right),$$

with B an $(n-1) \times (p-1)$ matrix.

Row and column operations performed on B can also be performed on A without disturbing its block form, so by the induction hypothesis we can further reduce to the case

$$A = \mathrm{diag}(a_1, b_2, \ldots, b_r, 0, \ldots, 0) \quad \text{with } b_2 \| \cdots \| b_r.$$

<u>Step 3.</u>   If $a_1 \| b_2$, we are done. If, say, a_1 does not left divide db_2 for some d, add d times row 2 to row 1. Then, as in Step 1, addition of a suitable right multiple of the first column to the second reduces $N(A)$, allowing the induction to go through. $\square$

3.3.3 Rank and invariant factors

The integer r in the above theorem is the *rank* of A and $e_1, \ldots, e_r$ are the *invariant factors* of A; the diagonal matrix $\mathrm{diag}(e_1, \ldots, e_r, 0, \ldots, 0)$ is the *invariant factor form* or *Smith normal form* of A. The rank of A is unique and the invariant factors of A are essentially unique, so that we can speak unambiguously of the invariant factor form for A.

When R is commutative, the uniqueness of the rank is a consequence of the fact that a commutative domain can be embedded in a field (1.1.12), since the rank of a matrix over a field is unique. This argument also works in the noncommutative case, because a right principal ideal domain can be embedded in a division ring – see [BK: CM], (6.2.16).

Again when R is commutative, the uniqueness result for invariant factors is that, if $\{e_1, \ldots, e_r\}$ and $\{f_1, \ldots, f_r\}$ are two sets of invariant factors, then, for each i, the factors e_i and f_i are associated, that is, $e_i = f_i u_i$ for some unit u_i. A proof is outlined in Exercise 3.3.2 below. In particular, in a commutative polynomial ring $\mathcal{K}[T]$, the invariant factors are uniquely determined by the requirement that they be monic polynomials. In the ring of integers $\mathbb{Z}$, the invariant factors are uniquely specified by the requirement that they should be positive.

When R is not commutative, the invariant factors are unique up to similarity (3.2.22). A proof of this claim in general would take us too far afield; details can be found in [Cohn 1985], §8, where the results are extended to noncommutative (twosided) principal ideal domains.

3.3.4 The structure of modules

For the moment, let R be an arbitrary ring. Recall that a right R-module M is cyclic if it can be generated by a single element, that is, $M = mR$ for some m in M. If we write $\mathfrak{a} = \{r \in R \mid mr = 0\}$, then $\mathfrak{a}$ is a right ideal of R and $M \cong R/\mathfrak{a}$ as a right R-module. Conversely, such a quotient module of R is obviously cyclic.

We now return to the situation that R is a Euclidean domain. Our aim is to apply the Diagonal Reduction Theorem to describe R-modules in terms of cyclic modules; for this we need a preliminary lemma. Recall from (3.2.10) that every right Euclidean domain is a principal right ideal domain.

3.3.5 Lemma

Let K be a submodule of the free right R-module R^n of finite rank n, where R is a principal right ideal domain. Then K is free, with $\operatorname{rank}(K) \leq n$.

Proof

We argue by induction on n. If $n = 1$, K is a right ideal in R and so is either the zero ideal (free of rank zero) or a nonzero principal right ideal (a free right module of rank one).

For $n > 1$, let π be the homomorphism from R^n to R given by projection to the nth term. Thus $\operatorname{Ker} \pi$ is isomorphic to the free submodule R^{n-1}. If $\pi K = 0$, we are done; if not, $\pi K = aR$ is a principal right ideal. Choose any k in K with $\pi k = a$; then $K \cong K \cap (\operatorname{Ker} \pi) \oplus kR$ and $\operatorname{Ker} \pi$ has rank $n - 1$.

$\square$

We come to the fundamental result, whose title will be justified in (3.3.7).

3.3.6 The Invariant Factor Theorem

Let M be a finitely generated right R-module, with R a Euclidean domain. Then

$$M \cong R/e_1 R \oplus \cdots \oplus R/e_\ell R \oplus R^s,$$

where $e_1, \ldots, e_\ell$ are nonunits in R, $e_1 \| e_2 \| \cdots \| e_\ell$ and $s \geq 0$.

Proof

Suppose that M has n generators. Then there is a surjection π from R^n to M, whose kernel K is free, of rank $p \leq n$ by the lemma. Choose a basis for K, say $\{f_1, \ldots, f_p\}$, and let $\{c_1, \ldots, c_n\}$ be the standard basis of R^n. Then we can describe the inclusion map ι from K to R^n by an $n \times p$ matrix $A = (a_{ij})$, where

$$\iota f_j = e_1 a_{1j} + \cdots + e_n a_{nj} = \begin{pmatrix} a_{1j} \\ \vdots \\ a_{nj} \end{pmatrix},$$

the jth column of A. (Recall that we regard the standard free module as consisting of 'column' vectors.)

By the Diagonal Reduction Theorem, there exist invertible matrices P and Q and nonzero elements $e_1, e_2, \ldots, e_r$ of R with

$$PAQ = D = \mathrm{diag}(e_1, \ldots, e_r, 0, \ldots, 0)$$

and

$$e_1 \| e_2 \| \cdots \| e_r.$$

As we can see from (2.2.11), this means that we can choose new bases of K and R^n so that ι has matrix D. Explicitly, there are bases $\{g_1, \ldots, g_n\}$ of R^n and $\{h_1, \ldots, h_p\}$ of K such that

$$\iota h_j = g_j e_j \text{ for } j = 1, \ldots, r$$

and

$$\iota h_j = 0 \text{ for } j = r+1, \ldots, p.$$

However, the inclusion map ι is tautologously an injection, so we must have $r = p$.

Since

$$R^n = g_1 R \oplus \cdots \oplus g_p R \oplus g_{p+1} R \oplus \cdots \oplus g_n R$$

and

$$K = g_1 e_1 R \oplus \cdots \oplus g_p e_p R$$

we see that

$$M = \pi g_1 R \oplus \cdots \oplus \pi g_n R$$

with

$$\pi g_i R \cong R/e_i R \text{ for } i = 1, \dots, p$$

and

$$\pi g_i R \cong R \text{ for } i = p + 1, \dots, n.$$

It may happen that some of the invariant factors of A, say $e_1, \dots, e_q$, are units. The corresponding summands M are then zero modules, which we omit from the expression for M after renumbering the remaining $\ell = p - q$ nonunit invariant factors. Collecting together the $s = n - p$ summands that are isomorphic to R, we obtain the desired expression. $\qquad\square$

3.3.7 Rank and invariant factors for modules

Extending the definitions given in (3.3.3) from matrices to modules, we say that the elements $e_1, \dots, e_\ell$ of R occurring in the above theorem are the *invariant factors* of M, and that the integer s is the *rank* of M. A direct sum decomposition of M as in the theorem is called an *invariant factor decomposition* of M. The rank and the number ℓ of invariant factors of M are unique, and if $\{e_1, \dots, e_\ell\}$ and $\{f_1, \dots, f_\ell\}$ are sets of invariant factors for M, then e_i and f_i are similar for each i. In the commutative case, this reduces to the more familiar result that e_i and f_i are associated, that is, $e_i = f_i u_i$ for some unit u_i.

We do not prove the general result in this text. The uniqueness of the rank and invariant factors in the commutative case will emerge as a consequence of our discussion of module theory over Dedekind domains – see Exercise 6.3.5. Proofs of uniqueness can be found in [Cohn 1982], §10.6, for commutative principal ideal domains, and in [Cohn 1985], §8.2, for the general case.

Notice that the uniqueness results for modules are not immediate consequences of those for matrices, since a given module may be described in many different ways as a homomorphic image of a free module, and there is no obvious connection between the corresponding matrices.

The terminology introduced above explains why we have called Theorem 3.3.6 the 'Invariant Factor Theorem'. The literature contains a number of similar results that have a variety of names. In the commutative case, the uniqueness assertions are usually part of the statement of the theorem. Some authors confine the use of the title 'Invariant Factor Theorem' to the case that there is no free component of the module, that is, $s = 0$.

3.3.8 Non-cancellation

Skew polynomial rings are an excellent source of examples for the phenomenon of *non-cancellation*: this happens for a ring R when there are (projective) right R-modules P and Q so that $P \oplus R^n \cong Q \oplus R^n$ but $P \not\cong Q$.

More precisely, a module P is *stably free* if $P \oplus R^n \cong R^m$ for some integers m and n. We give some criteria which allow the construction of stably free but non-free ideals in certain rings.

Our treatment follows [Rowen 1988], 8.4.40ff. We need a preliminary result.

3.3.9 Lemma

Let A be a right Noetherian ring, let α be an automorphism of A and let $R = A[T, \alpha]$ be the skew polynomial ring.

Given an element $r \in R$ and a monic (right) polynomial g in R, then there is a monic polynomial f with $rf \in gR$.

Proof

Let $\rho : R \to (rR + gR)/gR$ be left multiplication by r, so that $\operatorname{Ker} \rho = \{p \in R \mid rp \in gR\}$. Suppose that g has degree $n + 1$. Then R/gR is generated as a right A-module by $\overline{1}, \overline{T}, \ldots, \overline{T}^n$. Since A is right Noetherian, R/gR is Noetherian by (3.1.4), and hence the submodule $\operatorname{Im} \rho$ has a finite set of generators of the form $\overline{r}, \overline{rT}, \ldots, \overline{rT}^m$ for some m. Thus there is a relation

$$\overline{r}f_0 + \overline{rT}f_1 + \cdots + \overline{rT^m}f_m + \overline{rT^{m+1}} = 0, \text{ with } f_i \in A,$$

which gives

$$f_0 + Tf_1 + \cdots + T^m f_m + T^{m+1}$$

as the desired monic polynomial in $\operatorname{Ker} \rho$.   □

Note that without the key conclusion that f is monic, the above result would be a consequence of the right-handed version of (3.1.13).

This brings us to the main result.

3.3.10 Theorem

Let A be a right Noetherian domain with an automorphism α and let $R = A[T, \alpha]$ be the skew polynomial ring.

Suppose that there are elements a and b in A such that

(i) *a is not invertible,*

(ii) *$aR + (T + b)R = R$.*

Then the module $P = aR \cap (T + b)R$ has the following properties.

(a) *P is a projective right ideal in R,*

(b) $P \oplus R \cong R^2$,

(c) P *is not isomorphic to* R,

(d) P *is not free.*

Proof

Define $\pi : R^2 \to R$ by $(r, s) \mapsto ar - (T + b)s$. By (ii), π is surjective, and it is split since R is free. Further, since R is a domain by (3.2.5), the surjection $\operatorname{Ker} \pi \to P$, $r \mapsto ar$, is an isomorphism. Thus (b) is true, and (a) follows. The final assertion will follow from (c), since R has invariant basis number by (3.1.9).

We prove (c) by contradiction. Suppose that $P \cong R$ as a right R-module, so that $aR \cap (T + b)R = xR$ for some x in R. We argue in three steps to contradict (i) above, as follows.

Step 1. $\deg(x) = 1$.

Since A is right Noetherian, we can find nonzero u and v in A with $au = b(\alpha a)v$ (3.1.13). This gives

$$a(Tv + u) = (T + b)\alpha(a)v \in aR \cap (T + b)R.$$

Then $(T + b)\alpha(a)v = xs$ for some s in R; since R is a domain, this can happen only if $\deg(x) = 1$ and $s \in A$ or vice versa. However, $x \in (T + b)R$, so x must have degree 1.

Step 2. $x = a(Tc + d)$, where c is a unit in A and $d \in A$.

Since x has degree 1, we certainly have an equality $x = a(Tc + d)$ with $c, d \in A$. On the other hand, the lemma above shows that there is a monic polynomial f with $af \in (T + b)R$. This means that $af \in P$ and so $af = xh$ for some h; thus $f = (Tc + d)h$ and c is indeed a unit.

Step 3.   The contradiction.

Using (ii) again, we have $1 = ay + (T + b)z$ for some y and z in R. Since c is a unit, we can write $y = (Tc + d)q + e$ with e in A. (This is easily seen by induction on $\deg(y)$.) Then, because $a(Tc + d) = x \in (T + b)R$, we have

$$1 = a(Tc + d)q + ae + (T + b)z \in ae + (T + b)R.$$

A direct calculation now forces a to be a unit, contrary to (i). $\square$

As an application, we have

3.3.11 Theorem

Let $\mathcal{D}$ be a noncommutative division ring. Then there is a stably free, non-free, right ideal in the polynomial ring $\mathcal{D}[X_1, X_2]$.

Proof

Choose two elements e and f of $\mathcal{D}$ which do not commute, and set $A = \mathcal{D}[X_1]$, $T = X_2$, $a = X_1 + e$ and $b = f$ in the preceding theorem (here α is the identity). Then

$$a(T + b) - (T + b)a = ab - ba = ef - fe$$

is a unit in $\mathcal{D}$ and belongs to $a\mathcal{D}[X_1, X_2] + (T + b)\mathcal{D}[X_1, X_2]$. $\square$

3.3.12 Serre's Conjecture

This celebrated conjecture asserts that if $\mathcal{K}$ is a (commutative) field, then any stably free module over the (commutative) polynomial ring $\mathcal{K}[T_1, \ldots, T_n]$ must be free, whatever the value of $n \geq 1$. It was proved independently by Quillen and Suslin; an account is given in [Lam 1978]. Thus, for the above corollary, it is essential that the division ring $\mathcal{D}$ is not a field.

3.3.13 Background and developments

As we have indicated previously, these results can be extended to noncommutative (twosided) principal ideal domains, but with somewhat different arguments. A comprehensive account is given in [Cohn 1985], §8, and that author also analyses the structure of a skew polynomial ring $\mathcal{K}[T, \alpha, \delta]$ having a nontrivial derivation δ. A discussion of the uniqueness of the invariant factors of matrices and modules over noncommutative principal ideal domains can also be found in [Guralnick, Levy & Odenthal 1988]; these authors use methods from algebraic K-theory, which is one reason why we do not attempt to cover the general uniqueness results in this text.

When R is an arbitrary right Noetherian ring and M is a finitely generated right R-module, there is an exact sequence

$$R^m \xrightarrow{\alpha} R^n \to M \to 0;$$

such a sequence is called a *free presentation* of M. Since the homomorphism α is given by left multiplication by a matrix A, one might hope to describe M in terms of a standard form for A. However, this approach appears to be very difficult to implement when R is not a principal ideal domain. The problem is discussed in [Guralnick & Levy 1988].

Exercises

3.3.1   Let A be an $n \times n$ square matrix over a Euclidean domain R. Show that A is invertible (over R) if and only if the invariant factor form of A is $\operatorname{diag}(u_1,\ldots,u_n)$ with each u_i a unit of R.

3.3.2   Uniqueness of rank and invariant factors

Let R be a commutative Euclidean domain and let A be an $n \times p$ matrix over R. A $k \times k$ *minor* of A is, by definition, the determinant of a $k \times k$ submatrix of A, where $1 \le k \le \min(n,p)$. For fixed k, the kth *Fitting ideal* $\operatorname{Fit}_k(A)$ of A is the ideal of R generated by the set of all $k \times k$ minors of A.

Show that for any $p \times q$ matrix B, the columns of AB are R-linear combinations of those of A. Deduce that a $k \times k$ minor of AB is an R-linear combination of $k \times k$ minors of A and hence that $\operatorname{Fit}_k(AB) \subseteq \operatorname{Fit}_k(A)$. Show also that $\operatorname{Fit}_k(AB) \subseteq \operatorname{Fit}_k(B)$.

Prove that for an $n \times n$ invertible matrix P and a $p \times p$ invertible matrix Q, $\operatorname{Fit}_k(PAQ) = \operatorname{Fit}_k(A)$ for all k.

Let $E = \operatorname{diag}(e_1,\ldots,e_r,0,\ldots,0)$ be in invariant factor form. Show that $\operatorname{Fit}_k(E) = e_1 \cdots e_k R$ for $k = 1,\ldots,r$, and $\operatorname{Fit}_k(E) = 0$ for $k > r$.

Deduce that if E and $F = \operatorname{diag}(f_1,\ldots,f_s,0,\ldots,0)$ are both invariant factor forms of the same matrix A, then $r = s$ and, for $i = 1,\ldots,r$, $f_i = e_i u_i$ for some unit u_i of R.

3.3.3   A right R-module M, for R an arbitrary ring, is said to be *bounded* if $\operatorname{Ann}(M) \ne 0$, where the twosided ideal $\operatorname{Ann}(M) = \{r \in R \mid Mr = 0\}$ is the annihilator of M.

Let $\mathfrak{b}$ be a right ideal of R. Show that the cyclic module $R/\mathfrak{b}$ is bounded if and only if $\mathfrak{b}$ contains a nonzero twosided ideal of R.

Suppose now that M is a finitely generated right $\mathcal{D}[T,\alpha]$-module, where α is an automorphism of the division ring $\mathcal{D}$, and that the rank of M is 0. Show that there is at most one unbounded term in the invariant factor decomposition of M.

3.3.4   Let α be an automorphism of infinite inner order of the division ring $\mathcal{D}$. Find the unbounded cyclic $\mathcal{D}[T,\alpha]$-modules, and list the possible invariant factor decompositions of finitely generated $\mathcal{D}[T,\alpha]$-modules.

3.3.5   Show that, for $n \ge 2$, $\mathcal{D}[X_1,\ldots,X_n]$ has a stably free, non-free right ideal.

3.3.6   Let m be an even (nonzero) *squarefree* integer (that is, in the prime factorization $m = 2p_1 \cdots p_s$, the primes $2, p_1, \ldots, p_s$ are all distinct).

Put $\mathcal{Q} = \mathbb{Z}[m^{-1}]$, the subring of $\mathbb{Q}$ consisting of those fractions that can be written a/m^t with $a \in \mathbb{Z}$ and $t \ge 0$, and let $A = \mathcal{Q}[\sqrt{m}]$. Then A has an automorphism α given by

$$\alpha(u + v\sqrt{m}) = u - v\sqrt{m}.$$

Let $a = (1+\sqrt{m})/2$. Show that the ideal $A[T,\alpha]a \cap A[T,\alpha](T+\alpha a)$ is stably free but not free.

3.3.7   A converse to (3.2.10), based on (3.3.10) ([McConnell & Robson 1987]), §11.2.12).

Suppose that α is an automorphism of the ring A and that $R = A[T,\alpha]$ is a principal right ideal domain. Show that the following hold.

   (i) A is a domain.

  (ii) There is an order preserving injective map from the set of right ideals of A to those of R, namely $a \mapsto aR$; thus A is right Noetherian.

 (iii) Take $a \neq 0$ in A. Then $aR + TR$ is principal, generated by some b in A.

 (iv) We have that b has a right inverse, and so $aR + TR = R$.

  (v) If a is not a unit of A, then $aR \cap TR$ is not principal.

 (vi) Hence A is a division ring.

The remaining exercises apply the structure theory of modules over an ordinary polynomial ring $\mathcal{K}[T]$, $\mathcal{K}$ a (commutative) field, to determine normal forms of matrices over $\mathcal{K}$. There is a similar discussion for proper skew polynomial rings which we do not give in this text, apart from some special results in the one-dimensional case where a division ring $\mathcal{D}$ is considered as a $\mathcal{D}[T,\alpha]$-module (Exercise 4.2.7). Full details can be found in [Cohn 1985], §8.5.

3.3.8   Let $\mathcal{K}$ be a field and let $h(T) = T^n - h_{n-1}T^{n-1} - \cdots - h_1T - h_0$ be a polynomial in the non-skew polynomial ring $\mathcal{K}[T]$.

Write $M = \mathcal{K}[T]/h(T)\mathcal{K}[T]$ and put $f_0 = \bar{1} \in M$ and $f_i = f_0T^i$ for $i = 1,\ldots,n-1$. Verify that $\{f_0,\ldots,f_{n-1}\}$ is a $\mathcal{K}$-basis for M and that the action of T on M is described by the $n \times n$ matrix

$$B = \begin{pmatrix} 0 & 0 & 0 & \ldots & 0 & h_0 \\ 1 & 0 & 0 & \ldots & 0 & h_1 \\ 0 & 1 & 0 & \ldots & 0 & h_2 \\ \vdots & \vdots & \vdots & \ddots & \vdots & \vdots \\ 0 & 0 & 0 & \ldots & 0 & h_{n-2} \\ 0 & 0 & 0 & \ldots & 1 & h_{n-1} \end{pmatrix}$$

(in the terminology of (1.2.2), (iv), T acts as B).

The matrix B is called the *companion matrix* of $h(T)$.

3.3.9   Using the above result together with (3.3.6) and Exercise 2.1.1, prove the following.

Given an $n \times n$ matrix A over a field $\mathcal{K}$, there is an invertible $n \times n$ matrix Γ such that $\Gamma^{-1}A\Gamma = \mathrm{diag}(B_1, \ldots, B_k)$, a diagonal block matrix in which each block B_i is the companion matrix of some polynomial h_i over $\mathcal{K}$.

The matrix $\Gamma^{-1}A\Gamma$ is known as the *rational canonical form* of A.

Show further that the polynomials h_i are the invariant factors of the matrix $T \cdot I_n - A$ as computed in (3.3.2).

3.3.10   Elements a and b of a commutative ring R are said to be *coprime* if the ideals aR and bR are comaximal, that is, $aR + bR = R$ (2.6.5).

Given that a, b are coprime, show that $R/abR \cong R/aR \oplus R/bR$ as (right) R-modules.

Extend the definition and result to a finite set $\{a_1, \ldots, a_k\}$ of *pairwise coprime* elements of R – see Exercise 3.2.3.

3.3.11   Let $\mathcal{K}$ be a field and let $M = \mathcal{K}[T]/(T - \alpha)^n\mathcal{K}[T]$, where $\alpha \in \mathcal{K}$. Put $x_0 = \bar{1} \in M$ and $x_i = x_0(T - \alpha)^i$ for $i = 1, \ldots, n - 1$. Verify that $\{x_0, \ldots, x_{n-1}\}$ is a $\mathcal{K}$-basis of M and that the matrix giving the action of T with respect to this basis is

$$
J = \begin{pmatrix}
\alpha & 0 & 0 & \cdots & 0 & 0 \\
1 & \alpha & 0 & \cdots & 0 & 0 \\
0 & 1 & \alpha & \cdots & 0 & 0 \\
\vdots & \vdots & \vdots & \ddots & \vdots & \vdots \\
0 & 0 & 0 & \cdots & \alpha & 0 \\
0 & 0 & 0 & \cdots & 1 & \alpha
\end{pmatrix}.
$$

Such a matrix J is called a *Jordan block matrix*.

3.3.12   Suppose that the polynomial $h \in \mathcal{K}[T]$ splits into linear factors: $h = (T - \alpha_1)^{n(1)} \cdots (T - \alpha_k)^{n(k)}$ with $\alpha_i \neq \alpha_j$ for $i \neq j$. Let B be the companion matrix of h.

Show that there is an invertible matrix Γ such that

$$
\Gamma^{-1}B\Gamma = \mathrm{diag}(J_1, \ldots, J_k),
$$

each J_i being a Jordan block matrix.

Let A be any square matrix. Deduce that, if all the invariant factors of $T \cdot I_n - A$ split into linear factors, then, for some invertible matrix Γ, $\Gamma^{-1}A\Gamma = \mathrm{diag}(J_1, \ldots, J_k)$, where each J_i is a Jordan block matrix.

This form is called the *Jordan normal form* of A. It always exists if every polynomial over the field $\mathcal{K}$ splits into linear factors, that is, if $\mathcal{K}$ is *algebraically closed*, for example, $\mathbb{C}$.

4
ARTINIAN RINGS AND MODULES

The heart of this chapter is the Wedderburn–Artin Theorem (4.2.3), which shows that a certain type of ring, namely an Artinian semisimple ring, can be described very explicitly as a direct product of matrix rings over division rings. This result permits a complete description of all the finitely generated modules over such a ring (4.2.7), which in turn is the key to the classification of the finitely generated projective modules for several other kinds of ring. One of the basic techniques of module theory (and of K-theory) is to compare a given ring with one or more Artinian semisimple rings which are derived from the original ring; it is then often possible to piece together the transparent module theories of these ancillary rings to obtain information about modules over the original ring. We return to this theme at the end of Chapter 7 of [BK: CM], where we discuss methods of identifying modules over orders.

As preparation for the proof of the Wedderburn–Artin Theorem, we give an account of the theory of Artinian modules in section 4.1, while for the applications we are obliged to develop the theory of Artinian rings in general in section 4.3. As usual, we digress from the pursuit of our main objective here and there; in particular, we establish the Hopkins–Levitzki Theorem (4.3.21), which gives a description of right Artinian rings in general.

4.1 ARTINIAN MODULES

In this section, we make the definitions and derive the basic structure theory for Artinian modules, that is, modules which have no infinite descending chains of submodules. We obtain two important descriptions of Artinian modules in terms of irreducible modules. For the first, we look at composition series, which are chains of submodules of a module in which successive quotients are irreducible modules. In the second description, we are concerned with the circumstances when a module is the direct sum of its irreducible submodules.

As the terminology suggests, our treatment of the subject matter follows the axiomatic approach introduced in [Artin 1927]. Before then, the descending chain condition was used implicitly in the context of algebras over fields – see (4.1.2)(ii) below.

We concentrate on right modules; clearly, there is a parallel discussion for left modules. The ambient ring R is arbitrary. Many of our statements have trivial exceptional cases when the module (or ring) is 0; we do not state these separately.

4.1.1 The definition

Let M be a right R-module, where R is an arbitrary ring. A *descending chain* of submodules of M is a sequence

$$\cdots \subseteq M_k \subseteq \cdots \subseteq M_{i+1} \subseteq M_i \subseteq \cdots \subseteq M_1 \subseteq M,$$

each M_i being an R-submodule of M. Such a sequence is called *finite* if the set of indices is finite. Otherwise either it *terminates*, that is, there is an index h with $M_h = M_{h+j}$ for all $j \geq 1$, or it is *infinite* in the sense that it contains infinitely many distinct terms. The right R-module M is *Artinian* if M satisfies the *descending chain condition*, by which we mean that every descending chain of submodules of M either terminates or is finite.

The ring R is said to be *right Artinian* if it is Artinian when considered as a right module over itself. Put more directly, every descending chain of right ideals in R must terminate or be finite. A ring is left Artinian if it is Artinian as a left module. A right Artinian ring need not be left Artinian – see Exercise 4.3.8.

4.1.2 Examples

(i) The zero module is obviously Artinian. Next in complexity, we have the irreducible modules. Recall from (1.2.16) that an R-module M is irreducible if M is nonzero and M has no proper submodules, that is, the only submodules of M are 0 and M. Clearly, an irreducible module is Artinian.

(ii) A field or division ring is evidently both right and left Artinian. If R is an algebra over a field $\mathcal{K}$, that is, $\mathcal{K}$ is a subring of the centre of R, and R has finite dimension as a $\mathcal{K}$-space, then R is right and left Artinian since any ideal of R is also a subspace of R.

More generally, any finitely generated right R-module is Artinian for the same reason. Thus it is possible to avoid a formal statement of the descending chain condition if one works only with algebras.

(iii) For a nonzero integer d, the residue ring $\mathbb{Z}/d\mathbb{Z}$ is finite and so Artinian, both as a ring and as a $\mathbb{Z}$-module. More generally, if d is a nonzero element of a right Euclidean domain, then R/dR is an Artinian R-module. To see this, recall that any right ideal of R is principal (3.2.10), and so each proper submodule of R/dR has the form xR/dR, with x a proper left divisor of d. By (3.2.9), $\varphi(x) < \varphi(d)$. Thus any descending chain in R/dR must terminate in at most $\varphi(d)$ steps.

4.1.3 Fundamental properties

Our first general results are counterparts to those obtained for Noetherian modules in (3.1.2)ff.

4.1.4 Proposition

Let M be a right R-module. Then M is Artinian if and only if any nonempty set of submodules of M has a minimal member.

Proof

We argue by contradiction. If there is an infinite descending chain in M, then the set of distinct members of the chain does not have a minimal member. Conversely, if there is a set of submodules without a minimal member, it is clear that we can construct an infinite descending chain in M. □

4.1.5 Proposition

Let

$$0 \longrightarrow M' \longrightarrow M \longrightarrow M'' \longrightarrow 0$$

be an exact sequence of right R-modules. Then M is Artinian if and only if both M' and M'' are Artinian.

Proof

Suppose that M is Artinian. It is immediate that M' is Artinian. If

$$\cdots \subseteq M''_{i+1} \subseteq M''_i \subseteq \cdots \subseteq M''_1 \subseteq M''$$

is a descending chain in M'' (with infinite index set), define

$$M_i = \pi^{-1}(M''_i) = \{m \in M \mid \pi m \in M''_i\},$$

where π is the given homomorphism from M to M''. Then $\{M_i\}$ is a descending chain in M which terminates; thus its image in M'' terminates too.

Conversely, consider an (infinitely indexed) descending chain in M:

$$\cdots \subseteq M_{i+1} \subseteq M_i \subseteq \cdots \subseteq M_1 \subseteq M.$$

The chain $\{\pi M_i\}$ in M'' terminates, so, for some k, $\pi M_k = \pi M_{k+j}$ for all $j > 0$. Thus $M_k + M' = M_{k+j} + M'$ for all j. The chain $\{M_i \cap M'\}$ in M' terminates, hence we can choose k so that we also have $M_k \cap M' = M_{k+j} \cap M'$ for all j.

Let $x \in M_k$. Then, for any j, $x = y + m'$ for some $y \in M_{k+j}$ and $m' \in M'$, which forces m' to be in $M_k \cap M'$ and hence in $M_{k+j} \cap M'$. Thus $x \in M_{k+j}$ already and the original chain terminates at k. $\qquad\square$

4.1.6 Corollary

Let $\{M_1, \ldots, M_s\}$ be a finite set of Artinian right R-modules. Then the direct sum $M_1 \oplus \cdots \oplus M_s$ is also Artinian. $\qquad\square$

4.1.7 Theorem

Suppose that R is a right Artinian ring. Then

(i) *each free right R-module R^k of finite rank is Artinian, and*

(ii) *every finitely generated right R-module is Artinian.* $\qquad\square$

4.1.8 Composition series

For practical purposes, it is often more convenient to work with proper chains. A descending chain

$$0 \subset M_k \subset \cdots \subset M_{i+1} \subset M_i \subset \cdots \subset M_1 \subset M,$$

is *proper* if each inclusion is an inequality (as the notation suggests). The (proper) descending chain

$$0 \subset N_\ell \subset \cdots \subset N_{j+1} \subset N_j \subset \cdots \subset N_1 \subset M$$

is a *refinement* of the one above if $\{M_k, \ldots, M_1\}$ is a subset of $\{N_\ell, \ldots, N_1\}$; we allow a chain to be a refinement of itself.

A *composition series* for M is a proper chain

$$0 = M_{k+1} \subset M_k \subset \cdots \subset M_{i+1} \subset M_i \subset \cdots \subset M_1 \subset M_0 = M$$

in which every factor M_i/M_{i+1} is an irreducible module. It is easy to see that a composition series can have no refinement (other than itself). The *length* of this composition series is $k + 1$ and the *composition factors* arising from this series are the ireducible modules M_i/M_{i+1}, $i = 0, \ldots, k$. Note that M_i/M_{i+1} is irreducible precisely when M_{i+1} is a maximal submodule of M_i, that is, there is no module L with $M_{i+1} \subset L \subset M_i$ (1.2.13).

We can now characterize modules that have a composition series.

4.1.9 Lemma

Let M be a right R-module. The following statements are equivalent.

(i) *M is both Artinian and Noetherian.*

(ii) *Every proper descending chain in M can be refined to a composition series.*

(iii) *M has a composition series.*

Proof

(i) $\Rightarrow$ (ii)   Suppose that we have a descending chain

$$\mathcal{N}: \quad 0 \subset N_\ell \subset \cdots \subset N_{j+1} \subset N_j \subset \cdots \subset N_1 \subset N_0 = M$$

which is not a composition series, and choose the smallest index j such that N_j/N_{j+1} is not irreducible. Since M is Noetherian, so is N, and hence N_j and N_j/N_{j+1} are also Noetherian (3.1.2). By (3.1.6), N_j/N_{j+1} has a maximal submodule $\overline{L}$, whose inverse image is a maximal submodule L of N_j with $N_{j+1} \subset L \subset N_j$. Thus we have a refinement $\mathcal{N}^1$ of $\mathcal{N}$.

Proceeding in this way, we obtain a succession of refinements

$$\mathcal{N}^1, \ldots, \mathcal{N}^i, \mathcal{N}^{i+1}, \ldots$$

of $\mathcal{N}$. If we do not reach a composition series, this sequence is infinite. But then we can form an infinite descending chain in M.

(ii) $\Rightarrow$ (iii)   This is obvious.

(iii) $\Rightarrow$ (i)   Argue by induction on the length $k + 1$ of the shortest composition series. If $k = 0$, M is irreducible and there is only one (ascending or descending) chain.

In general, suppose that

$$0 = M_{k+1} \subset M_k \subset \cdots \subset M_{i+1} \subset M_i \subset \cdots \subset M_1 \subset M_0 = M$$

is a composition series of minimal length. Striking off the top term, we have a composition series for M_1 which must also be minimal. Thus both M_1 and M/M_1 are simultaneously Artinian and Noetherian. The assertion now follows from (4.1.5) above and (3.1.2). $\square$

An example of an Artinian module without a composition series is given in Exercise 4.1.6.

We now consider the uniqueness of the composition series. The following basic result originated in the study of finite groups in the mid 19th century, and was recast in terms of module theory in [Noether 1929]. It can be viewed as a far-reaching generalization of unique factorization (Exercises 4.1.2, 4.1.3).

4.1.10 The Jordan–Hölder Theorem

Let M be an Artinian right R-module and let

(I) $0 = M_{k+1} \subset M_k \subset \cdots \subset M_{i+1} \subset M_i \subset \cdots \subset M_1 \subset M_0 = M$

and

(II) $0 = N_{\ell+1} \subset N_\ell \subset \cdots \subset N_{j+1} \subset N_j \subset \cdots \subset N_1 \subset N_0 = M$

be two composition series for M.

Then $k = \ell$, and there is a bijective correspondence between the sets of composition factors

$$\{M_i/M_{i+1} \mid i = 0, \ldots, k\}$$

and

$$\{N_j/N_{j+1} \mid j = 0, \ldots, k\}$$

so that corresponding factors are isomorphic.

Proof

We proceed by induction on the length of the shortest composition series, noting that a submodule of an Artinian module is again Artinian. We may suppose that the series (I) has minimal length among all composition series of M. If $k = 0$, then M is irreducible and there is nothing to show. Suppose then that $k > 0$ and that the theorem is true for all modules that have a composition series of length less than $k + 1$.

If $M_1 = N_1$, the result is immediate, so we assume that $M_1 \neq N_1$. Then $M_1 + N_1 = M$, which implies that the natural homomorphism from M_1 to $(M_1 + N_1)/N_1$ induces an isomorphism from $M_1/(M_1 \cap N_1)$ to M/N_1 by the Second Isomorphism Theorem (Exercise 1.2.2). In particular, $M_1/(M_1 \cap N_1)$ is irreducible. Similarly, there is an isomorphism $N_1/(M_1 \cap N_1) \cong M/M_1$.

Now refine the chain

$$M_1 \cap N_1 \subset M_1 \subset M$$

to a composition series

$$0 = L_{s+1} \subset L_s \subset \cdots \subset L_2 = M_1 \cap N_1 \subset M_1 \subset M_0 = M$$

of M. Since

$$0 = M_{k+1} \subset M_k \subset \cdots \subset M_2 \subset M_1$$

and

$$0 = L_{s+1} \subset L_s \subset \cdots \subset L_2 = M_1 \cap N_1 \subset M_1$$

are both composition series for M_1, the induction hypothesis shows that $k = s$ and that there is a bijective correspondence between their sets of composition factors under which corresponding modules are isomorphic.

Similarly, as

$$0 = L_{k+1} \subset L_k \subset \cdots \subset L_2 = M_1 \cap N_1 \subset N_1$$

and

$$0 = N_{\ell+1} \subset N_\ell \subset \cdots \subset N_2 \subset N_1$$

are both composition series for N_1, we find that $k = \ell$ and there is again a bijection between the sets of composition factors of these series with corresponding modules isomorphic. The theorem follows. $\qquad\square$

4.1.11 Multiplicity

The Jordan–Hölder Theorem allows us to define an important numerical invariant for an R-module which is both Artinian and Noetherian, namely, its multiplicity type.

First, we define a *representative set*

$$\mathcal{I}(R) = \{I_\lambda \mid \lambda \in \Lambda\}$$

of irreducible right R-modules as follows. We require that if I is any irreducible right R-module, then $I \cong I_\lambda$ for some λ, and that $I_\lambda \not\cong I_\mu$ when $\lambda \neq \mu$; the set Λ is an ordered set but the choice of ordering is not important. In general, the set Λ need not be finite (consider $\mathbb{Z}$; the irreducible modules have the form $\mathbb{Z}/p\mathbb{Z}$, p prime). If Λ is finite, we write it as $\{1, \ldots, k\}$ for some integer k, in accordance with our usual practice.

Given an Artinian and Noetherian R-module M and an index λ, the *multiplicity* h_λ of I_λ in M is defined by choosing a composition series for M and counting the number h_λ of composition factors that are isomorphic to I_λ. The *multiplicity type* of M (sometimes called the dimension type) $\mathrm{mult}(M)$ is then the sequence

$$\mathrm{mult}(M) = (h_\lambda \mid \lambda \in \Lambda) \in \mathbb{Z}^\Lambda.$$

The Jordan–Hölder Theorem shows that the sequence $\mathrm{mult}(M)$ is independent of the choice of composition series; of course, its appearance may change if we decide to label the representative set $\mathcal{I}(R)$ differently.

An important property of multiplicity types is that they are *additive on short exact sequences*. This property is of significance in K-theory, which is concerned with the general theory of functions that are additive on short exact sequences.

4.1.12 Theorem

Let

$$0 \longrightarrow M' \longrightarrow M \longrightarrow M'' \longrightarrow 0$$

be an exact sequence of right R-modules.

Then the following hold.

 (i) *M has a composition series if and only if both M' and M'' have composition series.*

 (ii) *If the modules M, M' and M'' all have composition series, then*

$$\mathrm{mult}(M) = \mathrm{mult}(M') + \mathrm{mult}(M'').$$

Proof

Assertion (i) follows from criterion (i) of (4.1.9) above, together with the corresponding results for Noetherian and Artinian modules, (3.1.2) and (4.1.5) respectively. For (ii), we appeal to the implication (iii) $\Rightarrow$ (ii) of (4.1.9) to see that the chain $0 \subset M' \subset M$ can be refined to a composition series of M. Clearly, the set of composition factors of M arising from this series is the union of the set of composition factors of M' with the set of composition factors of $M/M' \cong M''$, provided we count multiplicities. $\square$

4.1.13 Reducibility

It is clear from the definition that an irreducible submodule I of an R-module M is a *minimal submodule*, in that there can be no other submodule J with $0 \subset J \subset I$; conversely, a minimal submodule must be irreducible.

We define the *socle* $\mathrm{soc}(M)$ of an arbitrary R-module M to be the sum of its minimal submodules:

$$\mathrm{soc}(M) = \sum \{I \mid I \subseteq M, I \text{ irreducible}\}.$$

If M has no minimal submodules (for example, if $M = 0$), we put $\mathrm{soc}(M) = 0$. We say that M is *semisimple* if $M = \mathrm{soc}(M)$; clearly, $\mathrm{soc}(M)$ is itself semisimple.

Suppose now that M is nonzero and Artinian. Then M must have at least one minimal submodule, and so $soc(M)$ is nonzero, and Artinian by (4.1.5). However, it may well be that the set of minimal submodules of a given Artinian semisimple module is infinite – an example is provided by the vector space $\mathcal{K}^2$ over an infinite field $\mathcal{K}$. Our aim now is to show that their sum will be given by a finite subset.

4.1.14 The Complementation Lemma

Let M be a semisimple R-module and let M' be any submodule of M. Let $\{I_\sigma \mid \sigma \in \Sigma\}$ be the set of minimal submodules of M, indexed by some ordered set Σ.

Then there is a subset Λ of Σ such that

(i) *the sum $S(\Lambda) = \sum\{I_\lambda \mid \lambda \in \Lambda\}$ is direct, and*
(ii) *$M' \oplus S(\Lambda) = M$ (internal direct sum).*

Conversely, if every submodule of a module M is a direct summand of M, then M is semisimple.

Proof

If M' is M, take $\Lambda = \emptyset$. Suppose that $M' \neq M$, and let X be the set of all subsets Λ of Σ with the properties that (i) holds and that $S(\Lambda) \cap M' = 0$. To see that X is not empty, consider any $m \notin M'$. By semisimplicity, there exist $m_\sigma \in I_\sigma$ with $m = \sum m_\sigma$. Then $m_\sigma \notin M'$ for some σ, and so the irreducible module $I_\sigma = m_\sigma R$ has trivial intersection with M'. It is straightforward to verify that X is an inductive set, ordered by inclusion of sets, and therefore has a maximal element, Λ say, by Zorn's Lemma (1.2.19).

We need only show that (ii) holds, that is, $M' + S(\Lambda) = M$. If not, we can argue as before to find an index σ for which

$$I_\sigma \cap (M' + S(\Lambda)) = 0.$$

But then the sum $S(\Lambda \cup \{\sigma\})$ is direct, contradicting the maximality of Λ.

The converse requires a preliminary result, that each nonzero submodule M' of M contains an irreducible submodule. Take a nonzero element x of M' and consider the set of submodules of M' that exclude x. Using Zorn's Lemma again, we find that this set has a maximal element N, and our hypothesis shows that $M' = N \oplus N'$ for some submodule N'. If N' is not irreducible, we have a further decomposition $N' = P \oplus P'$. But then one of P and P' does not contain x, and so N is not maximal after all.

Now consider the (nonempty) set of all semisimple submodules of M. Yet again, Zorn's Lemma shows that this set has a maximal member, say L. If

$M \neq L$, then $M = L \oplus L'$. But L' has an irreducible submodule, which gives another contradiction. Thus $M = L$ is semisimple. $\square$

4.1.15 Complete reducibility
A module M is said to be *completely reducible* if
$$M = I_1 \oplus \cdots \oplus I_k$$
for a finite set $\{I_1, \ldots, I_k\}$ of minimal submodules of M. Observe that M then has a composition series
$$0 \subset I_1 \subset I_1 \oplus I_2 \subset \cdots \subset I_1 \oplus \cdots \oplus I_{k-1} \subset I_1 \oplus \cdots \oplus I_k = M$$
with composition factors $\{I_1, \ldots, I_k\}$.

This definition leads to a series of results that have many applications.

4.1.16 The Complete Reducibility Theorem
Let M be an Artinian semisimple right R-module. Then M is completely reducible.

Proof
Applying the Complementation Lemma to the zero submodule, we find a set $\{I_\lambda \mid \lambda \in \Lambda\}$ of minimal submodules such that M is the direct sum $M = \bigoplus_\Lambda I_\lambda$. But then the index set Λ must be finite, since otherwise we could construct an infinite descending chain by omitting terms successively. $\square$

4.1.17 The Artinian Splitting Theorem
Let M' be a (nonzero) submodule of an Artinian semisimple module M and let $M'' = M/M'$. Then

(i) $M \cong M' \oplus M''$,
(ii) *both M' and M'' are Artinian semisimple, and*
(iii) *there is a direct sum decomposition*
$$M = I_1 \oplus \cdots \oplus I_k$$

of M into minimal submodules such that, for some h with $1 \leq h \leq k$,
$$M' = I_1 \oplus \cdots \oplus I_h,$$

and
$$M'' \cong I_{h+1} \oplus \cdots \oplus I_k.$$

Proof

The Complementation Lemma (4.1.14) shows that

$$M' \oplus \bigoplus \{I_\lambda \mid \lambda \in \Lambda\} = M$$

for some set $\{I_\lambda\}$ of minimal submodules of M. But then $M'' \cong \bigoplus I_\lambda$, giving (i). This also shows that M'' is semisimple, while by (4.1.5) M'' is Artinian. Likewise, the isomorphism $M' \cong M/\bigoplus I_\lambda$ shows M' to be Artinian and semisimple, which proves (ii).

Assertion (iii) follows by applying the Complete Reducibility Theorem to M' and M'' and labelling minimal submodules appropriately. $\qquad\square$

Here is another useful consequence of the Complete Reducibility Theorem.

4.1.18 Corollary

Let M be a semisimple module. Then the following assertions are equivalent.

(i) *M is Noetherian.*

(ii) *M is Artinian.*

(iii) *M has a composition series.*

(iv) *M is completely reducible.*

Proof

By (4.1.9), (iii) implies both (i) and (ii). The Complete Reducibility Theorem (4.1.16) shows that (ii) implies (iv), and clearly (iv) implies (iii). Finally, suppose that (i) holds. By the Complementation Lemma (4.1.14), $M = \bigoplus I_\lambda$ for some set $\{I_\lambda \mid \lambda \in \Lambda\}$ of minimal submodules of M. It is easy to see that, if Λ is infinite, then we can construct an infinite ascending chain in M, contrary to the implication (i) $\Rightarrow$ (ii) of (3.1.6). Thus Λ must be finite, which establishes that (iv) is a consequence of (i). $\qquad\square$

Next, we consider a direct product $R = R_1 \times \cdots \times R_k$ of rings. For future applications, we need some information on the relationships between irreducible, semisimple and Artinian R-modules and the corresponding modules over the component rings R_i.

4.1.19 Theorem

Suppose that as rings $R = R_1 \times \cdots \times R_k$ and let M be an R-module. Then M can be written in the form $M = M_1 \oplus \cdots \oplus M_k$ where each component M_i is an R_i-module, and the following relationships hold between the properties of M and those of its components.

(a) *The following are equivalent:*

   (i) *M is an irreducible R-module,*

   (ii) M_i *is an irreducible R_i-module for one value of $i \in \{1, \ldots, k\}$, and $M_j = 0$ for all $j \neq i$.*

(b) *If N is another irreducible R-module, then $M \cong N$ as an R-module if and only if $N_i \neq 0$ for the same i, and further $M_i \cong N_i$ as an R_i-module.*

(c) *M is a semisimple R-module if and only if every M_i is a semisimple R_i-module.*

(d) *M is Artinian as an R-module if and only if each summand M_i is Artinian as an R_i-module.*

(e) *M has a composition series if and only if each component M_i has a composition series, in which case*

$$\mathrm{mult}(M) = (\mathrm{mult}_1(M_1), \ldots, \mathrm{mult}_k(M_k)),$$

where $\mathrm{mult}_i(-)$ denotes the R_i-module multiplicity. This relation holds in particular when M is Artinian and semisimple.

Proof

First, we recall that the discussion in (2.6.6) shows that an R-module M can be written as a direct sum

$$M = M_1 \oplus \cdots \oplus M_k$$

with each component M_i an R_i-module, and that any R_i-module is an R-module by restriction of scalars.

(a)   Clearly, if one component of M is not irreducible as an R_i-module, or if two components are nonzero, then M cannot be irreducible, so (i) implies (ii). The converse is clear.

(b)   Immediate from (a) and (2.6.8).

(c)   By (a), the set of minimal R-submodules of M is the disjoint union of the sets of minimal R_i-submodules of each M_i, so M is the sum of its minimal submodules precisely when the same holds for each of its components.

(d)   This follows inductively from (4.1.5), since by our initial observation descending chains of R-modules in M_i coincide with descending chains of R_i-submodules of M_i.

(e)   It is now clear that a composition series for M_i as an R_i-module is the same as a composition series for M_i as an R-module, so the assertions follow from (4.1.12) and (4.1.18).   □

We next look at the effect of a homomorphism of modules on minimal (that is, irreducible) submodules and on the socle.

4.1.20 Lemma

Let $\alpha : M' \to M$ be a homomorphism of right R-modules. Then, when restricted to any irreducible submodule I of M', α is either the zero map or an isomorphism from I to the minimal submodule αI of M.

Proof

Let α' be the restriction of α to I. Then $\operatorname{Ker}\alpha'$ is $I \cap \operatorname{Ker}\alpha$, which is a submodule of I, and so it is either I or zero. When α' is injective, each submodule of αI gives a submodule of I, hence αI must also be irreducible. $\qquad\square$

In the case where $M' = M$ is itself irreducible, the preceding result (4.1.20) (or a variant thereof) is known as Schur's Lemma.

4.1.21 Schur's Lemma – First Version

Any R-module endomorphism of an irreducible right R-module is either the zero map or an automorphism. $\qquad\square$

4.1.22 Fully invariant submodules

To record the implications of these results for the socle, we make two definitions.

A submodule M' of a right R-module M is *fully invariant* if $\alpha M' \subseteq M'$ for every R-module endomorphism α of M.

It is clear that such a submodule is necessarily an $\operatorname{End}(M_R)$-R-bimodule. In particular, a right ideal $\mathfrak{a}$ of R is fully invariant precisely when it is a twosided ideal of R, since $\operatorname{End}(R_R)$ can be identified with R itself acting through left multiplication.

A weaker property that is sometimes useful is the following. A submodule M' of an R-module M is said to be *characteristic* if $\alpha M' = M'$ for every R-module automorphism α of M.

The next result is immediate from the definitions and (4.1.20), on noting that left multiplication by an element r of the ring R gives a homomorphism from a right ideal to R.

4.1.23 Proposition

Let M be a right R-module, with a minimal submodule N. Then

$$\sum \{M' \mid M' \text{ minimal submodule}, M' \cong N\}$$

and $\operatorname{soc}(M)$ are fully invariant. In particular, for each minimal right ideal J

of R,

$$\sum \{I \mid I \text{ minimal right ideal, } I_R \cong J_R\}$$

and soc(R) *are twosided ideals of R.*   □

4.1.24 The socle series

Let M be a right R-module. A chain

$$0 \subseteq \cdots \subseteq M_k \subseteq \cdots \subseteq M_{i+1} \subseteq M_i \subseteq \cdots \subseteq M_1 \subseteq M$$

of submodules of M is called a *fully invariant series* for M if every term M_i is a fully invariant submodule of M. It is called a *characteristic series* for M if every term M_i is a characteristic submodule of M. The same terminology appplies to infinite chains, whether ascending or descending.

Any module M has a fully invariant series, the *socle series* or *Loewy series*

$$0 \subseteq \text{soc}(M) = \text{soc}^1(M) \subseteq \cdots \subseteq \text{soc}^i(M) \subseteq \text{soc}^{i+1}(M) \subseteq \cdots \subseteq M,$$

whose terms are defined inductively as follows.

We start with $\text{soc}(M) = \text{soc}^1(M)$. Suppose that we have defined $\text{soc}^i(M)$ for some $i \geq 1$. Then we put

$$\text{soc}^{i+1}(M) = \{m \in M \mid \overline{m} \in \text{soc}(M/\text{soc}^i(M))\},$$

the inverse image of the socle of the quotient module $M/\text{soc}^i(M)$.

The fact that each $\text{soc}^{i+1}(M)$ is a fully invariant submodule of M follows by induction on observing that an endomorphism α of M induces an endomorphism $\overline{\alpha}$ of $M/\text{soc}^i(M)$.

Note that each factor $\text{soc}^{i+1}(M)/\text{soc}^i(M)$ of the socle series is, by design, a semisimple module.

In general, the socle series may never escape from 0, as is the case with the integers $\mathbb{Z}$ considered as a $\mathbb{Z}$-module. However, when M is Artinian, any nonzero quotient module of M has a nonzero socle, and thus the series either reaches M, that is $\text{soc}^k(M) = M$ for some k, or is infinite.

4.1.25 Theorem

Let M be an Artinian R-module. Then the socle series reaches M if and only if M is Noetherian.

Proof

By repeated applications of (4.1.5), each factor $\text{soc}^{i+1}(M)/\text{soc}^i(M)$ is Artinian; it is certainly semisimple, and so Noetherian by (4.1.18). Thus (3.1.2) shows that each term in the socle series is Noetherian, as must M be if the series reaches it.

On the other hand, if M is Noetherian, the ascending chain condition, together with the fact that each factor in the series is nontrivial, guarantees that the series reaches M. $\square$

Exercises

4.1.1   Let $\mathcal{D}[T, \alpha]$ be a skew polynomial ring over a division ring $\mathcal{D}$, with α an automorphism of $\mathcal{D}$. Show that a finitely generated right $\mathcal{D}[T, \alpha]$-module M is Artinian if and only if the rank of M is 0. (On the other hand, any finitely generated $\mathcal{D}[T, \alpha]$-module is Noetherian.)

Extend this result to arbitrary (twosided) Euclidean domains.

4.1.2   Show that a $\mathbb{Z}$-module S is irreducible if and only if $S \cong \mathbb{Z}/p\mathbb{Z}$ for some (positive) prime number p, and that $\mathbb{Z}/p\mathbb{Z} \cong \mathbb{Z}/q\mathbb{Z}$ if and only if $p = q$. Describe the multiplicity type of a cyclic module $\mathbb{Z}/d\mathbb{Z}$ in terms of the prime factorization of d, and hence find a connection between the Jordan–Hölder Theorem and unique prime factorization in $\mathbb{Z}$.

4.1.3   **Unique factorization**

The observation made in the preceding exercise can be used to formulate a unique factorization in an arbitrary (right) Euclidean domain R, as promised in (3.2.22). Prove the following assertions.

(i)   An element $u \in R$ has a right inverse if and only if u is a unit – note (3.2.9).

(ii)   A right R-module S is irreducible if and only if $S \cong R/pR$ with p irreducible in R.

(iii)   If $d = ab$ in R, then $aR/abR \cong R/bR$.

Deduce that any nonzero, nonunit element d of R has a factorization $d = p_1 \cdots p_k$ into irreducible elements of R, and that the set of isomorphism classes of irreducible modules $R/p_i R$, counting multiplicities, is independent of the factorization of d into irreducibles.

Recall from (3.2.22) that elements p, q of R are right similar if $R/pR \cong R/qR$. Prove that if $d = q_1 \cdots q_\ell$ is another irreducible factorization of d, then $k = \ell$ and the sets $\{p_1, \ldots, p_k\}$ and $\{q_1, \ldots, q_k\}$ of irreducible factors of d can be arranged into right similar pairs.

4.1.4   **Similar elements**

Continuing the theme of the preceding exercise, let R be a (possibly noncommutative) domain, and let d and e be elements of R. Show that the following are equivalent.

(i) $R/dR \cong R/eR$ as right R-modules (that is, d and e are right similar).

(ii) There are elements a, b of R satisfying the following four requirements:

(a) $R = aR + eR$,

(b) $ad = eb$,

(c) $aR \cap eR = adR$,

(d) if $r \in R$ and $ar \in eR$, then $r \in dR$.

Hint. An isomorphism $\theta : R/dR \to R/eR$ must be induced by a homomorphism $\alpha : R \to R$, which in turn is given by left multiplication by an element a of R.

(iii) There is an exact sequence

$$0 \longrightarrow R \xrightarrow{\binom{d}{b}} R^2 \xrightarrow{(a-e)} R \longrightarrow 0$$

of right R-modules, where the matrices act by left multiplication. Now assume that R is commutative. Show that (ii)(a) and (ii)(c) together imply that e and d are associates. Part (vi) of Exercise 4.2.7 shows that similar elements need not be associates when R is not commutative.

4.1.5   For this exercise, we need to assume the *elementary divisor theorem* for $\mathbb{Z}$-modules: if M is a finitely generated $\mathbb{Z}$-module, then

$$M \cong \mathbb{Z}^h \oplus \mathbb{Z}/q_1\mathbb{Z} \oplus \cdots \oplus \mathbb{Z}/q_s\mathbb{Z}, \quad \text{where } q_i = p_i^{n(i)}, \ p_i \text{ prime};$$

the primes $p_1, \ldots, p_s$ are unique up to order and the rank h and the integers $n(1), \ldots, n(s)$ are unique. (We eventually derive this result as a special case of our discussion of modules over Dedekind domains – see (6.3.24).)

Find $\mathrm{soc}(\mathbb{Z}/q\mathbb{Z})$, q a prime power, and hence give the socle series of a finitely generated Artinian $\mathbb{Z}$-module.

4.1.6   Let p be a fixed prime. The *quasicyclic p-group* (or *Prüfer group of type* p^∞) G is defined as follows. For each $i > 0$, let C_i be the (additive) subgroup of $\mathbb{Q}/\mathbb{Z}$ consisting of all those elements which have order dividing p^i. Then each C_i is cyclic, and $C_i \subset C_{i+1}$. Put $G = \bigcup C_i$.

Show that G is Artinian as a $\mathbb{Z}$-module, but not Noetherian.

4.1.7   The $\mathbb{Z}$-module G in the previous exercise is not finitely generated. Our next example, due to P. M. Cohn (private communication), is cyclic and Artinian, but not Noetherian.

Let V be a vector space over a field $\mathcal{K}$, with a countably infinite basis $\{u, v_1, v_2, \ldots\}$. Let $R = \mathcal{K}\langle T_1, T_2, \ldots \rangle$ be the noncommutative polynomial ring in a countable set of variables. Then V becomes an R-module by the rules

$$uT_n = v_n \text{ for } n \geq 1 \text{ and } v_i T_n = \left\{ \begin{array}{ll} v_{i-1} & i > 1, \\ 0 & i = 1. \end{array} \right.$$

Show

(i) V is generated by u as an R-module,

(ii) $M_n = v_1 \mathcal{K} + \cdots + v_n \mathcal{K}$ is an R-submodule of V for each $n \geq 1$, and V has no other nonzero submodules,

(iii) V is Artinian but not Noetherian.

4.1.8   Let $f : R \to S$ be a ring homomorphism and let M be a right S-module. Show that, if M is Artinian as a right R-module via f, then M is Artinian as a right S-module.

Show that the converse also holds when f is surjective.

4.2 ARTINIAN SEMISIMPLE RINGS

This section is devoted to a full exposition of the structure theory of Artinian semisimple rings, that is, rings which are Artinian and semisimple when considered as right modules over themselves. The key result is the Wedderburn–Artin Theorem, which gives the structure of Artinian semisimple rings very explicitly. The result was first given by [Wedderburn 1908] in the context of algebras, and subsequently extended by [Artin 1927]. From it, we obtain a listing of the modules over such a ring which is of considerable importance, both in K-theory and elsewhere.

One consequence of the Wedderburn–Artin Theorem is that there is no difference between a left and a right Artinian semisimple ring, which is not obvious from the definition. On the other hand, a right Artinian ring need not be left Artinian (see Exercise 4.3.8)).

4.2.1 Definitions and the statement of the Wedderburn–Artin Theorem

Recall that a ring R is right Artinian if R is itself Artinian when considered as a right R-module; it is *(right) Artinian semisimple* if it is also semisimple as a right R-module.

A right submodule of a ring R being the same thing as a right ideal of R, such a submodule $\mathfrak{m}$ is irreducible precisely when it is minimal as a right ideal, that is, there are no right ideals $\mathfrak{a}$ with $0 \subset \mathfrak{a} \subset \mathfrak{m}$. (To avoid a multitude of trivial exceptional cases, the zero ideal is regarded as neither irreducible nor minimal.) Thus a right Artinian ring is one in which every descending chain of right ideals terminates, while from the Complete Reducibility Theorem (4.1.16) a right Artinian semisimple ring is a ring which is the direct sum of a finite number of its minimal right ideals.

In anticipation of the fact that a right Artinian semisimple ring is the same thing as a left Artinian semisimple ring, we sometimes omit to mention the chirality.

Recall that a ring R is simple (as a ring) if it has no twosided ideals other than itself and the zero ideal. We say that R is *Artinian simple* if it is both Artinian, as a right module, and simple, as a ring. The theory of non-Artinian simple rings is very deep and far from complete; for more information, see [McConnell & Robson 1987] for example.

The following result illustrates the relationship between these notions.

4.2.2 Proposition

Let R be a right Artinian simple ring. Then R is semisimple as a right R-module.

Proof

By (4.1.23), the socle $\mathrm{soc}(R)$ of R is a twosided ideal of R. Since R is Artinian, $\mathrm{soc}(R) \neq 0$; thus $\mathrm{soc}(R) = R$. $\square$

We now state the crucial structure theorem for Artinian semisimple rings, which contains a much more precise version of the preceding proposition.

4.2.3 The Wedderburn–Artin Theorem

(i) *A ring R is a (right) Artinian simple ring if and only if*

$$R \cong M_n(\mathcal{D})$$

where $\mathcal{D}$ is a division ring. The natural number n is uniquely determined by R, and $\mathcal{D}$ is unique up to ring isomorphism.

(ii) *A ring R is a (right) Artinian semisimple ring if and only if*

$$R \cong M_{n_1}(\mathcal{D}_1) \times \cdots \times M_{n_k}(\mathcal{D}_k),$$

a direct product of matrix rings, where $\mathcal{D}_1, \ldots, \mathcal{D}_k$ are division rings. Apart from permuting the order of the components, the natural numbers $n_1, \ldots, n_k$ are uniquely determined by R, and the division rings are unique up to ring isomorphism.

The proof of the theorem is fairly intricate, so we break it up into several steps. We commence with the easier argument, that the explicit rings mentioned in the theorem do have the desired properties. We then classify the modules over these rings. Finally, we present the harder argument that shows that the abstract definitions lead to explicit matrix rings as promised.

Notice that the Wedderburn–Artin Theorem does show the promised left–right symmetry of Artinian semisimple rings. This is because the matrix rings themselves have left–right symmetric properties, which will be evident from the arguments we give. Alternatively, observe that the opposite ring to $M_n(\mathcal{D})$ is $M_n(\mathcal{D}^\circ)$, and that $\mathcal{D}^\circ$ is again a division ring, so we can translate from right to left using the principle set out in (1.2.6).

4.2.4 Division rings

We start by recollecting some facts about modules over a division ring $\mathcal{D}$. It is obvious that $\mathcal{D}$ is an Artinian simple ring.

A (right) $\mathcal{D}$-module M is, by definition, the same thing as a (right) vector space over $\mathcal{D}$; if M is finitely generated, it has a finite basis, and any two bases of M have the same number of elements. Thus $M \cong \mathcal{D}^h$, where the integer h is unique; we write $h = \dim(M_{\mathcal{D}})$, the *dimension* of M over $\mathcal{D}$. The dimension is the same as the multiplicity type $\text{mult}(M)$, which in this case is a single number rather than a vector.

These results are very familiar when $\mathcal{D}$ is a (commutative) field, and a careful examination of their usual proofs reveals that they are true for an arbitrary division ring. Details can be found in [Cohn 1979], §1.4, for instance.

4.2.5 Matrix rings

The fact that a matrix ring $M_n(\mathcal{D})$, where $\mathcal{D}$ is a division ring, is Artinian simple is easily verified. It is Artinian because any right ideal of $M_n(\mathcal{D})$ is also a finite-dimensional right vector space over $\mathcal{D}$. Alternatively, we can appeal to (4.1.7). Simplicity follows from a manipulation with matrix units which is spelt out in Exercise 1.1.4.

From (4.2.2) it follows that $M_n(\mathcal{D})$ is Artinian semisimple, and so has a decomposition as a direct sum of finitely many irreducible right $M_n(\mathcal{D})$-modules. Let us see how these modules are realized explicitly as minimal right ideals. Let $\{e_{ij}\}$ be the standard set of matrix units in $M_n(\mathcal{D})$, so that

$$e_{hi}e_{jk} = \begin{cases} e_{hk} & \text{if } i = j, \\ 0 & \text{if } i \neq j, \end{cases}$$

and a matrix $A = (a_{ij})$ can be written

$$A = \sum_{i,j} a_{ij}e_{ij} \text{ with each } a_{ij} \in \mathcal{D}.$$

Then, for each i, the right $M_n(\mathcal{D})$-module $e_{ii}M_n(\mathcal{D})$ consists of the 'ith' row of $M_n(\mathcal{D})$, that is, it comprises those matrices A in $M_n(\mathcal{D})$ whose entries are all 0 except possibly in the ith row. Clearly,

$$M_n(\mathcal{D}) = e_{11}M_n(\mathcal{D}) \oplus \cdots \oplus e_{nn}M_n(\mathcal{D}).$$

Right multiplication by matrix units shows that each right ideal $e_{ii}M_n(\mathcal{D})$ is an irreducible module, being generated by any of its nonzero members.

Notice also that for each pair of indices i, j, left multiplication by e_{ij} defines an $M_n(\mathcal{D})$-module isomorphism from $e_{jj}M_n(\mathcal{D})$ to $e_{ii}M_n(\mathcal{D})$. Now, as we saw in (2.2.14), $M_n(\mathcal{D})$ can be identified as the endomorphism ring $\text{End}(^n\mathcal{D})$, so that $^n\mathcal{D}$ is a $\mathcal{D}$-$M_n(\mathcal{D})$-bimodule. Then each $e_{ii}M_n(\mathcal{D})$ is isomorphic as a $\mathcal{D}$-$M_n(\mathcal{D})$-bimodule to the free left $\mathcal{D}$-module $^n\mathcal{D}$ of rank n, which we regard as the row-space (of dimension n) over $\mathcal{D}$. Thus we can rewrite the right $M_n(\mathcal{D})$-module decomposition of the matrix ring in the form

$$M_n(\mathcal{D}) \cong (^n\mathcal{D})^n.$$

4.2.6 Products of matrix rings

Next, we consider a direct product of rings

$$R = M_{n_1}(\mathcal{D}_1) \times \cdots \times M_{n_k}(\mathcal{D}_k)$$

with each $\mathcal{D}_i$ a division ring. Since each $M_{n_i}(\mathcal{D}_i)$ is Artinian semisimple, it follows from (4.1.19)(c),(d) that R is also.

By our previous analysis, the decomposition of R as a direct sum of irreducible right modules is given by

$$R \cong (^{n_1}\mathcal{D}_1)^{n_1} \oplus \cdots \oplus (^{n_k}\mathcal{D}_k)^{n_k}.$$

We now obtain the structure theorem for modules over direct products of matrix rings, which, by the Wedderburn–Artin Theorem, will include all Artinian semisimple rings.

4.2.7 Theorem

Let $R = M_{n_1}(\mathcal{D}_1) \times \cdots \times M_{n_k}(\mathcal{D}_k)$, a direct product of rings, with $\mathcal{D}_1, \ldots, \mathcal{D}_k$ division rings, and let M be a finitely generated right R-module.
Then the following hold.

(i) M is both projective and semisimple as a right R-module. In particular, R is semisimple as a right R-module.

(ii) A representative set of irreducible R-modules is given by

$$\mathcal{I}(R) = \{(^{n_1}\mathcal{D}_1), \ldots, (^{n_k}\mathcal{D}_k)\}$$

where each $^{n_i}\mathcal{D}_i$ is a right R-module by change of rings.

(iii) As a direct sum of right R-modules

$$M \cong (^{n_1}\mathcal{D}_1)^{h_1} \oplus \cdots \oplus (^{n_k}\mathcal{D}_k)^{h_k}$$

for unique nonnegative integers $h_1, \ldots, h_k$ (some of which may be 0).

(iv) There are orthogonal idempotents $\eta_1, \ldots, \eta_k$ of R with right R-module isomorphisms

$$\eta_1 R \cong {}^{n_1}\mathcal{D}_1, \ \ldots, \eta_k R \cong {}^{n_k}\mathcal{D}_k.$$

Proof

Since M is finitely generated, there is a surjective homomorphism from R^s to M for some integer s. As R^s is both projective and semisimple, (i) follows from (2.5.5) and (4.1.17). Clearly R^s has a series with factors all isomorphic to R, so R^s has a composition series all of whose composition factors belong to the set

$$\{^{n_1}\mathcal{D}_1, \ldots, {}^{n_k}\mathcal{D}_k\}.$$

In particular, any irreducible R-module must be isomorphic to a member of this set, by the Jordan–Hölder Theorem (4.1.10). Thus (ii) holds.

Part (iii) follows from (i), (ii) and the Jordan–Hölder Theorem again. For part (iv), take $\eta_i = (0, \ldots, 0, e_{11}, 0, \ldots, 0)$, that is, the member of R which has zero entry in each summand except the ith, where the entry is a matrix unit of $M_{n_i}(\mathcal{D}_i)$. $\qquad\square$

4.2.8 Corollary

The finitely generated R-module M is determined up to isomorphism by its multiplicity type

$$\text{mult}(M) = (h_1, \ldots, h_k). \qquad \square$$

In the case when M is not necessarily finitely generated, one can still say the following.

4.2.9 Proposition

Let R be an Artinian semisimple ring and let M be a right R-module. Then M is semisimple.

Proof

By the Complete Reducibility Theorem (4.1.16) we have $R = \mathfrak{a}_1 \oplus \cdots \oplus \mathfrak{a}_k$ for some minimal right ideals $\mathfrak{a}_i$. Let x be in M. Then each nonzero term $x\mathfrak{a}_i$ must be a minimal submodule of M by (4.1.20). Hence $x \in x\mathfrak{a}_1 \oplus \cdots \oplus x\mathfrak{a}_k \subseteq \text{soc}(M)$. $\qquad \square$

By combining this result with (4.1.18), we obtain the following useful result.

4.2.10 Corollary

Let R be an Artinian semisimple ring and let M be a right R-module. Then the following are equivalent.

(i) *M is Noetherian.*
(ii) *M is Artinian.*
(iii) *M has a composition series.*
(iv) *M is completely reducible.* $\qquad \square$

4.2.11 Finishing the proof of the Wedderburn–Artin Theorem

We now turn to the proof of the stronger assertion in the Wedderburn–Artin Theorem, that Artinian simple and semisimple rings can be described in terms of matrix rings.

Our first result can be viewed as a refinement of the observation in (4.2.1) that an Artinian semisimple ring is a finite direct sum of some of its minimal right ideals.

4.2.12 Proposition

Let R be a (right) Artinian semisimple ring, for which a set of distinct representatives of isomorphism classes of minimal right ideals is $\{J_1, \ldots, J_k\}$. Then the following hold.

(i) *There is an internal direct product decomposition $R = R_1 \times \cdots \times R_k$ of R as a ring, where each component R_i is given by*

$$R_i = \sum \{I \mid I \cong J_i \text{ as right ideals}\}.$$

(ii) *Each component R_i is a simple (right) Artinian ring.*

(iii) *The components $R_1, \ldots, R_k$ are unique up to re-ordering.*

Proof

(i)   It is clear that, as a $\mathbb{Z}$-module, $R = R_1 + \cdots + R_k$, and, by (4.1.23), each R_i is a twosided ideal of R. Thus the conclusion follows from (2.6.3) provided that we can show that R is the direct sum of $R_1, \ldots, R_k$, which in turn requires us to prove that

$$R_i \cap (R_1 + \cdots + R_{i-1} + R_{i+1} + \cdots + R_k) = 0$$

for each i. However, if the intersection is not 0, it contains some minimal right ideal J and J is isomorphic to both J_i and some J_h, $h \neq i$, a contradiction. Thus the intersection is zero, and the result follows.

(ii)   By (i) and (4.1.19), each R_i is an Artinian semisimple ring whose minimal right ideals are all isomorphic to J_i. So the result is a consequence of the next lemma.

(iii)   This is clear from the method of construction of the components. $\square$

4.2.13 Lemma

Suppose that R is an Artinian semisimple ring. Then R is a simple ring if and only if all the minimal right ideals of R are isomorphic as R-modules.

Proof

Suppose first that R is a simple ring. Since R is Artinian, it has a minimal right ideal J, say (4.1.4). By (4.1.23),

$$\sum \{I \mid I \cong J \text{ as right ideals}\}$$

is a twosided ideal of R and so must be R itself.

For the converse, suppose that all minimal right ideals of R are isomorphic, and, by way of contradiction, assume that R is not a simple ring. Then R has a proper nonzero twosided ideal $\mathfrak{a}$. Since R is Artinian, by (4.1.4) we can choose $\mathfrak{a}$ to be minimal among the twosided proper ideals of R.

By the Complete Reducibility Theorem (4.1.16), we can find a direct decomposition $R = I_1 \oplus \cdots \oplus I_k$ of R (as a right R-module) into minimal ideals such that $\mathfrak{a} = I_1 \oplus \cdots \oplus I_h$ for some $h < k$. Put $\mathfrak{b} = I_{h+1} \oplus \cdots \oplus I_k$, a right ideal of R.

The main point in the argument is that this decomposition $R = \mathfrak{a} \oplus \mathfrak{b}$ into right ideals with $\mathfrak{a}$ twosided and minimal is in fact a decomposition of R into a direct product of rings.

First, observe that $\mathfrak{b}\mathfrak{a} \subseteq \mathfrak{a} \cap \mathfrak{b} = 0$. Next, $\mathfrak{a}\mathfrak{b}$ is a twosided ideal contained in $\mathfrak{a}$, so either $\mathfrak{a}\mathfrak{b} = 0$ or $\mathfrak{a}\mathfrak{b} = \mathfrak{a}$. If the latter holds, then

$$\mathfrak{a} = R\mathfrak{a} = (\mathfrak{a}\mathfrak{b} + \mathfrak{b})\mathfrak{a} = \mathfrak{b}\mathfrak{a} = 0,$$

a contradiction; hence $\mathfrak{a}\mathfrak{b} = 0$ also. Finally, $R\mathfrak{b} = (\mathfrak{a} + \mathfrak{b})\mathfrak{b} \subseteq \mathfrak{b}$, which shows that $\mathfrak{b}$ is also twosided.

It follows from (2.6.3) that $R = \mathfrak{a} \oplus \mathfrak{b}$ is a decomposition of R into rings. Thus we obtain a contradiction to the hypothesis that all minimal right R-modules are isomorphic, since I_1 is an $\mathfrak{a}$-module, I_k is a $\mathfrak{b}$-module, and so there can be no R-module isomorphism between I_1 and I_k, by (4.1.19). □

Next, we record a reformulation of (4.1.21).

4.2.14 Schur's Lemma – Second Version
Let I be an irreducible module over a ring R. Then $\mathrm{End}_R(I)$ is a division ring. □

Our next result provides the final link in the proof of the Wedderburn–Artin theorem.

4.2.15 Theorem
Let R be an Artinian simple ring. Then $R \cong M_n(\mathcal{D})$ for some division ring $\mathcal{D}$ and natural number n.

The division ring $\mathcal{D}$ is unique up to ring isomorphism, and the natural number n is unique.

Proof

Choose any minimal right ideal I of R. By (4.2.13) and complete reducibility, we have, for some n, $R \cong I^n$ as a right R-module. Then it is easy to verify that

$$R = \mathrm{End}(R_R) \cong M_n(\mathrm{End}(I_R))$$

(see Exercise 2.1.6). We put $\mathcal{D} = \mathrm{End}(I_R)$, which is a division ring by Schur's Lemma.

If also $R \cong M_m(\mathcal{D}')$, then, as in (4.2.5), the first row of the matrix ring is (isomorphic to) a minimal right ideal J of R, and $\mathcal{D}' \cong \mathrm{End}_R(J)$. But $I \cong J$ as a right R-module, so $\mathcal{D} \cong \mathcal{D}'$ as a ring. Further, we can view J as a left

$\mathcal{D}$-module, and then

$$^{n}\mathcal{D} \cong I \cong J \cong {}^{m}\mathcal{D},$$

which shows that $m = n$. $\qquad\qquad\qquad\qquad\qquad\qquad\qquad\qquad$ □

4.2.16 Recapitulation of the argument

As our proof of the Wedderburn–Artin Theorem has been spread over many separate results, we give a brief summary of the arguments.

The first part of the theorem asserts that an Artinian simple ring must have the form $R \cong M_n(\mathcal{D})$ for a division ring $\mathcal{D}$ and natural number n which are (essentially) unique. The discussion in subsection 4.2.5 establishes that such a matrix ring is Artinian simple. In the reverse direction, Proposition 4.2.2 tells us that a (right) Artinian simple ring R is necessarily semisimple as a right R-module, so the existence and uniqueness of the matrix form are given by Theorem 4.2.15.

The second part of the theorem says that a (right) Artinian semisimple ring has the form

$$R \cong M_{n_1}(\mathcal{D}_1) \times \cdots \times M_{n_k}(\mathcal{D}_k)$$

where the division rings $\mathcal{D}_1, \ldots, \mathcal{D}_k$ are unique up to isomorphism and the integers $n_1, \ldots, n_k$ are unique, up to renumbering the components.

Again, the discussion in subsection 4.2.5 shows that such a direct product of matrix rings is Artinian semisimple. Conversely, if R is Artinian semisimple, then Proposition 4.2.12 tells us that R decomposes into a direct product of simple Artinian rings, which are unique, and so the result is a consequence of the first part of the theorem.

To complete the circle of ideas, we record a converse to part (i) of Theorem 4.2.7.

4.2.17 Theorem

Suppose that R is a ring such that every right R-module is projective. Then R is an Artinian semisimple ring.

Proof

The hypothesis ensures that every quotient module of R, and hence every right ideal in R, is a direct summand of R, so R is semisimple by the converse assertion in the Complementation Lemma 4.1.14. But the identity element of R is a sum $1 = \sum_i e_i$ with only finitely many nonzero components, and so any element $r = 1 \cdot r$ involves only the same set of nonzero components.

Thus R is the direct sum of finitely many minimal right ideals and hence R is Artinian. ☐

Exercises

4.2.1   Let R be a right Artinian ring. Show that, if $a \in R$, then there are an integer k and an element b in R with $a^k = a^{k+1}b$.

Deduce that, if R is a domain, then it is a division ring.

4.2.2   Prove that a ring is Artinian semisimple if and only if it is equal to the sum of its minimal ideals.

Hint. Writing 1 as a finite sum of elements from minimal ideals, show that the ring itself is also the sum of the same minimal ideals. Thereby obtain a composition series.

4.2.3   Let $f : R \twoheadrightarrow S$ be a surjective ring homomorphism. Show that, if R is an Artinian semisimple ring, then so is S.

Hint. Exercise 4.1.8.

4.2.4   Let $T = T_2(\mathcal{K}) = \begin{pmatrix} \mathcal{K} & \mathcal{K} \\ 0 & \mathcal{K} \end{pmatrix}$ be the ring of upper triangular matrices over a field, and let $\mathfrak{a} = \begin{pmatrix} \mathcal{K} & \mathcal{K} \\ 0 & 0 \end{pmatrix}$, $\mathfrak{b} = \begin{pmatrix} 0 & \mathcal{K} \\ 0 & 0 \end{pmatrix}$ and $\mathfrak{c} = \begin{pmatrix} 0 & 0 \\ 0 & \mathcal{K} \end{pmatrix}$.

Show that $\mathfrak{a}$ and $\mathfrak{b}$ are twosided ideals of T while $\mathfrak{c}$ is a right ideal, and that $\mathfrak{b}$ and $\mathfrak{c}$ are isomorphic minimal right ideals of T.

Find a composition series for T as a right T-module, and show that any minimal right T-ideal is isomorphic to $\mathfrak{b}$. Show further that there are two isomorphism classes of irreducible right T-modules, corresponding to the two isomorphism classes of irreducible right $(T/\mathfrak{b})$-modules (4.1.19)(b). Deduce also that T is not semisimple. (This illustrates that the ring R in (4.2.13) must be assumed to be semisimple.)

4.2.5   This example shows that the results of (4.2.5) do not hold for infinite matrix rings, which may fail to be simple even if the coefficient ring is simple. For an application, see Exercise 2.3.1.

Let R be a ring. Recall that $M_\omega^{cf}(R)$ is the ring of column-finite square matrices with rows and columns indexed by the natural numbers, that is, each column of a matrix has only finitely many nonzero entries. Recall also that the cone CR of R is the subring of $M_\omega^{cf}(R)$ comprising those matrices with only finitely many nonzero entries in any row.

Let $\mathfrak{M}$ be the subset of $M_\omega^{cf}(R)$ consisting of all those matrices with a finite set of nonzero rows, and define $mR = \mathfrak{M} \cap CR$. (Thus, $a \in mR \Leftrightarrow a = \begin{pmatrix} b & 0 \\ 0 & 0 \end{pmatrix}$ with b finite.)

Show that $\mathfrak{M}$ and mR are twosided ideals, in $M_\omega^{cf}(R)$ and CR respectively.

Now take the coefficient ring to be a field $\mathcal{K}$, and recall that there is a ring isomorphism $M_\omega^{cf}(\mathcal{K}) \cong \mathrm{End}(V)$, where $V = \mathcal{K}^\omega$, a $\mathcal{K}$-space of countably infinite dimension (2.2.13).

Show that $\mathfrak{M}$ corresponds to the set $\mathrm{Endf}(V)$ of endomorphisms f of V of *finite rank*, that is, $\dim(fV) < \infty$. Verify directly that $\mathrm{Endf}(V)$ is a maximal twosided ideal of $\mathrm{End}(V)$, so that $\mathfrak{M}$ is maximal in $M_\omega^{cf}(\mathcal{K})$.

Hint. By the method of (1.2.20), any linearly independent set of vectors in V is part of a basis for V.

Show also that $\mathfrak{M}$ is the only nontrivial proper twosided ideal of $M_\omega^{cf}(\mathcal{K})$.

Deduce that $m\mathcal{K}$ is the only nontrivial proper twosided ideal of $C\mathcal{K}$.

4.2.6 Let $\mathcal{K}$ be a field. Show that maximal ideals of the polynomial ring $\mathcal{K}[T]$ correspond to the irreducible polynomials $p(T)$ in $\mathcal{K}[T]$. Show that the simple rings $\mathcal{L} = \mathcal{K}[T]/p(T)\mathcal{K}[T]$ are fields, and that there is a natural injective ring homomorphism ι from $\mathcal{K}$ into $\mathcal{L}$.

Regarding ι as an identification, show that $\mathcal{L}$ is a finite extension of $\mathcal{K}$ (that is, $\dim_\mathcal{K} \mathcal{L}$ is finite.)

4.2.7 **$\mathcal{D}$ as a $\mathcal{D}[T, \alpha]$-module**

Let $\mathcal{D}$ be a division ring, α an automorphism of $\mathcal{D}$.

Given any element a of $\mathcal{D}$, we turn $\mathcal{D}$ into a $\mathcal{D}[T, \alpha]$-module $M(a)$ as follows: write m_0 for the identity element of $\mathcal{D}$, and define

$$m_0 \cdot f(T) = m_0 f_0 + m_0 a f_1 + \cdots + m_0 a^n f_n$$

where $f(T) = f_0 + T f_1 + \cdots + T^n f_n$.

Then $\mathcal{D}$ is a $\mathcal{D}[T, \alpha]$-module with T 'acting as a' as in (iv) of (1.2.2).

Caution: be careful to put polynomials into the 'right' form before calculating $m_0 f$: thus $m_0(xT) = m_0 a \alpha(x)$ for $x \in \mathcal{D}$.

(i) Show that $M(a) \cong \mathcal{D}[T, \alpha]/(T - a)\mathcal{D}[T, \alpha]$ as a $\mathcal{D}[T, \alpha]$-module.

(ii) Let $\mathfrak{a}$ be a twosided ideal of $\mathcal{D}[T, \alpha]$. Show that $M(a)\mathfrak{a} = 0$ if and only if $\mathfrak{a} \subseteq (T - a)\mathcal{D}[T, \alpha]$.

(iii) Suppose α has infinite inner order. Deduce that, for $a \neq 0$, $M(a)\mathfrak{a} \neq 0$ for any nonzero twosided ideal $\mathfrak{a}$ of $\mathcal{D}[T, \alpha]$ (see (3.2.18).)

(Thus we have a supply of Artinian $\mathcal{D}[T,\alpha]$-modules which are not modules over an Artinian residue ring of $\mathcal{D}[T,\alpha]$ in any natural manner.)

(iv) Let α again be an arbitrary automorphism, and define

$$C(a,\alpha) = \{x \in \mathcal{D} \mid xa = a\alpha(x)\},$$

the *skew centralizer* of a. Verify that $C(a,\alpha)$ is a division subring of $\mathcal{D}$, always nontrivial. (When α is the identity, this is the usual centralizer of a.)

(v) Show that $\operatorname{End}_{\mathcal{D}[T,\alpha]}(M(a)) \cong C(a,\alpha)$ as rings.
Hint. An endomorphism of $M(a)$ is defined by a map $m_0 \mapsto m_0 x$ for some x in $\mathcal{D}$.

(vi) Show further that, for elements a,b of $\mathcal{D}$,

$$\operatorname{Hom}_{\mathcal{D}[T,\alpha]}(M(a), M(b)) = \{x \in \mathcal{D} \mid xa = b\alpha(x)\}.$$

In particular, show that $M(a) \cong M(1)$ if and only if $a = x^{-1}\alpha(x)$ for some nonzero x. (This can happen nontrivially when α is not the identity automorphism.)

(vii) Suppose that α is the identity automorphism of $\mathcal{D}$. Deduce that $M(a) \cong M(b)$ if and only if a, b are *conjugate*, that is, $a = x^{-1}bx$ for some nonzero x in $\mathcal{D}$.

4.2.8 Let $\mathcal{K}$ be a (commutative) field and let α be an automorphism of $\mathcal{K}$ of finite order n. Write $\mathcal{Q} = \mathcal{K}^\alpha$, the fixed field of α.

Show that the ideals of $\mathcal{K}[T,\alpha]$ have the form $T^t g(T^n)\mathcal{K}[T,\alpha]$ where $g(T) \in \mathcal{Q}[T]$, $t \geq 0$ (see (3.2.14) and (3.2.16)).

Deduce that the maximal ideals of $\mathcal{K}[T,\alpha]$ correspond to the (monic) irreducible polynomials $p(T)$ over $\mathcal{Q}$. Show further that each ring $\mathcal{K}[T,\alpha]/p(T^n)\mathcal{K}[T,\alpha]$ is Artinian simple and contains $\mathcal{Q}$ in its centre.

Remark. The rings given by linear polynomials $T - a$, $a \in \mathcal{Q}$, are called *cyclic algebras*, and play a crucial role in the theory of division rings; for more details, see [Cohn 1979], §10.9, for example. A very special case is explored in the next exercise.

4.2.9 **Quaternions again**

This exercise generalizes Exercise 2.2.3. Suppose that (in the notation of the preceding exercise) $n = 2$ and that the characteristic of $\mathcal{K}$ is not 2.

Show that $\mathcal{K}$ has dimension 2 over $\mathcal{Q}$.

Let $a \neq 0$ and write $S = \mathcal{K}[T,\alpha]/(T^2 - a)\mathcal{K}[T,\alpha]$. Prove that S

has dimension 4 over $\mathcal{Q}$, and hence that S is either a division ring or the matrix ring $M_2(\mathcal{Q})$.

Suppose $a = 1$. Show that S is a division ring if and only if the equation

$$x_0^2 + x_1^2 + x_2^2 + x_3^2 = 0$$

has no nonzero solution in $\mathcal{Q}$. Give examples to show that both possibilities can occur.

Generalize this criterion to arbitrary nonzero a.

(When S is a division ring, it is known as a *generalized quaternion ring*.)

4.3 ARTINIAN RINGS

We can now give some structure theory for Artinian rings in general. In contrast to the semisimple case, this structure theory is far from complete, as can be seen from the citations in (4.3.27). Our main result is the Hopkins–Levitzki Theorem (4.3.21), which characterizes Artinian rings by the fact that they are right Noetherian, with nilpotent Jacobson radical rad(R) and Artinian semisimple residue ring $R/\mathrm{rad}(R)$. In Chapter 7 of [BK: CM], this result helps us to classify the projective modules over an Artinian ring, which is an important foundation for further developments in K-theory. We also give some results about a closely related class of rings, the semilocal rings.

We commence with a discussion of the Jacobson radical for rings and modules in general.

4.3.1 The Jacobson radical

For the moment, we take R to be an arbitrary ring.

Let M be an arbitrary right R-module M, and recall from (1.2.16) that a maximal submodule of M is a proper submodule M' of M such that if N is a submodule of M with $M' \subseteq N \subseteq M$, then either $M' = N$ or $N = M$. The *radical* rad(M) of M is defined to be the intersection

$$\mathrm{rad}(M) = \bigcap \{M' \mid M' \text{ maximal in } M\}.$$

(By convention, rad(M) = M if M has no maximal submodules.)

When M is finitely generated, any submodule of M is contained in a maximal submodule (1.2.21), so that the definition of rad(M) is not vacuous in the case of greatest interest to us.

After [Jacobson 1945], the radical of the ring R itself is called the *Jacobson radical* of R. At first sight, it appears that a ring will have both a left and a right Jacobson radical, but these in fact coincide (4.3.11).

4.3.2 Examples

(i) If I is an irreducible module, then 0 is its unique maximal submodule and so $\text{rad}(I) = 0$. If $S = \bigoplus_{\lambda \in \Lambda} I_\lambda$ is semisimple, Λ an arbitrary index set, then $\text{rad}(S) = 0$ also.

(ii) Let p be a prime number. Then $\text{rad}(\mathbb{Z}/p^r\mathbb{Z}) = p\mathbb{Z}/p^r\mathbb{Z}$, both as ring and as $\mathbb{Z}$-module. If $a = p_1^{r(1)} \cdots p_n^{r(n)}$ is the factorization of an integer a into powers of distinct primes, then $\text{rad}(\mathbb{Z}/a\mathbb{Z}) = p_1 \cdots p_n\mathbb{Z}/a\mathbb{Z}$.

(iii) Let $\mathcal{K}[\epsilon]$ be the ring of dual numbers over a field $\mathcal{K}$. Then $\text{rad}(\mathcal{K}[\epsilon]) = \epsilon\mathcal{K}[\epsilon]$.

4.3.3 Basic properties

The first result follows easily from the definition.

4.3.4 Lemma

For any module M, $\text{rad}(M/\text{rad}(M)) = 0$. $\square$

If M' is a maximal submodule of M, then the quotient module M/M' is irreducible. Conversely, if I is an irreducible module and there is a nontrivial R-module homomorphism β from M to I, then $\text{Ker}\,\beta$ is a maximal submodule of M. This leads to a fundamental result.

4.3.5 Lemma

The radical of a module M is given by

$$\text{rad}(M) = \bigcap\{\text{Ker}\,\beta \mid \beta : M \to I, \ I \ \text{irreducible}\},$$

where the intersection is taken over all homomorphisms β and irreducible modules I. $\square$

4.3.6 Proposition

Let $\alpha : M \to N$ be a homomorphism of right R-modules. Then

$$\alpha(\text{rad}(M)) \subseteq \text{rad}(N).$$

Proof

If β is a homomorphism from N to I with I irreducible, then $\beta\alpha$ is a homomorphism from M to I. Hence $\beta\alpha(\text{rad}(M)) = 0$ for all such β. $\square$

4.3.7 Corollary

For any R-module M, rad(M) is a fully invariant submodule of M.
In particular, the Jacobson radical rad(R) is a twosided ideal of R. □

Next, we investigate the relation between the radical of an R-module and the Jacobson radical of the ring R. Recall that, given a right R-module M and a right ideal $\mathfrak{a}$ of R, the submodule $M\mathfrak{a}$ of M is

$$M\mathfrak{a} = \{m_1 a_1 + \cdots + m_h a_h \mid m_i \in M, \ a_i \in \mathfrak{a}, \ h \in \mathbb{N}\}.$$

4.3.8 Lemma

Let M be a right R-module. Then $M \cdot \mathrm{rad}(R) \subseteq \mathrm{rad}(M)$.

Proof

Let I be an irreducible module and choose any x in I. Define $\beta : R \to I$ by $r \mapsto xr$. Then $\beta(\mathrm{rad}(R)) = 0$, so $I \cdot \mathrm{rad}(R) = 0$. Now if $\gamma : M \to I$ is any homomorphism, $\gamma(M \cdot \mathrm{rad}(R)) \subseteq I \cdot \mathrm{rad}(R)$, so $M \cdot \mathrm{rad}(R) \subseteq \mathrm{Ker}\,\gamma$. □

4.3.9 Lemma

Let M be a finitely generated right R-module and let M' be a submodule of M. Then the following are equivalent.

(i) *$M' \subseteq \mathrm{rad}(M)$.*
(ii) *If L is a submodule of M with $M' + L = M$, then $L = M$.*

Proof

(i) $\Rightarrow$ (ii):   Suppose that $M' + L = M$. If L is proper, then by (1.2.21) there is a maximal submodule that contains both L and M'.
(ii) $\Rightarrow$ (i):   If N is maximal, then (ii) forces $M' + N = N$. So $M' \subseteq N$, as required. □

For the next result, recall that the group of units $U(R)$ of an arbitrary ring R is the set of elements of R that have twosided inverses under multiplication; thus $x \in U(R)$ precisely when there is some $y \in R$ with $xy = 1 = yx$.

The following is one of the more important theorems in the subject, despite its traditional title as a lemma.

4.3.10 Nakayama's Lemma

Let j be a right ideal in R. Then the following statements are equivalent:

(i) *$\mathfrak{j} \subseteq \mathrm{rad}(R)$;*
(ii) *if M is a finitely generated right R-module and L is a submodule of M with $M = L + M\mathfrak{j}$, then $M = L$;*

(iii) *if M is a finitely generated right R-module with $Mj = M$, then $M = 0$;*
(iv) $1 + j = \{1 + x \mid x \in j\}$ *is a subgroup of $U(R)$.*

Proof

By Lemmas (4.3.8) and (4.3.9), (i) implies (ii). Taking L to be 0, we see that (ii) implies (iii) and the converse follows from considering M/L.

(ii) $\Rightarrow$ (iv): Let $x \in j$ and put $u = 1 + x$. Then $R = j + uR$, hence by (ii) $R = uR$ and $uv = 1$ for some v. But $v = 1 - xv \in 1 + j$, so v also has a right inverse which is necessarily u. Moreover, the set $1 + j$ is easily seen to be closed under both multiplication and taking inverses.

(iv) $\Rightarrow$ (i): Suppose that j is not contained in some maximal right ideal $\mathfrak{a}$. Then $R = \mathfrak{a} + j$, so $1 = a + x$, with a in $\mathfrak{a}$, x in j. But then, by (iv), a is invertible, so $\mathfrak{a} = R$. $\square$

4.3.11 Alternative descriptions of the Jacobson radical

Since $\mathrm{rad}(R)$ is a twosided ideal and condition (iv) of Nakayama's Lemma is left–right symmetric, it follows from the left R-module version of the above that the intersection of all maximal left ideals of R is a twosided ideal satisfying (iv) and hence (i) above. Thus by symmetry we see that the radical of a ring could equally well be defined in terms of maximal left ideals. Moreover, the radical can also be described as the set of those elements for which every right (or left) multiple is quasi-invertible, in other words, the set of all x such that, for all $r \in R$, $1 + xr$ is in $U(R)$.

To validate this description, write $\mathfrak{a}$ for the set of elements x such that $1 + xr \in U(R)$ for all r. By Nakayama's Lemma, $\mathrm{rad}(R) \subseteq \mathfrak{a}$, and equality follows once we know that $\mathfrak{a}$ is a right ideal. To establish this, it suffices to show that, if x and y are in $\mathfrak{a}$, then $x + y$ is in $\mathfrak{a}$ also. But, if

$$(1 + xr)(1 + x') = 1 \text{ for some } r \text{ and } x',$$

then

$$(1 + (x + y)r)(1 + x') = 1 + yr(1 + x') \in U(R).$$

We now have a series of results about the radicals of rings and modules.

4.3.12 Lemma

Let j be a twosided ideal of R, with $\pi : R \to R/j$ the canonical ring surjection. If $j \subseteq \mathrm{rad}(R)$, then $\mathrm{rad}(R/j) = \pi(\mathrm{rad}(R))$.

Proof

If $\mathfrak{n}$ is a maximal right ideal of $R/\mathfrak{j}$, then its inverse image $\pi^{-1}\mathfrak{n} = \{r \in R \mid \pi r \in \mathfrak{n}\}$ is a maximal ideal in R. Conversely, if $\mathfrak{m}$ is maximal in R, then $\pi\mathfrak{m}$ is maximal in $R/\mathfrak{j}$ since $\mathfrak{j} \subseteq \operatorname{rad}(R) \subseteq \mathfrak{m}$. Thus

$$\operatorname{rad}(R/\mathfrak{j}) = \bigcap\{\mathfrak{n} \mid \mathfrak{n} \text{ maximal in } R/\mathfrak{j}\} = \pi(\operatorname{rad}(R)). \qquad \square$$

The next result uses the elementary observations on 'change of rings' made in (1.2.14).

4.3.13 Proposition

Let $\mathfrak{j}$ be a twosided ideal of R, with $\mathfrak{j} \subseteq \operatorname{rad}(R)$. Then a representative set of irreducible right R-modules is given by

$$\mathcal{I}(R) = \mathcal{I}(R/\mathfrak{j}) = \mathcal{I}(R/\operatorname{rad}(R)).$$

Proof

For both R and $R/\operatorname{rad}(R)$, representative irreducible modules are of the form $R/\mathfrak{m}$ where $\mathfrak{m}$ is a maximal right ideal (and so $\mathfrak{j} \subseteq \operatorname{rad}(R) \subseteq \mathfrak{m}$). $\qquad \square$

We now return to the study of right Artinian rings. First, we record two preliminary results, which may be stated for Artinian modules in general.

Note that Lemma 4.3.5 may be paraphrased to say that for arbitrary M there is an embedding of $M/\operatorname{rad}(M)$ in the direct product $\prod_\alpha(M/M_\alpha)$, where $\{M_\alpha\}_\alpha$ is the set of maximal submodules of M. For Artinian modules, this product can be taken to be finite, as follows.

4.3.14 Lemma

Let M be an Artinian right R-module. Then

$$\operatorname{rad}(M) = M_1 \cap \cdots \cap M_k$$

for a finite set $\{M_1, \ldots, M_k\}$ of maximal submodules of M.

Proof

Suppose that we have found maximal submodules $M_1, \ldots, M_h$ such that the intersections $L_i = M_1 \cap \cdots \cap M_i$, $i \le h$, form a descending chain

$$\operatorname{rad}(M) \subseteq L_h \subset \cdots \subset L_{i+1} \subset L_i \subset \cdots \subset L_1 = M_1 \subset M.$$

If $\operatorname{rad}(M) \ne L_h$, there must be some maximal submodule M_{h+1} not containing L_h, so we can extend the chain downwards by defining $L_{h+1} = L_h \cap M_{h+1} \subset L_h$. By the descending chain condition, we must reach $\operatorname{rad}(M)$ in a finite number of steps. $\qquad \square$

4.3.15 Theorem

Let M be an Artinian right R-module. Then M is semisimple if and only if $\operatorname{rad}(M) = 0$.

Proof

We take maximal submodules $\{M_1, \ldots, M_k\}$ of M such that $\operatorname{rad}(M) = M_1 \cap \cdots \cap M_k$, as in the preceding lemma. Then $M/\operatorname{rad}(M)$ is a submodule of the semisimple module $M/M_1 \oplus \cdots \oplus M/M_k$, hence semisimple. In particular, if $\operatorname{rad}(M) = 0$, then M is semisimple.

The converse follows from the Complete Reducibility Theorem (4.1.16) together with (4.3.5).   □

We now come to the main results.

4.3.16 Theorem

Let R be a right Artinian ring. Then $R/\operatorname{rad}(R)$ is a semisimple Artinian ring.

Proof

First, note that $R/\operatorname{rad}(R)$ is Artinian as a module over itself since by (4.1.5) it is already Artinian as an R-module (see Exercise 4.1.8). By (4.3.12), $\operatorname{rad}(R/\operatorname{rad}(R)) = 0$, and hence $R/\operatorname{rad}(R)$ is semisimple by the previous result (4.3.15).   □

4.3.17 Corollary

Let M be a right module over a right Artinian ring R. Then

$$\operatorname{rad}(M) = M \cdot \operatorname{rad}(R).$$

In particular, M is semisimple if and only if $M \cdot \operatorname{rad}(R) = 0$.

Proof

We know that $M \cdot \operatorname{rad}(R) \subseteq \operatorname{rad}(M)$ by (4.3.8). Conversely, $M/(M \cdot \operatorname{rad}(R))$ is a module over the semisimple ring $R/\operatorname{rad}(R)$ and so is semisimple by (4.2.9), which gives the reverse inequality.   □

We can now strengthen (4.3.13).

4.3.18 Corollary

Let R be right Artinian. Then R has a finite representative set of irreducible right R-modules, given by

$$\mathcal{I}(R) = \mathcal{I}(R/\operatorname{rad}(R)).$$ □

4.3.19 Nilpotent ideals and a characterization of Artinian rings

Recall that the product of two right ideals $\mathfrak{a}$ and $\mathfrak{b}$ of a ring R is

$$\mathfrak{a}\mathfrak{b} = \{a_1b_1 + \cdots + a_kb_k \mid a_i \in \mathfrak{a},\ b_i \in \mathfrak{b},\ 1 \leq i \leq k,\ k > 0\},$$

which is again a right ideal of R. This is a special case of the product $M\mathfrak{b}$ for a right module M. We write $\mathfrak{a}^2 = \mathfrak{a}\mathfrak{a}$ and inductively $\mathfrak{a}^k = \mathfrak{a}^{k-1}\mathfrak{a}$. (There is an ambiguity in notation, since $\mathfrak{a}^k$ could also denote the direct sum of k copies of $\mathfrak{a}$, but this will not cause any confusion in practice.) The definition of the product for left or twosided ideals is identical to that for right ideals.

A right, left or twosided ideal $\mathfrak{a}$ of a ring R is *nilpotent* if $\mathfrak{a}^k = 0$ for some natural number k.

4.3.20 Proposition

Let R be a right Artinian ring. Then the following hold.

(i) $\mathrm{rad}(R)$ *is nilpotent.*

(ii) *If $\mathfrak{j}$ is any nilpotent (right or left) ideal of R, then $\mathfrak{j} \subseteq \mathrm{rad}(R)$.*

Proof

By the descending chain condition, $(\mathrm{rad}(R))^k = (\mathrm{rad}(R))^{k+1}$ for some k. Suppose that $(\mathrm{rad}(R))^k \neq 0$. Then there is some nonzero right ideal $\mathfrak{a}$ of R with $\mathfrak{a}(\mathrm{rad}(R))^k \neq 0$, (for example, R itself), and, as R is right Artinian, there must be a right ideal $\mathfrak{a}$ which is minimal among such ideals by (4.1.4). Choose some $a \in \mathfrak{a}$ with $a(\mathrm{rad}(R))^k \neq 0$. Then

$$a(\mathrm{rad}(R)) \subseteq aR \subseteq \mathfrak{a}$$

and

$$a(\mathrm{rad}(R)) \cdot (\mathrm{rad}(R))^k = a(\mathrm{rad}(R))^{k+1} = a(\mathrm{rad}(R))^k \neq 0.$$

By the minimality of $\mathfrak{a}$, we have $a(\mathrm{rad}(R)) = aR = \mathfrak{a}$. The implication (i) $\Rightarrow$ (iii) of Nakayama's Lemma (4.3.10) now shows that $aR = 0$, a contradiction. Thus (i) holds.

If $x \in \mathfrak{j}$, then x itself is nilpotent and $1 + x$ is a unit with inverse $1 - x + \cdots \pm x^{k-1}$, which gives (ii) by the implication (iv) $\Rightarrow$ (i) of Nakayama's Lemma. $\square$

We now have a complete characterization of right Artinian rings.

4.3.21 The Hopkins–Levitzki Theorem

A ring R is right Artinian if and only if R is right Noetherian, $\mathrm{rad}(R)$ is nilpotent and $R/\mathrm{rad}(R)$ is an Artinian semisimple ring.

Proof

Necessity of the second and third conditions has already been proved in (4.3.20) and (4.3.16) respectively. We therefore suppose that $\mathrm{rad}(R)$ is nilpotent and that $R/\mathrm{rad}(R)$ is Artinian semisimple, and show that R is right Artinian if and only if R is right Noetherian.

We argue by induction on the smallest exponent k with $(\mathrm{rad}(R))^k = 0$. If $k = 0$, then $R = 0$ and there is nothing to show. In general, put $S = R/(\mathrm{rad}(R))^{k-1}$.

By (4.3.12), $\mathrm{rad}(S) = \mathrm{Im}(\mathrm{rad}(R))$, so that $(\mathrm{rad}(S))^{k-1} = 0$. Moreover, $R/\mathrm{rad}(R)$ maps onto $S/\mathrm{rad}(S)$, making $S/\mathrm{rad}(S)$ an Artinian semisimple ring as indicated in Exercise 4.2.3. So, by the induction hypothesis, S is a right Artinian ring precisely when it is a right Noetherian ring. Using a standard change of rings argument, this assertion is equivalent to the assertion that S is Artinian as a right R-module precisely when it is Noetherian as a right R-module (see Exercises 4.1.8 and 3.1.5). Thus, by (4.1.5) and (3.1.2), it suffices to show that the kernel $(\mathrm{rad}(R))^{k-1}$ of the canonical map from R to S is right Artinian precisely when it is right Noetherian.

Since $(\mathrm{rad}(R))^{k-1} \cdot \mathrm{rad}(R) = 0$, $(\mathrm{rad}(R))^{k-1}$ is naturally an $(R/\mathrm{rad}(R))$-module. By the usual change of rings argument, it is enough to verify the assertion for $(\mathrm{rad}(R))^{k-1}$ considered as an $(R/\mathrm{rad}(R))$-module. But $R/\mathrm{rad}(R)$ is an Artinian semisimple ring, by hypothesis, so the result follows from (4.2.10).   □

4.3.22 Semilocal rings

A ring R is called *semilocal* if $R/\mathrm{rad}(R)$ is a semisimple Artinian ring. Clearly, a right Artinian ring is semilocal, and it turns out that many results concerning Artinian rings hold also for semilocal rings. For example, we record the following straightforward generalization of (4.3.18).

4.3.23 Proposition

Let R be a semilocal ring. Then R has a finite representative set of irreducible right R-modules, given by $\mathcal{I}(R) = \mathcal{I}(R/\mathrm{rad}(R))$.   □

Remark. When R is Artinian, the indecomposable projective R-modules correspond bijectively to the irreducible $(R/\mathrm{rad}(R))$-modules under the correspondence

$$P \longleftrightarrow P/P\,\mathrm{rad}(R).$$

4.3.24 Local rings

An important special case of a semilocal ring is a *local ring*. By definition, R is local if $R/\operatorname{rad}(R)$ is a division ring. It is clear that such a ring R is one with a unique irreducible module, namely $R/\operatorname{rad}(R)$ itself. Equivalently (see (1.2.16)), a local ring is a ring R with a unique maximal twosided ideal which is also simultaneously the unique maximal left ideal and the unique maximal right ideal of R. Since this unique maximal ideal is necessarily the Jacobson radical, a further characterization is that R is partitioned as the disjoint union

$$R = U(R) \sqcup \operatorname{rad}(R).$$

This partition is equivalent to the information that every multiple of a non-invertible element is quasi-invertible. Or again, from the fact that here the sum of two nonunits is a nonunit, one concludes that R is local if and only if the equation $r + s = 1$ forces either r or s to be a unit.

Finally, we note two technical lemmas that are very useful in K-theory.

4.3.25 Lemma

Let R be a ring, with a twosided ideal $\mathfrak{j} \subseteq \operatorname{rad}(R)$. Let x be an element of R, and suppose that the image of x is invertible in $R/\mathfrak{j}$. Then x is invertible in R.

Proof

Since x is invertible in $R/\mathfrak{j}$, $xR + \mathfrak{j} = R$, and so, by the implication (i) $\Rightarrow$ (ii) of Nakayama's Lemma (4.3.10), $xR = R$. Hence x has a right inverse. Similarly x has a left inverse. $\qquad\square$

4.3.26 Lemma ([Bass 1968], Chapter III, §2.)

Let R be a semilocal ring, let $\mathfrak{a}$ be a right ideal of R and let b in R be such that $\mathfrak{a} + bR = R$. Then the coset

$$\mathfrak{a} + b = \{a + b \mid a \in \mathfrak{a}\}$$

contains a unit of R.

Proof

By the previous lemma, we can assume that $\operatorname{rad}(R) = 0$ and so that R is itself an Artinian semisimple ring. By the Wedderburn–Artin Theorem (4.2.3), we know that R decomposes into a direct product of Artinian simple rings and it is clearly enough to verify the assertion in each component.

Suppose then that $R = M_n(\mathcal{D})$, where $\mathcal{D}$ is a division ring. We regard R as the ring of endomorphisms of the right $\mathcal{D}$-space $V = \mathcal{D}^n$. Since R is completely reducible as a right R-module, the Artinian Splitting Theorem (4.1.17) shows that there is a right ideal $\mathfrak{a}'$ of R with $\mathfrak{a} \oplus \mathfrak{a}' = R$ as a right R-module. Then $\mathfrak{a}V \oplus \mathfrak{a}'V = V$ as a $\mathcal{D}$-space, from which we see that $\mathfrak{a} = \{a \in R \mid aV \subseteq \mathfrak{a}V\}$.

Since $\mathfrak{a} + bR = R$, we have $\mathfrak{a}V + \mathrm{Im}(b) = V$, and so there is a subspace W of $\mathfrak{a}V$ with $W \oplus \mathrm{Im}(b) = V$. Write $V = \mathrm{Ker}(b) \oplus U$ for some U. Then (as in the proof of the 'rank and nullity theorem' in undergraduate linear algebra) b induces an isomorphism from U to $\mathrm{Im}(b)$ and so $\mathrm{Ker}(b) \cong W$.

We can now choose an element a in $R = \mathrm{End}(V)$ with the properties that $aU = 0$ and that a induces an isomorphism from $\mathrm{Ker}(b)$ to W. Then $aV = W \subseteq \mathfrak{a}V$, so $a \in \mathfrak{a}$, and

$$(a + b)V = (a + b)(\mathrm{Ker}(b) + U) = W \oplus \mathrm{Im}(b) = V,$$

which shows that $a + b$ induces an automorphism of V and hence is a unit of R. □

4.3.27 Further reading

The structure theory for modules over an arbitrary Artinian ring is an active topic of research. More extensive expositions of the basic results can be found in [Cohn 1979] and [Rowen 1988], and some recent developments are given in [Auslander, Reiten & Smalø 1995].

Exercises

4.3.1   Show that a representative set of irreducible modules for $\mathbb{Z}$ is

$$\mathcal{I}(\mathbb{Z}) = \{\mathbb{Z}/p\mathbb{Z} \mid p \text{ prime}\}$$

and that $\mathrm{rad}(\mathbb{Z}) = 0$. Why is $\mathbb{Z}$ not isomorphic to $\bigoplus_p \mathbb{Z}/p\mathbb{Z}$?

4.3.2   Let $T = T_2(\mathcal{K}) = \begin{pmatrix} \mathcal{K} & \mathcal{K} \\ 0 & \mathcal{K} \end{pmatrix}$ be the ring of upper triangular matrices over a field $\mathcal{K}$. Find $\mathrm{rad}(T)$.

4.3.3   Let R and S be (right Artinian) rings and let M be an R-S-bimodule. Let $T = \begin{pmatrix} R & M \\ 0 & S \end{pmatrix}$ be the generalized triangular matrix ring defined by the triple (R, S, M); recall that T is the set of all matrices $\begin{pmatrix} r & m \\ 0 & s \end{pmatrix}$ with $r \in R$, $s \in S$ and $m \in M$, with addition and multiplication given in the obvious way.

Find $\mathrm{rad}(T)$.

Hence or otherwise find $\mathrm{rad}(T_k(\mathcal{K}))$, where $T_k(\mathcal{K})$ is the ring of $k \times k$ upper triangular matrices over a field $\mathcal{K}$.

4.3.4   Let M be an Artinian module. The *radical series* of M is defined by

$$\text{rad}^0(M) = M \text{ and } \text{rad}^{i+1}(M) = \text{rad}(\text{rad}^i(M)) \text{ for } i \geq 1.$$

Show that this is a characteristic series for M.

  Show that the following are equivalent.

  (i) The radical series of M reaches 0 (that is, $\text{rad}^i(M) = 0$ for some i).
  (ii) The socle series of M reaches M.
  (iii) M is Noetherian.

  Hint. (4.1.25) is useful; Exercise 4.1.6 shows that the radical series need not reach 0.

  Verify that the radical series and the socle series are the same if (and only if) these conditions hold.

4.3.5   Find the radical series of a finitely generated Artinian $\mathbb{Z}$-module (see Exercise 4.1.5).

4.3.6   Find the socle series of the ring $T_k(\mathcal{K})$ of $k \times k$ upper triangular matrices over a field $\mathcal{K}$.

4.3.7   Let R be a right Artinian ring. Show that the following are equivalent.

  (i) The exact sequence

$$0 \longrightarrow \text{rad}(R) \longrightarrow R \longrightarrow R/\text{rad}(R) \longrightarrow 0$$

  of right R-modules is split.
  (ii) $\text{rad}(R)$ is generated by an idempotent.
  (iii) $\text{rad}(R) = 0$.
  (iv) R is semisimple.

4.3.8   Let $\mathcal{K}$ be an extension field of a field $\mathcal{Q}$, of infinite dimension over $\mathcal{Q}$. (For example, take $\mathcal{K} = \mathcal{Q}(X_1, X_2, \ldots)$, the field of rational functions in a countably infinite set of variables.) Using (1.2.20), show that there is an infinite descending chain of $\mathcal{Q}$-subspaces in $\mathcal{K}$.

  Let $T = \begin{pmatrix} \mathcal{Q} & \mathcal{K} \\ 0 & \mathcal{K} \end{pmatrix}$. Show that, for any $\mathcal{Q}$-subspace V of $\mathcal{K}$, $\begin{pmatrix} 0 & V \\ 0 & 0 \end{pmatrix}$ is a left ideal in T. Deduce that T is neither left Artinian nor left Noetherian.

  Find $\text{rad}(T)$, and show that T is right Artinian and right Noetherian.

4.3.9   Maschke's Theorem

Let G be a finite group of order n. Given a field $\mathcal{K}$, the group ring $\mathcal{K}G$ is called the *group algebra* of G over $\mathcal{K}$ (see Exercise 2.6.7). Let p be the characteristic of $\mathcal{K}$; p is a prime by (1.1.10).

Maschke's Theorem states that, if p does not divide n, then $\mathcal{K}G$ is an Artinian semisimple ring.

By (ii) of (4.1.2), $\mathcal{K}G$ is Artinian. To prove semisimplicity, consider any surjective (right) $\mathcal{K}G$-module homomorphism $\pi : L \to M$. As a $\mathcal{K}$-space homomorphism, π has a splitting σ. Define τ by

$$\tau(m) = \Big(\sum_{g \in G} g \sigma g^{-1}(m) \Big)/n \text{ for each } m \text{ in } M;$$

we can divide by n since $n \neq 0$ in $\mathcal{K}$ (more properly, the image of n under the canonical homomorphism χ from $\mathbb{Z}$ to $\mathcal{K}$ is nonzero (1.1.10)). Then τ is a $\mathcal{K}G$-module splitting of π. Thus $\mathcal{K}G$ is semisimple by, say, Exercise 4.3.7 above.

4.3.10   Maschke's Theorem is false if the order n of G is 0 in $\mathcal{K}$, that is, if the characteristic p of $\mathcal{K}$ divides n.

Let $\Sigma = \sum \{g \mid g \in G\}$. Then $\Sigma^2 = n\Sigma = 0$ in $\mathcal{K}G$ and Σ is in the centre of $\mathcal{K}G$. Thus the principal ideal $\Sigma \mathcal{K}G$ is twosided and nilpotent, hence contained in the radical of $\mathcal{K}G$.

4.3.11   Let C be a cyclic group of order p with generator γ, and let $\mathcal{K} = \mathbb{Z}/p\mathbb{Z}$. Show that $\mathrm{rad}(\mathcal{K}C) = (1 - \gamma)\mathcal{K}C$.

This result can be extended to arbitrary finite p-groups: if G is such a group,

$$\mathrm{rad}(\mathcal{K}G) = \sum_{g \neq 1} \mathcal{K}(g - 1).$$

Details are given in [Curtis & Reiner 1981], (5.24).

4.3.12   Let R be an arbitrary ring. Prove the following assertions.

(a) $\mathrm{rad}(M_n(R)) = M_n(\mathrm{rad}(R))$.

(b) If $\mathfrak{j}$ is a twosided ideal of R with $\mathfrak{j} \subseteq \mathrm{rad}(R)$, then the canonical homomorphism from $\mathrm{GL}_n(R)$ to $\mathrm{GL}_n(R/\mathfrak{j})$ is surjective.

Suppose further that R is semilocal. Show

(c) $M_n(R)$ is also semilocal,

(d) assertion (b) holds for any ideal $\mathfrak{j}$.

Hints. Use (iii) $\Rightarrow$ (i) of Nakayama's Lemma (4.3.10) to show

$$\mathrm{rad}(M_n(R)) \supseteq M_n(\mathrm{rad}(R)),$$

noting that any $M_n(R)$ module is an R-module. In the opposite direction, recall from Exercise 1.1.4 that any twosided ideal $\mathfrak{A}$ in $M_n(R)$ can be written as $M_n(\mathfrak{a})$ for some ideal $\mathfrak{a}$ of R, and use (i) $\Rightarrow$ (iv) of Nakayama. The same exercise helps with (c).

Parts (a) and (c) can also be derived from the Morita theory ([BK: CM], Chapter 4).

4.3.13   [Berrick & Keating 1997]

      Show that the rings R and $R/\operatorname{rad}(R)$ have the same type.

      Hint. Nakayama's Lemma.

5

DEDEKIND DOMAINS

Dedekind domains occupy a pivotal position at the interface of algebra and number theory. From the algebraic point of view, Dedekind domains are important since they are commutative rings whose properties are direct extensions of those established for commutative Euclidean domains in Chapter 3. The main distinction between the two types of ring lies in the fact that the ideals of a Dedekind domain $\mathcal{O}$ need not be principal. The extent to which $\mathcal{O}$ fails to be a principal ideal domain is measured by the class group $\mathrm{Cl}(\mathcal{O})$ of $\mathcal{O}$.

In number theory, the study of Dedekind domains originates with rings of integers in algebraic number fields, that is, finite extensions of the field of rational numbers $\mathbb{Q}$. An example is the ring $\mathbb{Z}[\sqrt{-5}]$ in $\mathbb{Q}(\sqrt{-5})$, which turns out to have class group of order 2. Thus the class group contains information about the solubility, in $\mathbb{Z}$, of polynomial equations of the type $x^2 + 5y^2 = 0$. We are able only to touch on this topic in this text.

Another reason for paying detailed attention to Dedekind domains arises from K-theory. The class group can be interpreted in terms of K-theory, and analogues of the class group can thereby be found for more general types of ring.

In this chapter, we examine rings of integers, and we show how this leads to the basic ring-theoretic properties of Dedekind domains. We make a detailed analysis of the integers in the quadratic case, which allows some elementary calculations of class groups.

Our development of the theory follows the axiomatic method introduced in [Noether 1926]. The connection between Dedekind's original work and its modern version can be found in the introductory commentary by its translator, John Stillwell, to [Dedekind 1996]. This makes clear that much of our modern terminology (ideals, modules, ...) was introduced by Dedekind in the context of algebraic number fields.

All rings in this chapter are commutative, save for a few examples. Thus any module can be regarded as a balanced bimodule (1.2.7) and any ideal is twosided.

5.1 DEDEKIND DOMAINS AND INVERTIBLE IDEALS

In this section we give the definition of a Dedekind domain and the fundamental results on the factorization of ideals in a Dedekind domain. After a preliminary discussion of ideal theory in a general commutative domain, we define a Dedekind domain as a domain $\mathcal{O}$ all of whose nonzero ideals are 'invertible'. This definition leads directly to the definition of the ideal class group $\mathrm{Cl}(\mathcal{O})$ of $\mathcal{O}$, and is well suited to the development of the ideal theory of $\mathcal{O}$, culminating in the Unique Factorization Theorem for ideals (5.1.19).

Since a (commutative) principal ideal domain is a special type of Dedekind domain, we can reinterpret the unique factorization of ideals to obtain the unique factorization of elements which was promised in (3.2.22).

Some results about invertible ideals in (noncommutative) orders are given in the exercises.

5.1.1 Prime ideals

Before we can define a Dedekind domain, we need a few preliminary concepts from commutative algebra.

Let $\mathcal{O}$ be any commutative ring. An ideal $\mathfrak{p}$ of $\mathcal{O}$ is *prime* if it is a proper ideal with the following property:

(P) if $x \in \mathcal{O}$ and $y \in \mathcal{O}$ and $xy \in \mathfrak{p}$, then either $x \in \mathfrak{p}$ or $y \in \mathfrak{p}$.

5.1.2 Lemma

An ideal $\mathfrak{p}$ is prime if and only if the residue ring $\mathcal{O}/\mathfrak{p}$ is a domain.

In particular, a maximal ideal of $\mathcal{O}$ is prime.

Proof

The first assertion is a matter of checking definitions. For the second, note that, for any ideal $\mathfrak{i}$ of $\mathcal{O}$, there is a bijective correspondence between ideals $\mathfrak{a}$ of $\mathcal{O}$ with $\mathfrak{i} \subseteq \mathfrak{a} \subseteq \mathcal{O}$ and ideals $\mathfrak{b}$ of the residue ring $\mathcal{O}/\mathfrak{i}$, given by $\mathfrak{b} = \pi \mathfrak{a}$, where π is the canonical surjection. $\square$

An element p of $\mathcal{O}$ is *prime* if the following holds:

(PE) if $p \mid xy$, then either $p \mid x$ or $p \mid y$.

Recall that an element z of $\mathcal{O}$ is irreducible if it has no factorizations $z = uv$ except those trivial factorizations in which either u or v is a unit of $\mathcal{O}$.

The general relationship between prime elements, prime ideals and irreducible elements is given by the next, easily verified, result.

5.1.3 Lemma

(i) *The element p is prime if and only if the principal ideal $p\mathcal{O}$ is prime.*

(ii) *Suppose that $\mathcal{O}$ is a domain. If p is a prime element of $\mathcal{O}$, then p is irreducible.* $\square$

In most contexts of interest to us, a prime element is the same thing as an irreducible element. However, in general rings, prime elements and irreducible elements are different – see Exercise 5.1.7. The principal ideal given by an irreducible element need not be prime (Exercise 5.3.3).

5.1.4 Coprime ideals

Two ideals $\mathfrak{a}$ and $\mathfrak{b}$ in a commutative ring $\mathcal{O}$ are *coprime* if

$$\mathfrak{a} + \mathfrak{b} = \mathcal{O}.$$

(Sometimes, the term *comaximal* is used.)

Both the following results have a claim to be known as the Chinese Remainder Theorem, so we name them accordingly; however, the second version, in terms of congruences, is more often given the appellation. (In fact, the original problem was posed about 1600 years ago by Sun Zi in terms of integral remainders, and solved by Qin Jiushao in 1247 – see [Lam & Ang 1992], (7.5).)

5.1.5 Chinese Remainder Theorem – I

Suppose that $\mathfrak{a}$ and $\mathfrak{b}$ are coprime ideals of a commutative ring $\mathcal{O}$. Then

(i) $\mathfrak{a} \cap \mathfrak{b} = \mathfrak{ab}$,

(ii) $\mathcal{O}/\mathfrak{ab} \cong \mathcal{O}/\mathfrak{a} \times \mathcal{O}/\mathfrak{b}$ *as rings.*

Proof

(i)   We can write $1 = a + b$ with a in $\mathfrak{a}$ and b in $\mathfrak{b}$. If $x \in \mathfrak{a} \cap \mathfrak{b}$, then $x = x1$ is in $\mathfrak{ab}$. The reverse inclusion is clear.

(ii)   Let θ be the ring homomorphism from $\mathcal{O}$ to $\mathcal{O}/\mathfrak{a} \times \mathcal{O}/\mathfrak{b}$ given by $\theta r = (\bar{r}, \bar{r})$, where $\bar{r}$ indicates the residue of r both in $\mathcal{O}/\mathfrak{a}$ and in $\mathcal{O}/\mathfrak{b}$. The kernel of θ is $\mathfrak{a} \cap \mathfrak{b}$, so, by the Induced Mapping Theorem for rings (1.1.9), there is an injective ring homomorphism $\bar{\theta}$ from $\mathcal{O}/\mathfrak{ab}$ to $\mathcal{O}/\mathfrak{a} \times \mathcal{O}/\mathfrak{b}$ induced by θ.

Given an element $(\bar{r}, \bar{s})$ in the direct product, write $x = sa + rb$ in $\mathcal{O}$. Then $\theta x = (\bar{r}, \bar{s})$, so $\bar{\theta}$ is also surjective. $\qquad\square$

5.1.6 Chinese Remainder Theorem – II

Suppose that $\mathfrak{a}$ and $\mathfrak{b}$ are coprime ideals of $\mathcal{O}$ and that elements r and s of $\mathcal{O}$ are given. Then there is an element x of $\mathcal{O}$ which satisfies the congruences

$$x \equiv r \pmod{\mathfrak{a}} \quad and \quad x \equiv s \pmod{\mathfrak{b}}.$$

More generally, suppose that $\{\mathfrak{a}_1, \ldots, \mathfrak{a}_k\}$ is a finite set of ideals in $\mathcal{O}$, any two of which are coprime, and let $r_1, \ldots, r_k$ be any elements of $\mathcal{O}$. Then there is an element x in $\mathcal{O}$ with

$$x \equiv r_i \pmod{\mathfrak{a}_i} \quad for \quad i = 1, \ldots, k.$$

Proof

The first assertion is a restatement of the preceding theorem, and the second follows by induction, on noting that $\mathfrak{a}_1$ and the product $\mathfrak{a}_2 \cdots \mathfrak{a}_k$ are coprime (also by induction). $\qquad\square$

5.1.7 Fractional ideals

We prepare the way for the definition of a Dedekind domain by extending the notion of an ideal. Let $\mathcal{O}$ be a commutative Noetherian domain with field of fractions $\mathcal{K}$. Evidently, $\mathcal{K}$ is an $\mathcal{O}$-module by restriction of scalars.

A *fractional ideal* of $\mathcal{O}$ is a nonzero finitely generated $\mathcal{O}$-submodule of $\mathcal{K}$. Among these are the principal fractional ideals $\mathcal{O}x$ where x can be any nonzero element of $\mathcal{K}$, and, of course, all the nonzero ideals of $\mathcal{O}$. If we need to emphasize that a particular fractional ideal $\mathfrak{a}$ is actually an ideal in $\mathcal{O}$, we say that $\mathfrak{a}$ is an *integral* ideal.

Let $\mathrm{Frac}(\mathcal{O})$ denote the set of all fractional ideals, $\mathrm{Pr}(\mathcal{O})$ the subset consisting of principal ideals. We define a product in $\mathrm{Frac}(\mathcal{O})$ by extending the usual product of ideals:

$$\mathfrak{a}\mathfrak{b} = \{a_1 b_1 + \cdots + a_k b_k \mid a_i \in \mathfrak{a}, b_i \in \mathfrak{b}, k \geq 0\}.$$

The identification of a fractional ideal is facilitated by an elementary but useful manoeuvre, that of *clearing denominators*.

Suppose that $\{x_1, \ldots, x_\ell\}$ is a finite set of elements in $\mathcal{K}$. For each i, we have $x_i = a_i/b_i$ with a_i, b_i in $\mathcal{O}$ (1.1.12). Put $d = b_1 \cdots b_\ell$; then $x_i = c_i/d$ for suitable $c_i \in \mathcal{O}$ and clearly dx_i is in $\mathcal{O}$ for all i. Thus d is the *common denominator* of the set $\{x_1, \ldots, x_\ell\}$, which is cleared by cross-multiplication.

5.1.8 Lemma

Suppose that $\mathcal{O}$ is a commutative Noetherian domain with field of fractions $\mathcal{K}$ and let $\mathfrak{a}$ be an $\mathcal{O}$-submodule of $\mathcal{K}$. Then the following statements are equivalent:

(i) *$\mathfrak{a}$ is a fractional ideal of $\mathcal{O}$;*

(ii) *$\mathcal{O}c \subseteq \mathfrak{a} \subseteq \mathcal{O}d^{-1}$ where c and d are nonzero elements of $\mathcal{O}$;*

(iii) *$\mathfrak{a}$ is nonzero and $\mathfrak{a} \subseteq \mathcal{O}d^{-1}$ where d is a nonzero element of $\mathcal{O}$.*

Proof

(i) $\Rightarrow$ (ii)   By definition, $\mathfrak{a}$ has a finite set of generators $\{x_1, \ldots, x_\ell\}$. Place these over a common denominator d; then $\mathfrak{a} \subseteq \mathcal{O}d^{-1}$. Since $\mathfrak{a}$ is nonzero, it contains a nonzero element y of $\mathcal{K}$, and some multiple of y will be an element $c \in \mathfrak{a} \cap \mathcal{O}$.

(ii) $\Rightarrow$ (iii)   Immediate.

(iii) $\Rightarrow$ (i)   Since $\mathcal{O} \cong \mathcal{O}d^{-1}$ as an $\mathcal{O}$-module, $\mathcal{O}d^{-1}$ is Noetherian and so $\mathfrak{a}$ is finitely generated. $\square$

Our next result states that $\mathrm{Frac}(\mathcal{O})$ is an abelian monoid under multiplication of fractional ideals, with identity element the ring $\mathcal{O}$ itself.

5.1.9 Proposition

Let $\mathcal{O}$ be a commutative Noetherian domain and let $\mathfrak{a}$, $\mathfrak{b}$ and $\mathfrak{c}$ be in $\mathrm{Frac}(\mathcal{O})$. Then

(i) $\mathfrak{a}\mathfrak{b} \in \mathrm{Frac}(\mathcal{O})$,

(ii) $(\mathfrak{a}\mathfrak{b})\mathfrak{c} = \mathfrak{a}(\mathfrak{b}\mathfrak{c})$,

(iii) $\mathfrak{a}\mathfrak{b} = \mathfrak{b}\mathfrak{a}$,

(iv) $\mathcal{O}\mathfrak{a} = \mathfrak{a} = \mathfrak{a}\mathcal{O}$.

Proof

By the above lemma, $\mathfrak{a} \subseteq \mathcal{O}x$ and $\mathfrak{b} \subseteq \mathcal{O}y$ for some nonzero x, y in $\mathcal{K}$. Clearly, $0 \neq \mathfrak{a}\mathfrak{b} \subseteq \mathcal{O}xy$, which gives (i) using the lemma again. The remaining assertions are easily checked. $\square$

5.1.10 Dedekind domains – the definition

We say that a fractional ideal $\mathfrak{a}$ of a commutative domain $\mathcal{O}$ is *invertible* if for some fractional ideal $\mathfrak{a}^{-1}$ of $\mathcal{O}$ we have

$$\mathfrak{a}\mathfrak{a}^{-1} = \mathcal{O}.$$

The set of invertible fractional ideals forms an abelian group under multiplication, which is contained in Frac($\mathcal{O}$), and, obviously, every nonzero principal fractional ideal belongs to this group. An easy standard argument shows that the inverse of $\mathfrak{a}$ is unique if it exists.

We can now make the definition that we have been working towards.

The commutative Noetherian domain $\mathcal{O}$ is a *Dedekind domain* if every fractional ideal of $\mathcal{O}$ is invertible.

Thus a field is a trivial example of a Dedekind domain. Most of our results have uninteresting special cases for fields, so we usually assume that a Dedekind domain $\mathcal{O}$ is different from its field of fractions $\mathcal{K}$.

A slight modification of the definitions and arguments given here allows one to deduce that a Dedekind domain is a Noetherian ring, rather than including this fact as an axiom. The modification is indicated in Exercise 5.1.2 below.

5.1.11 The class group

The set Frac($\mathcal{O}$) of fractional ideals of a Dedekind domain is thus a multiplicative abelian group with Pr($\mathcal{O}$) as a subgroup. The *class* of the fractional ideal $\mathfrak{a}$ is the set of ideals of the form $\mathfrak{a}x$ where x runs through the nonzero elements of $\mathcal{K}$. The *ideal class group* Cl($\mathcal{O}$) of $\mathcal{O}$ is the set of all such ideal classes, with multiplication induced by the multiplication of fractional ideals. Thus, Cl($\mathcal{O}$) is the quotient group

$$\mathrm{Cl}(\mathcal{O}) = \mathrm{Frac}(\mathcal{O})/\mathrm{Pr}(\mathcal{O}).$$

We indicate the class of $\mathfrak{a}$ by $\{\mathfrak{a}\} \in \mathrm{Cl}(\mathcal{O})$.

The ideal class group is often referred to simply as the *class group*. In number theory texts, it is very often called the class group of the field $\mathcal{K}$ rather than $\mathcal{O}$, since, in that context, the ring $\mathcal{O}$ is almost invariably the ring of algebraic integers in $\mathcal{K}$.

The determination of the class group of any particular Dedekind domain is usually a difficult problem which requires some number theory or algebraic geometry for its solution, if indeed a solution is known. Some examples of nontrivial class groups are given in section 5.3 – see (5.3.17) and the following results, and the exercises.

On the other hand, any given abelian group can be realized as the ideal class group of some Dedekind domain, as shown by [Claborn 1966]. An exposition is given in [Fossum 1973], Theorem 14.10.

Our first result about class groups is a characterization of principal ideal domains that follows directly from the definitions.

5.1.12 Proposition

Let $\mathcal{O}$ be a commutative domain. Then $\mathcal{O}$ is a principal ideal domain if and only if $\mathcal{O}$ is a Dedekind domain for which $\mathrm{Cl}(\mathcal{O}) = 1$.   □

5.1.13 An exact sequence

The fact that the class group measures the failure of a Dedekind domain to be a principal ideal domain is also described explicitly by an exact sequence of abelian groups. (Since an abelian group is another name for a $\mathbb{Z}$-module, the definition of such an 'exact sequence' is as in (2.4.3). However, we are now writing our groups multiplicatively, so that identity elements and trivial groups are written as 1.)

Define a map $\partial : U(\mathcal{K}) \to \mathrm{Frac}(\mathcal{O})$ by $\partial(x) = \mathcal{O}x$, that is, we associate with each nonzero element x of the field of fractions the principal fractional ideal which it generates. It is clear that ∂ is a group homomorphism, and it is easy to verify that it fits into the following exact sequence, in which each unnamed arrow is a canonical homomorphism:

$$1 \longrightarrow U(\mathcal{O}) \longrightarrow U(\mathcal{K}) \overset{\partial}{\longrightarrow} \mathrm{Frac}(\mathcal{O}) \longrightarrow \mathrm{Cl}(\mathcal{O}) \longrightarrow 1.$$

The existence of the above exact sequence can also be established as a special case of a more general construction in algebraic K-theory, namely the localization sequence.

Next we look more closely at the circumstances when two ideals correspond to the same class.

5.1.14 Proposition

Let $\mathfrak{a}$ and $\mathfrak{b}$ be fractional ideals in a Dedekind domain $\mathcal{O}$. Then $\{\mathfrak{a}\} = \{\mathfrak{b}\}$ in $\mathrm{Cl}(\mathcal{O})$ precisely when there is an $\mathcal{O}$-module isomorphism $\mathfrak{a} \cong \mathfrak{b}$.

Moreover, given $\mathfrak{a}$, there is an integral ideal $\mathfrak{b}$ with $\{\mathfrak{a}\} = \{\mathfrak{b}\}$.

Proof

If $\{\mathfrak{a}\} = \{\mathfrak{b}\}$, then, by definition, $\mathfrak{a} = \mathfrak{b}x$ for some nonzero element x in the field of fractions $\mathcal{K}$ of $\mathcal{O}$, and the isomorphism is given simply by multiplication by x.

Conversely, suppose that $\alpha : \mathfrak{a} \to \mathfrak{b}$ is an $\mathcal{O}$-module isomorphism. Since $\mathcal{O}c \subseteq \mathfrak{a}$ for some c in $\mathcal{O}$, any element k of $\mathcal{K}$ can be written in the form $k = a/c$ with $a \in \mathfrak{a}$. We can therefore extend α to a mapping $\alpha' : \mathcal{K} \to \mathcal{K}$ by setting $\alpha'(k) = \alpha(a)/c$. A routine calculation, requiring only the definition of the field of fractions (1.1.12), shows that α' is a well-defined $\mathcal{K}$-module endomorphism of $\mathcal{K}$. Thus α' has the form $k \mapsto kx$ for some (nonzero) x in $\mathcal{K}$. Therefore α is also given by multiplication by x, and so $\mathfrak{a}x = \mathfrak{b}$.

Given $\mathfrak{a}$, we can clear denominators to find some $d \neq 0$ with $\mathfrak{a}d$ integral, which gives the final assertion. $\qquad\square$

5.1.15 Ideal theory in a Dedekind domain

We now examine the ideal theory of a Dedekind domain $\mathcal{O}$. First, we find the inverse of an ideal explicitly.

5.1.16 Lemma

Let $\mathfrak{a}$ be any fractional ideal of a Dedekind domain $\mathcal{O}$. Then

(i) $\mathfrak{a}^{-1} = \{x \in \mathcal{K} \mid x\mathfrak{a} \subseteq \mathcal{O}\}$,

(ii) *if $\mathfrak{b}$ is a fractional ideal with $\mathfrak{b} \subseteq \mathfrak{a}$, then $\mathfrak{a}^{-1} \subseteq \mathfrak{b}^{-1}$.*

Proof

(i)   Denote the right-hand set by $\mathfrak{a}^I$, which is clearly an $\mathcal{O}$-submodule of $\mathcal{K}$. If a is a nonzero element of $\mathfrak{a}$, we have $\mathfrak{a}^I \subseteq \mathcal{O}a^{-1}$, so $\mathfrak{a}^I$ is a fractional ideal. It is obvious that $\mathfrak{a}^{-1} \subseteq \mathfrak{a}^I$. On the other hand, $\mathfrak{a}^I\mathfrak{a} \subseteq \mathcal{O} = \mathfrak{a}^{-1}\mathfrak{a}$, so, multiplying on the right by $\mathfrak{a}^{-1}$, we obtain the equality.

(ii)   This follows readily from (i). $\qquad\square$

Given integral ideals $\mathfrak{a}$ and $\mathfrak{b}$ of any commutative domain $\mathcal{O}$, we say that $\mathfrak{a}$ *divides* $\mathfrak{b}$ if $\mathfrak{b} = \mathfrak{a}\mathfrak{c}$ for some integral ideal $\mathfrak{c}$; the notation is $\mathfrak{a} \mid \mathfrak{b}$.

5.1.17 Corollary

Let $\mathfrak{a}$ and $\mathfrak{b}$ be integral ideals in a Dedekind domain $\mathcal{O}$. Then $\mathfrak{a}$ divides $\mathfrak{b}$ if and only if $\mathfrak{b} \subseteq \mathfrak{a}$.

Proof

We have $\mathfrak{b} = \mathfrak{a}\mathfrak{c}$ where $\mathfrak{c} = \mathfrak{a}^{-1}\mathfrak{b}$ is a unique fractional ideal, and $\mathfrak{a}$ divides $\mathfrak{b}$ if and only if $\mathfrak{c}$ is integral. $\qquad\square$

The divisibility properties of ideals in a Dedekind domain lead to a result on prime ideals that has no counterpart for general domains (see Exercise 5.1.6).

5.1.18 Lemma

Let $\mathfrak{p}$ be a nonzero prime ideal of the Dedekind domain $\mathcal{O}$. Then $\mathfrak{p}$ is a maximal ideal.

Proof

Since $\mathcal{O}$ is Noetherian, there is a maximal ideal $\mathfrak{m}$ containing $\mathfrak{p}$, and $\mathfrak{p} = \mathfrak{mc}$ for the integral ideal $\mathfrak{c} = \mathfrak{pm}^{-1}$. Suppose that $\mathfrak{p} \neq \mathfrak{m}$, and choose some $m \in \mathfrak{m} \setminus \mathfrak{p}$. Then $mc \in \mathfrak{p}$ for all c in $\mathfrak{c}$ and so $\mathfrak{c} \subseteq \mathfrak{p}$. Therefore $\mathfrak{c} = \mathfrak{p\eth}$ for some integral ideal $\eth$. But then

$$\mathfrak{m}^{-1} = \mathfrak{p}^{-1}\mathfrak{c} = \eth \subseteq \mathcal{O},$$

hence $\mathcal{O} \subseteq (\mathfrak{m}^{-1})^{-1} = \mathfrak{m}$, a contradiction. $\qquad\qquad\square$

We can now obtain the unique factorization theorem for fractional ideals, which in essence tells us that $\mathrm{Frac}(\mathcal{O})$ is a free abelian group generated by the nonzero prime ideals of $\mathcal{O}$. We use the standard conventions that, in any multiplicative group G, $g^0 = 1$ and $g^{-i} = (g^{-1})^i$ for $g \in G$ and $i \in \mathbb{N}$.

5.1.19 The Unique Factorization Theorem for Ideals

Let $\mathfrak{a}$ be a fractional ideal of the Dedekind domain $\mathcal{O}$. Then $\mathfrak{a}$ is a unique product of nonzero prime ideals of $\mathcal{O}$.

More precisely, let $\mathbf{P}$ be the set of nonzero prime ideals in $\mathcal{O}$. Then there are unique integers $v(\mathfrak{p}, \mathfrak{a})$, $\mathfrak{p} \in \mathbf{P}$, almost all of which are 0, so that

$$\mathfrak{a} = \prod_{\mathfrak{p} \in \mathbf{P}} \mathfrak{p}^{v(\mathfrak{p}, \mathfrak{a})}.$$

Proof

We first establish the existence of a factorization when $\mathfrak{a}$ is integral. We suppose that some such $\mathfrak{a}$ has no factorization and obtain a contradiction. Clearly, $\mathfrak{a}$ is not maximal (nor is it $\mathcal{O}$ itself). Thus there is some maximal ideal $\mathfrak{p}_1$ containing $\mathfrak{a}$; put $\mathfrak{a}_1 = \mathfrak{ap}_1^{-1}$, which is contained in $\mathfrak{aa}^{-1}$ by (5.1.16), so that $\mathfrak{a}_1$ is integral and $\mathfrak{a} \subset \mathfrak{a}_1$ (since $\mathfrak{p}_1 \subset \mathcal{O}$). In turn, $\mathfrak{a}_1$ cannot be maximal (or equal to $\mathcal{O}$), for otherwise we have factored $\mathfrak{a}$. Hence we can find a maximal ideal $\mathfrak{p}_2$ with $\mathfrak{a}_1 \subset \mathfrak{a}_1\mathfrak{p}_2^{-1} = \mathfrak{a}_2$. Continuing this way, we obtain an infinite ascending chain of ideals in $\mathcal{O}$, which is impossible since $\mathcal{O}$ is Noetherian (3.1.6). Thus any integral fractional ideal does have a factorization.

By (5.1.8), given an arbitrary fractional ideal $\mathfrak{a}$, we have $\mathfrak{a}d \subseteq \mathcal{O}$ for some nonzero d in $\mathcal{O}$. Thus $\mathfrak{a} = \mathfrak{bc}^{-1}$ with both $\mathfrak{b}$ and $\mathfrak{c}$ integral, and so $\mathfrak{a}$ can also be expressed as a product of prime ideals. Collecting like terms gives the product expression.

Suppose that there are two different expressions for some fractional ideal $\mathfrak{a}$. Cross-multiplying, expanding and cancelling if possible, we obtain an equality of the form

$$\mathfrak{p}_1 \cdots \mathfrak{p}_k = \mathfrak{q}_1 \cdots \mathfrak{q}_\ell$$

with $\mathfrak{p}_1, \ldots, \mathfrak{p}_k, \mathfrak{q}_1, \ldots, \mathfrak{q}_\ell$ all prime ideals, $k \geq 1$, $\ell \geq 0$ and $\mathfrak{p}_i \neq \mathfrak{q}_j$ for any i and j.

Obviously, $\ell \geq 1$. Choose some $q \in \mathfrak{q}_1 \setminus \mathfrak{p}_1$. Then $q(\mathfrak{q}_2 \cdots \mathfrak{q}_\ell) \subseteq \mathfrak{p}_1$ and so $\mathfrak{q}_2 \cdots \mathfrak{q}_\ell \subseteq \mathfrak{p}_1$. Repeating this argument leads to the relation $\mathfrak{q}_\ell \subseteq \mathfrak{p}_1$. But $\mathfrak{q}_\ell$ is a maximal ideal by (5.1.18), so $\mathfrak{q}_\ell = \mathfrak{p}_1$, a contradiction. $\qquad\square$

We record two consequences of the theorem.

5.1.20 Corollary

Let $\mathfrak{a}$ be a fractional ideal of the Dedekind domain $\mathcal{O}$. Then $\mathfrak{a}$ is integral if and only if $v(\mathfrak{p}, \mathfrak{a}) \geq 0$ for all nonzero prime ideals $\mathfrak{p}$. $\qquad\square$

5.1.21 Corollary

Suppose that $\mathfrak{a}$ is an integral ideal of the Dedekind domain $\mathcal{O}$.

(i) *There is a finite set of nonzero prime ideals $\mathfrak{p}$ which contain $\mathfrak{a}$, namely, those with $v(\mathfrak{p}, \mathfrak{a}) \geq 1$.*

(ii) *Let $\mathfrak{p}_1, \ldots, \mathfrak{p}_k$ be the nonzero prime ideals containing $\mathfrak{a}$. Then the integral ideals which contain $\mathfrak{a}$ are those that can be written in the form*

$$\mathfrak{b} = \mathfrak{p}_1^{u(1)} \cdots \mathfrak{p}_k^{u(k)} \text{ with } 0 \leq u(i) \leq v(\mathfrak{p}_i, \mathfrak{a}) \text{ for all } i.$$

In particular, there is a finite set of such ideals. $\qquad\square$

The above corollaries lead to some results which are useful in our discussion of modules and in the calculation of class groups. First, we extend a definition which we made previously only for domains. A *principal ideal ring* is a ring in which every (left or right) ideal is a principal ideal.

5.1.22 Proposition

Let $\mathfrak{a}$ be a (nonzero) integral ideal in a Dedekind domain $\mathcal{O}$. Then the residue ring $\mathcal{O}/\mathfrak{a}$ is an Artinian principal ideal ring.

Proof

Let $\pi : \mathcal{O} \to \mathcal{O}/\mathfrak{a}$ be the canonical surjection. Any ideal $\mathfrak{b}'$ of the residue ring has the form $\pi\mathfrak{b}$ for some integral ideal $\mathfrak{b}$ of $\mathcal{O}$ which contains $\mathfrak{a}$; we can take $\mathfrak{b}$ to be the inverse image of $\mathfrak{b}'$ (1.2.12). As by the preceding corollary there are only finitely many such ideals, $\mathcal{O}/\mathfrak{a}$ is Artinian.

Let $\mathfrak{p}_1, \ldots, \mathfrak{p}_k$ be the maximal ideals containing $\mathfrak{a}$. If each ideal $\pi\mathfrak{p}_i$ is principal, then it is clear that every ideal in $\mathcal{O}/\mathfrak{a}$ is principal. To see that $\pi\mathfrak{p}_1$, say, is principal, choose an element x in $\mathcal{O}$ with $x \in \mathfrak{p}_1 \setminus \mathfrak{p}_1^2$ – note that $\mathfrak{b} \neq \mathfrak{b}^2$

for any proper integral ideal, since $\mathfrak{b}$ has an inverse in $\mathrm{Frac}(\mathcal{O})$. The Chinese Remainder Theorem (5.1.6) shows that there is an element y of $\mathcal{O}$ with

$$y \equiv x \,(\mathrm{mod}\ \mathfrak{p}_1^2) \ \text{and} \ y \equiv 1\,(\mathrm{mod}\ \mathfrak{p}_i) \ \text{for}\ i \geq 1.$$

Then $\mathfrak{a} + y\mathcal{O} = \mathfrak{p}_1$, so πy generates $\pi\mathfrak{p}_1$.   □

5.1.23 Corollary

Let $\mathfrak{a}$ be an integral ideal of a Dedekind domain $\mathcal{O}$. Then $\mathfrak{a}$ is generated by two elements, one of which can be chosen to be an arbitrary nonzero element of $\mathfrak{a}$.   □

5.1.24 Proposition

Let $\mathfrak{a}$ and $\mathfrak{b}$ be integral ideals of the Dedekind domain $\mathcal{O}$. Then there is an $\mathcal{O}$-module isomorphism

$$\mathcal{O}/\mathfrak{b} \cong \mathfrak{a}/\mathfrak{ab}.$$

Proof

Using the last corollary, we can choose any a in $\mathfrak{ab}$ and write $\mathfrak{a} = (a, y)$ for some y in $\mathfrak{a}$. Define $\theta : \mathcal{O} \to \mathfrak{a}/\mathfrak{ab}$ by $\theta x = yx + \mathfrak{ab}$. Clearly, θ is surjective, and

$$\mathrm{Ker}\,\theta = \{x \mid yx \in \mathfrak{ab}\} = \{x \mid \mathfrak{a}x \subseteq \mathfrak{ab}\} = \mathfrak{a}^{-1}\mathfrak{ab} = \mathfrak{b}.$$

The result follows by the Induced Mapping Theorem (1.2.11).   □

5.1.25 Principal ideal domains

Let $\mathcal{Z}$ be a principal ideal domain. Since $\mathcal{Z}$ is a special type of Dedekind domain (5.1.12), we have a roundabout proof of the unique factorization of elements in $\mathcal{Z}$ which we promised in (3.2.22).

By part (i) of (5.1.3), a nonzero prime ideal $\mathfrak{p}$ of $\mathcal{Z}$ must be of the form $p\mathcal{Z}$ for some prime element p. The generator p need not be unique, but it is unique up to multiplication by a unit of $\mathcal{Z}$ (that is, p is unique up to associates). Let **Pe** be a set of prime elements of $\mathcal{Z}$ which represents the set **P** of nonzero prime ideals, that is, each prime ideal $\mathfrak{p}$ is generated by a member p of **Pe**, and no two members of **Pe** generate the same prime ideal.

5.1.26 Theorem

Let $\mathcal{Z}$ be a principal ideal domain with field of fractions $\mathcal{K}$, and let x be a nonzero element of $\mathcal{K}$.

Then there are unique integers $v(p, x)$, one for each $p \in$ Pe, almost all of which are 0, such that $x = u \cdot \prod_{p \in \mathbf{Pe}} p^{v(p,x)}$ for some unit u of $\mathcal{Z}$.

Proof

Take $v(p, x) = v(p\mathcal{Z}, x\mathcal{Z})$ in (5.1.19). □

The next observation follows from the fact that an irreducible element cannot be factorized nontrivially.

5.1.27 Corollary

Let a be an irreducible element of the principal ideal domain $\mathcal{Z}$. Then a is prime. □

5.1.28 Primes versus irreducibles

As we noted in (5.1.3), a prime element of any domain is necessarily irreducible. The converse need not hold, even in a Dedekind domain, as we illustrate in Exercise 5.3.3.

When we are working with a principal ideal domain, the terms 'irreducible' and 'prime' can be used interchangeably, and tradition dictates which is to be used. When we are dealing with $\mathbb{Z}$ or, more generally, rings of algebraic integers which are principal ideal domains, 'prime' is the preferred term. In most other situations, particularly when we are working with polynomial rings, the term 'irreducible' is used. The choice depends on whether we are doing 'number theory' or 'geometry'.

Exercises

5.1.1   Let $\pi : R \to S$ be a surjective homomorphism of commutative rings. Recall that, given an ideal $\mathfrak{b}$ of S, the inverse image of $\mathfrak{b}$ is $\pi^{-1}\mathfrak{b} = \{r \in R \mid \pi r \in \mathfrak{b}\}$, an ideal of R. Show that, if $\mathfrak{b}$ is prime, then so is $\pi^{-1}\mathfrak{b}$.

Give an example to show that the converse is not true.

5.1.2   As promised in (5.1.10) above, we indicate how the hypothesis that $\mathcal{O}$ is Noetherian can be omitted from the definition of a Dedekind domain. For this purpose, we must change the definition of a fractional ideal. As usual, let $\mathcal{O}$ be a commutative domain with field of fractions $\mathcal{K}$.

Then a fractional ideal is a nonzero $\mathcal{O}$-submodule $\mathfrak{a}$ of $\mathcal{K}$ with the property that $d\mathfrak{a} \subseteq \mathcal{O}$ for some nonzero $d \in \mathcal{O}$. (If $\mathcal{O}$ is already known

to be Noetherian, this definition coincides with the previous one by
(5.1.8).)

(a) Show that if $\mathfrak{a}$ and $\mathfrak{b}$ are fractional ideals, so is $\mathfrak{ab}$.

(b) Suppose that the fractional ideal $\mathfrak{a}$ is invertible, that is $\mathfrak{ab} = \mathcal{O}$ for
some $\mathcal{O}$-submodule of $\mathcal{K}$. Show that $\mathfrak{b}$ is also a fractional ideal.

(c) Given that $\mathfrak{a}$ is invertible with inverse $\mathfrak{b}$, write

$$1 = x_1 y_1 + \cdots + x_k y_k \text{ with } x_1, \ldots, x_k \text{ in } \mathfrak{a} \text{ and } y_1, \ldots, y_k \text{ in } \mathfrak{b}.$$

Show that $x_1, \ldots, x_k$ generate $\mathfrak{a}$ as an $\mathcal{O}$-module.

(d) Deduce that if every integral ideal of $\mathcal{O}$ is invertible, then $\mathcal{O}$ is
Noetherian.

(e) Hence conclude that a Dedekind domain can be defined as a commutative domain for which every nonzero integral ideal is invertible.

5.1.3   Let $\mathfrak{a}$ be an ideal of a commutative ring R. The *radical* of $\mathfrak{a}$ is

$$\sqrt{\mathfrak{a}} = \{r \in R \mid r^k \in \mathfrak{a} \text{ for some } k \geq 0\}.$$

Verify that $\sqrt{\mathfrak{a}}$ is an ideal in R.

Suppose that R is Artinian. Verify that $\sqrt{0} = \mathrm{rad}(R)$, the Jacobson
radical of R.

Suppose now that $\mathfrak{a}$ is an integral ideal of a Dedekind domain $\mathcal{O}$
and let $\mathfrak{p}_1, \ldots, \mathfrak{p}_k$ be the nonzero prime ideals containing $\mathfrak{a}$.

Show that $\sqrt{\mathfrak{a}} = \mathfrak{p}_1 \cdots \mathfrak{p}_k$ and deduce that $\mathfrak{a} = \sqrt{\mathfrak{a}}$ if and only if $\mathfrak{a}$
is squarefree, that is, $v(\mathfrak{p}, \mathfrak{a}) \leq 1$ for all $\mathfrak{p}$ (see (5.1.19) and (5.1.21)).

Find all integral ideals $\mathfrak{a}$ with $\mathcal{O}/\mathfrak{a}$ semisimple.

(*Harder.* For arbitrary commutative R, $\sqrt{\mathfrak{a}}$ is the intersection of
the maximal ideals containing $\mathfrak{a}$.)

5.1.4   Let $\mathfrak{p}$ be a nonzero prime ideal of a Dedekind domain $\mathcal{O}$. Show that
$\bigcap_{i=1}^{\infty} \mathfrak{p}^i = 0$.

5.1.5   **Greatest common divisors for ideals**

Let $\mathfrak{a}$ and $\mathfrak{b}$ be nonzero integral ideals of a Dedekind domain $\mathcal{O}$,
with prime factorizations $\mathfrak{a} = \prod_{\mathfrak{p}} \mathfrak{p}^{v(\mathfrak{p}, \mathfrak{a})}$ and $\mathfrak{b} = \prod_{\mathfrak{p}} \mathfrak{p}^{v(\mathfrak{p}, \mathfrak{b})}$. Show
that

(a) $\mathfrak{a} + \mathfrak{b}$ divides both $\mathfrak{a}$ and $\mathfrak{b}$,

(b) if $\mathfrak{c}$ is any integral ideal which also divides both $\mathfrak{a}$ and $\mathfrak{b}$, then $\mathfrak{c}$
divides $\mathfrak{a} + \mathfrak{b}$,

(c) $\mathfrak{a} + \mathfrak{b} = \prod_{\mathfrak{p}} \mathfrak{p}^{\min(v(\mathfrak{p}, \mathfrak{a}), v(\mathfrak{p}, \mathfrak{b}))}$.

Thus $\mathfrak{a} + \mathfrak{b}$ can be regarded as the *greatest common divisor* of the ideals $\mathfrak{a}$ and $\mathfrak{b}$.

Show also that $\mathfrak{a} \cap \mathfrak{b}$ can be regarded as the *least common multiple* of $\mathfrak{a}$ and $\mathfrak{b}$, and find its prime factorization.

5.1.6   Let $\mathcal{K}$ be a field. Show that the ideal (X, Y) in the polynomial ring $\mathcal{K}[X, Y]$ is maximal and not invertible.

More generally, show that the ideal $(X_1, \ldots, X_k)$ of the polynomial ring $\mathcal{K}[X_1, \ldots, X_n]$ is prime, but not invertible, for $n \geq k \geq 2$.

5.1.7   Let $\mathcal{K}$ be a field and let $R = \mathcal{K}[\epsilon][X, Y]$ be the polynomial ring in two variables over the ring of dual numbers $\mathcal{K}[\epsilon]$.

Show that X is irreducible, but that X is not prime, since X does not divide either $X - \epsilon Y$ or $X + \epsilon Y$, but X does divide their product.

5.1.8   Let $\mathfrak{a}$ and $\mathfrak{b}$ be fractional ideals of a commutative Noetherian domain $\mathcal{O}$, with field of fractions $\mathcal{K}$. Using the identification of $\mathrm{End}_{\mathcal{K}}(\mathcal{K})$ with $\mathcal{K}$ itself in which elements of $\mathcal{K}$ operate on the right module $\mathcal{K}$ by left multiplication, prove that

$$\mathrm{Hom}_{\mathcal{O}}(\mathfrak{a}, \mathfrak{b}) = \{x \in \mathcal{K} \mid x\mathfrak{a} \subseteq \mathfrak{b}\}.$$

Verify that $\mathrm{Hom}_{\mathcal{O}}(\mathfrak{a}, \mathfrak{b})$ is also a fractional ideal of $\mathcal{O}$. In particular, show that $\mathrm{Hom}_{\mathcal{O}}(\mathfrak{a}, \mathcal{O}) = \mathfrak{a}^I$ as defined in the proof of (5.1.16).

Deduce that $\mathfrak{a}$ is invertible if and only if $\mathfrak{a}^I$ is the inverse for $\mathfrak{a}$ in the monoid $\mathrm{Frac}(\mathcal{O})$.

Suppose further that $\mathcal{O}$ is a Dedekind domain. Prove that

$$\mathrm{Hom}_{\mathcal{O}}(\mathfrak{a}, \mathfrak{b}) = \mathfrak{a}^{-1}\mathfrak{b}.$$

5.1.9   Let the ring $R = \mathcal{O}_1 \times \cdots \times \mathcal{O}_n$ be the direct product of commutative domains. Define 'fractional' and 'invertible' ideals of R, and characterize the invertible R-ideals in terms of the invertible ideals of the components $\mathcal{O}_i$.

Find the prime ideals of R.

5.1.10   Let $\mathcal{O}$ be a Dedekind domain with field of fractions $\mathcal{K}$ and let $\mathcal{O}'$ be a proper subring of $\mathcal{O}$ which also has field of fractions $\mathcal{K}$. Show that $\mathcal{O}$ is not an invertible $\mathcal{O}'$-module. Deduce that, if $\mathcal{O}$ is finitely generated over $\mathcal{O}'$, then $\mathcal{O}'$ cannot be a Dedekind domain.

5.1.11   Let $\mathcal{O}$ be a commutative Noetherian domain with field of fractions $\mathcal{K}$, and let R be an $\mathcal{O}$-order. Define a *fractional ideal* of R to be a twosided R-submodule $\mathfrak{A}$ of the Artinian $\mathcal{K}$-algebra $A = \mathcal{K}R$ such that

(a) $\mathfrak{A}$ is finitely generated over $\mathcal{O}$,

(b) $\mathcal{K}\mathfrak{A} = A$.

An *invertible* ideal is a fractional ideal $\mathfrak{A}$ such that there is another fractional ideal $\mathfrak{A}'$ with $\mathfrak{A}\mathfrak{A}' = R = \mathfrak{A}'\mathfrak{A}$.

Verify that the set $\text{In}(R)$ of invertible ideals of R is a group under multiplication of ideals.

Show that for any fractional ideal $\mathfrak{a}$ of $\mathcal{O}$, $\mathfrak{a}R$ is a fractional ideal of R, which is invertible if $\mathfrak{a}$ is invertible.

Now take $R = M_n(\mathcal{O})$, the ring of $n \times n$ matrices over $\mathcal{O}$. Using Exercise 1.1.4, show that the invertible fractional ideals are precisely those of the form $\mathfrak{a}R$ with $\mathfrak{a}$ an invertible ideal of $\mathcal{O}$.

Note. This topic is considered from another point of view in the discussion of Morita theory in [BK: CM].

5.1.12 Let $\mathcal{O}$ be a Dedekind domain and let $\mathfrak{a}$ be a proper integral ideal of $\mathcal{O}$. Put $R = \begin{pmatrix} \mathcal{O} & \mathfrak{a} \\ \mathcal{O} & \mathcal{O} \end{pmatrix}$, the tiled order consisting of those matrices in $M_2(\mathcal{O})$ with $(1,2)$ entry belonging to $\mathfrak{a}$.

Show that $\begin{pmatrix} \mathfrak{a} & \mathfrak{a} \\ \mathcal{O} & \mathfrak{a} \end{pmatrix}$ is an invertible ideal of R but that $\begin{pmatrix} \mathcal{O} & \mathfrak{a} \\ \mathcal{O} & \mathfrak{a} \end{pmatrix}$ and $\begin{pmatrix} \mathfrak{a} & \mathfrak{a} \\ \mathcal{O} & \mathcal{O} \end{pmatrix}$ are twosided ideals which are not invertible.

Check that all of these ideals are projective, both as left and as right R-modules.

5.2 ALGEBRAIC INTEGERS

In this section, we give the definition of an algebraic integer and we show that the ring of algebraic integers in a number field is a Dedekind domain. Our main interest lies in the properties of algebraic integers over the familiar ring of integers $\mathbb{Z}$, but we consider rings of integers over more general (commutative) principal ideal domains, as these also provide some interesting examples.

We illustrate the theory with some explicit calculations of integers in the quadratic case, which provide the foundation for the computations of factorizations and of class groups that we make in the next section.

5.2.1 Integers

Although we are mainly interested in integers that are defined over the ordinary integers $\mathbb{Z}$ or a polynomial ring $\mathcal{F}[X]$, it will be useful to make our initial definitions in a more general setting.

Let $\mathcal{Z}$ be a commutative domain with field of fractions $\mathcal{Q}$.

An *extension field* $\mathcal{K}$ of $\mathcal{Q}$ is any field that contains $\mathcal{Q}$ as a subfield. We say that $\mathcal{K}$ is a *finite extension* of $\mathcal{Q}$, or more simply that the extension $\mathcal{K}/\mathcal{Q}$ is *finite*, if $\mathcal{K}$ is a finite-dimensional vector space over $\mathcal{Q}$. When $\mathcal{K}/\mathcal{Q}$ is finite, the *degree* $[\mathcal{K} : \mathcal{Q}]$ is the dimension of $\mathcal{K}$ as a $\mathcal{Q}$-space.

Suppose that we have such a finite extension $\mathcal{K}/\mathcal{Q}$, and let x be any element of $\mathcal{K}$. Then powers $1, x, x^2, \ldots$ cannot be linearly independent, so there is some nonzero polynomial $g(X) \in \mathcal{Q}[X]$ with $g(x) = 0$. The set of all such g is an ideal in $\mathcal{Q}[X]$, and so has a unique monic generator $m(X)$, since the polynomial ring $\mathcal{Q}[X]$ is a principal ideal domain (3.2.10). We say that $m(X)$ is the *minimal polynomial* of x as it is the monic polynomial of smallest degree with $m(x) = 0$.

We note that $m(X)$ is irreducible, since a factorization $m = gh$ in $\mathcal{Q}[X]$ gives $0 = g(x)h(x)$ in the field $\mathcal{K}$, hence $g(x) = 0$ or $h(x) = 0$, hence $m \mid g$ or $m \mid h$.

A *$\mathcal{Z}$-integer* in $\mathcal{K}$ is defined to be an element a of $\mathcal{K}$ which is the root of a monic polynomial over $\mathcal{Z}$; that is, there is a polynomial

$$f(X) = X^n + f_{n-1}X^{n-1} + \cdots + f_0, \quad \text{with } f_{n-1}, \ldots, f_0 \in \mathcal{Z},$$

such that $f(a) = 0$.

If the ring $\mathcal{Z}$ can be taken as granted, a $\mathcal{Z}$-integer is sometimes referred to as an *integer* or an *algebraic integer*, although, strictly speaking, the latter term should be reserved for the case that the coefficient ring is $\mathbb{Z}$.

We now specialize to the case that $\mathcal{Z}$ is a commutative principal ideal domain, which will be the situation in our applications. The next result is the basic tool for handling integers.

5.2.2 Gauss' Lemma

Suppose that f is a monic polynomial over $\mathcal{Z}$ and that $f = gh$ where g and h are monic polynomials over $\mathcal{Q}$. Then g and h also have coefficients in $\mathcal{Z}$.

Proof

Let $\mathfrak{a} = \{r \in \mathcal{Z} \mid rg(X) \in \mathcal{Z}[X]\}$ and $\mathfrak{b} = \{r \in \mathcal{Z} \mid rh(X) \in \mathcal{Z}[X]\}$. Both $\mathfrak{a}$ and $\mathfrak{b}$ are nonzero ideals in the principal ideal domain $\mathcal{Z}$; let y and z be their respective generators.

Let d be the greatest common divisor of the coefficients of yg, that is, a generator of the ideal $(y, yg_{k-1}, \ldots, yg_0)$ of $\mathcal{Z}$, where

$$g(X) = X^k + g_{k-1}X^{k-1} + \cdots + g_0.$$

Then $y/d \in \mathcal{Z}$ and $(y/d)g(X) \in \mathcal{Z}[X]$, which means that y/d is also a generator of $\mathfrak{a}$. Thus d must be a unit in $\mathcal{Z}$, and we may take d to be 1. Similarly, we can assume that the greatest common divisor of the coefficients of $zh(X)$ is 1.

Now suppose that yz is not a unit in $\mathcal{Z}$. By the Unique Factorization Theorem (5.1.26), there is an irreducible element p of $\mathcal{Z}$ which divides yz. Since p and therefore $p\mathcal{Z}$ are also prime ((5.1.27) and (5.1.18)), the residue ring $\mathcal{Z}/p\mathcal{Z}$ is a field. Thus $(\mathcal{Z}/p\mathcal{Z})[X]$ is easily seen to be a domain. Taking residues, we have $(\overline{yg})(X) \cdot (\overline{zh})(X) = (\overline{yzf})(X) = 0$ in $(\mathcal{Z}/p\mathcal{Z})[X]$, so either $(\overline{yg})(X) = 0$ or $(\overline{zh})(X) = 0$. Thus p is a common divisor of the coefficients of $(yg)(X)$ or of $(zh)(X)$, a contradiction. Hence y and z must be units of $\mathcal{Z}$, and so g and h have coefficients in $\mathcal{Z}$. □

5.2.3 Corollary

Let $\mathcal{K}/\mathcal{Q}$ be a finite extension of fields, and let $a \in \mathcal{K}$. Then a is a $\mathcal{Z}$-integer if and only if the minimal polynomial $m(X)$ of a has coefficients in $\mathcal{Z}$.

Proof

By definition, a is a $\mathcal{Z}$-integer if and only if a is a root of some monic polynomial $f(X)$ over $\mathcal{Z}$, and m divides f. □

Next, we have some alternative criteria for an element x of $\mathcal{K}$ to be an integer. We write $\mathcal{Z}[x]$ for the $\mathcal{Z}$-submodule of $\mathcal{K}$ which is generated by the powers $\{x^i \mid i \geq 0\}$ of x. Then $\mathcal{Z}[x]$ is a subring of $\mathcal{K}$, and there is an obvious surjective ring homomorphism from the polynomial ring $\mathcal{Z}[X]$ to $\mathcal{Z}[x]$, given by evaluating X at x.

5.2.4 Proposition

For an element x of $\mathcal{K}$, the following statements are equivalent:

(i) *x is a $\mathcal{Z}$-integer;*

(ii) *$\mathcal{Z}[x]$ is a finitely generated $\mathcal{Z}$-module;*

(iii) *there is a subring R of $\mathcal{K}$ which contains both $\mathcal{Z}$ and x and which is finitely generated as a $\mathcal{Z}$-module;*

(iv) *$xM \subseteq M$ for some nonzero finitely generated $\mathcal{Z}$-submodule M of $\mathcal{K}$;*

(v) *x is an eigenvalue of some square matrix A with entries in $\mathcal{Z}$.*

Proof

(i) $\Rightarrow$ (ii): If x is a root of some monic polynomial over $\mathcal{Z}$ of degree n, say, then $\mathcal{Z}[x]$ is generated by $1, x, \ldots, x^{n-1}$ as a $\mathcal{Z}$-module.

(ii) $\Rightarrow$ (iii) $\Rightarrow$ (iv): These are clear since $\mathcal{Z}$ is Noetherian.

(iv) $\Rightarrow$ (v): Let $m_1, \ldots, m_k$ generate M as a $\mathcal{Z}$-module, and write

$$xm_i = a_{i1}m_1 + \cdots + a_{ik}m_k \text{ for } i = 1, \ldots, k.$$

Put $A = (a_{ij})$, a matrix with entries in $\mathcal{Z}$. Then

$$(xI - A) \begin{pmatrix} m_1 \\ \vdots \\ m_k \end{pmatrix} = 0 \quad \text{with} \quad \begin{pmatrix} m_1 \\ \vdots \\ m_k \end{pmatrix} \neq 0 \ \text{in} \ \mathcal{K}^k.$$

(v) $\Rightarrow$ (i):   Working over the field $\mathcal{K}$, we see that x is a root of the characteristic polynomial $\det(XI - A)$ of A, which has coefficients in $\mathcal{Z}$. $\square$

5.2.5 Theorem

Let $\mathcal{O}$ be the set of all $\mathcal{Z}$-integers in the finite extension field $\mathcal{K}$ of $\mathcal{Q}$, where $\mathcal{Z}$ is a principal ideal domain and $\mathcal{Q}$ is the field of fractions of $\mathcal{Z}$.

Then $\mathcal{O}$ is a ring.

Proof

Clearly $\mathcal{Z} \subseteq \mathcal{O}$. Suppose that $x, y \in \mathcal{O}$ satisfy monic polynomials over $\mathcal{Z}$ of degrees m and n respectively. The ring $\mathcal{Z}[x, y]$ is generated as a $\mathcal{Z}$-module by the finite set of products $x^i y^j$ where $i = 0, \dots, m - 1$ and $j = 0, \dots, n - 1$. Since $x + y$ and xy belong to $\mathcal{Z}[x, y]$, we see that $x + y$ and xy are in $\mathcal{O}$ by the implication (iv) $\Rightarrow$ (i) of the preceding result. $\square$

5.2.6 Quadratic fields

A *quadratic extension* of $\mathcal{Q}$ is an extension field $\mathcal{K}$ of $\mathcal{Q}$ which has degree 2; that is, $\mathcal{K}$ has dimension 2 as a vector space over $\mathcal{Q}$. As a first step to providing some explicit calculations of class groups, we describe the ring $\mathcal{O}$ of $\mathcal{Z}$-integers in such extensions.

To avoid undue complications, we restrict our attention to two cases. The first, and more important for us, is that where the principal ideal domain $\mathcal{Z}$ is $\mathbb{Z}$, the ring of ordinary integers; we call this the *number field* case. In this case, $\mathcal{Q}$ is the field of rational numbers $\mathbb{Q}$.

The second case is that in which $\mathcal{Z} = \mathcal{F}[X]$, the polynomial ring over a field $\mathcal{F}$; we call this the *function field* case. Here, $\mathcal{Q}$ is the field of rational functions $\mathcal{F}(X)$. In this case, we also assume that the characteristic of $\mathcal{F}$ is not equal to 2. Then 2 is a unit in $\mathcal{Z}$ and the extension is 'separable' (see (5.2.9)).

We say that a nonzero element d of $\mathcal{Z}$ is *squarefree* if its prime factors are all distinct, that is, in the Unique Factorization Theorem (5.1.26) each exponent satisfies $v(p, d) \leq 1$. (In the function field case, tradition demands that we say 'irreducible' rather than 'prime'; as our main interest is with number fields, we prefer to use 'prime' when considering both cases together.)

It is not hard to see that, in either case, a quadratic extension $\mathcal{K}$ of $\mathcal{Q}$ has basis $\{1, \sqrt{d}\}$ with $d \in \mathcal{Z}$ squarefree. Thus

$$\mathcal{K} = \mathcal{Q}(\sqrt{d}) = \{x_0 + x_1\sqrt{d} \mid x_0, x_1 \in \mathcal{Q}\},$$

which is the field of fractions of its subring

$$\mathcal{Z}[\sqrt{d}] = \{x_0 + x_1\sqrt{d} \mid x_0, x_1 \in \mathcal{Z}\}.$$

Let $x = x_0 + x_1\sqrt{d} \in \mathcal{K}$, where $x_0, x_1 \in \mathcal{Q}$ are uniquely determined by x. The *conjugate* of x is $\gamma x = x_0 - x_1\sqrt{d}$, the *trace* of x is $Tx = x + \gamma x$, and the *norm* of x is $Nx = x \cdot \gamma x$. Routine calculation gives the following result.

5.2.7 Lemma

The following hold:

 (i) γ *is a ring automorphism of* $\mathcal{K}$;

 (ii) $\gamma x = x \Leftrightarrow x \in \mathcal{Q}$;

 (iii) $T(x + y) = Tx + Ty$ *and* $N(xy) = Nx \cdot Ny$ *for any* $x, y \in \mathcal{K}$;

 (iv) x *is a root of the polynomial* $C_x(X) = X^2 - Tx \cdot X + Nx$, *which has coefficients in* $\mathcal{Q}$;

 (v) $C_x(X)$ *is the minimal polynomial of* x, *provided that* $x \in \mathcal{K} \setminus \mathcal{Q}$.   □

We can now describe the integers explicitly.

5.2.8 Theorem

Let $\mathcal{O}$ *be the ring of* $\mathcal{Z}$-*integers in the quadratic extension* $\mathcal{Q}(\sqrt{d})$ *of* $\mathcal{Q}$, *where* d *is a squarefree element of* $\mathcal{Z}$.

 (i) *In the function field case with characteristic other than* 2,

$$\mathcal{O} = \mathcal{Z}[\sqrt{d}].$$

 (ii) *In the number field case, either*

$$\mathcal{O} = \mathbb{Z}[\sqrt{d}] \quad if \ d \equiv 2, 3 \ (\mathrm{mod} \ 4),$$

or

$$\mathcal{O} = \mathbb{Z}[(1 + \sqrt{d})/2] \quad if \ d \equiv 1 \ (\mathrm{mod} \ 4).$$

In the latter case,

$$\mathcal{O} = \{z_0 + z_1(1 + \sqrt{d})/2 \mid z_0, z_1 \in \mathbb{Z}\}$$

and $(1 + \sqrt{d})/2$ *has minimal polynomial* $X^2 - X + (1 - d)/4$.

Proof

Let $x = x_0 + x_1\sqrt{d} \in \mathcal{K}$. We know from (5.2.3) that $x \in \mathcal{O}$ precisely when its minimal polynomial has coefficients in $\mathcal{Z}$. If $x \in \mathcal{Q}$, this polynomial is simply $X - x$, so $x \in \mathcal{Z}$. Assume that $x \notin \mathcal{Q}$, so that its minimal polynomial is $X^2 - Tx \cdot X + Nx$. We have to determine when $Tx = 2x_0$ and $Nx = x_0^2 - dx_1^2$ both belong to $\mathcal{Z}$. This is certainly the case if both x_0 and x_1 are already in $\mathcal{Z}$, but the converse need not hold.

We make repeated use of the fact that, if $w \in \mathcal{Q}$ and $dw^2 \in \mathcal{Z}$, then $w \in \mathcal{Z}$ already. This follows from Theorem 5.1.26 because d is squarefree by hypothesis. Now assume that both Tx and Nx lie in $\mathcal{Z}$.

In case (i), 2 is a unit, and we have immediately that $x_0 \in \mathcal{Z}$. But $Nx \in \mathcal{Z}$, which gives $dx_1^2 \in \mathcal{Z}$, and so $x_1 \in \mathcal{Z}$ by the fact stated above.

In case (ii) write $x_0 = y_0/2$ with $y_0 = Tx \in \mathbb{Z}$. Put $x_1 = y_1/2$, with $y_1 \in \mathbb{Q}$. Then $y_0^2 - dy_1^2 \in 4\mathbb{Z}$, hence $dy_1^2 \in \mathbb{Z}$, and, as above, $y_1 \in \mathbb{Z}$.

Suppose that we can find some $\mathbb{Z}$-integer x for which $x_0 \notin \mathbb{Z}$, that is, $y_0 \equiv 1 \pmod 2$. Then $y_0^2 \equiv 1 \pmod 4$ and so $dy_1^2 \equiv 1 \pmod 4$. It follows that $y_1 \equiv 1 \pmod 2$ and $y_1^2 \equiv 1 \pmod 4$ also, which forces the congruence $d \equiv 1 \pmod 4$. We have therefore established the first assertion of (ii).

Suppose finally that $d \equiv 1 \pmod 4$. Then $(1 + \sqrt{d})/2$ certainly has minimal polynomial as claimed and so belongs to $\mathcal{O}$. It is clear from our discussion that if x is in $\mathcal{O}$, then either x is in $\mathbb{Z}[\sqrt{d}]$ or $x = (y_0 + y_1\sqrt{d})/2$ with $y_0 \equiv y_1 \equiv 1 \pmod 2$, from which we see that $x - (1 + \sqrt{d})/2$ is in $\mathbb{Z}[\sqrt{d}]$. $\square$

5.2.9 Separability and integral closure

The argument which shows that the ring $\mathcal{O}$ of $\mathcal{Z}$-integers in a finite extension field $\mathcal{K}$ of $\mathcal{Q}$ is a Dedekind domain is indirect. We in fact show that $\mathcal{O}$ satisfies a set of conditions which provide an alternative characterization of Dedekind domains; we establish this characterization in the next section.

At this point we are obliged to invoke a result from number theory. First, some definitions.

An irreducible polynomial f over $\mathcal{Q}$ is *separable* if it has no repeated roots in any extension field of $\mathcal{Q}$. The finite extension $\mathcal{K}/\mathcal{Q}$ is *separable* if the minimal polynomial of any of its elements is separable.

We do not develop the theory of separable extensions here, as it would take too much space; some details are outlined in Exercise 5.2.3 below. We remark that any extension of a field of characteristic 0 is separable, as is any extension of a finite field. If $\mathcal{F}$ is a field of characteristic other than 2, then any quadratic extension of $\mathcal{F}(X)$ is separable.

We need to assume the following fact, which is clearly true for the quadratic extensions which we discussed above. Proofs can be found in [Cohn 1979], Theorem 4, 11.5, and [Fröhlich & Taylor 1991], Theorem 5, which texts also contain discussions of separability.

5.2.10 Proposition
Suppose that the finite extension $\mathcal{K}/\mathcal{Q}$ is separable. Then the ring $\mathcal{O}$ of $\mathcal{Z}$-integers in $\mathcal{K}$ is finitely generated as a $\mathcal{Z}$-module.   ○

A domain $\mathcal{O}$ is said to be *integrally closed* if the set of $\mathcal{O}$-integers in its field of fractions $\mathcal{K}$ is $\mathcal{O}$ itself. An easy application of Gauss' Lemma (5.2.2) shows that a principal ideal domain $\mathcal{Z}$ is integrally closed.

The fact that the ring $\mathcal{O}$ of $\mathcal{Z}$-integers in a finite extension field $\mathcal{K}$ of $\mathcal{Q}$ is a Dedekind domain stems from the following result.

5.2.11 Theorem
Let $\mathcal{Z}$ be a principal ideal domain with field of fractions $\mathcal{Q}$, and let $\mathcal{O}$ be the ring of $\mathcal{Z}$-integers in a separable finite extension field $\mathcal{K}$ of $\mathcal{Q}$. Then

(i) *$\mathcal{O}$ is Noetherian,*
(ii) *every nonzero prime ideal of $\mathcal{O}$ is a maximal ideal,*
(iii) *$\mathcal{O}$ is integrally closed.*

Proof
Since $\mathcal{O}$ is finitely generated as a $\mathcal{Z}$-module and $\mathcal{Z}$ itself is Noetherian, (i) follows from (3.1.12).

Let $\mathfrak{p}$ be a nonzero prime ideal of $\mathcal{O}$. Then $\mathfrak{q} = \mathcal{Z} \cap \mathfrak{p}$ is a nonzero prime ideal of $\mathcal{Z}$, hence a maximal ideal, and $\mathcal{Z}/\mathfrak{q}$ is a field. To establish (ii), we have to show that the domain $\mathcal{O}/\mathfrak{p}$ is also a field.

There is a canonical ring homomorphism from $\mathcal{Z}$ to $\mathcal{O}/\mathfrak{p}$ with kernel $\mathfrak{q}$, and so, by the Induced Mapping Theorem (1.1.9), $\mathcal{Z}/\mathfrak{q}$ is isomorphic to the image of $\mathcal{Z}$ in $\mathcal{O}/\mathfrak{p}$, which of course is a subring. Thus $\mathcal{O}/\mathfrak{p}$ is a vector space over $\mathcal{Z}/\mathfrak{q}$, of finite dimension. Choose $x \neq 0$ in $\mathcal{O}/\mathfrak{p}$. Then there must be a nontrivial identity of the form

$$y_0 + y_1 x + \cdots + y_k x^k = 0 \text{ with } y_0, y_1, \ldots, y_k \in \mathcal{Z}/\mathfrak{q} \text{ and } y_k \neq 0.$$

Let k be minimal. Then $y_0 \neq 0$ also, since otherwise we could cancel a factor x in the domain $\mathcal{O}/\mathfrak{p}$. Thus x has an inverse. This makes $\mathcal{O}/\mathfrak{p}$ a field and establishes (ii).

Finally, let x be any $\mathcal{O}$-integer in $\mathcal{K}$. The ring $\mathcal{O}[x]$ is then finitely generated as an $\mathcal{O}$-module, and hence it is also finitely generated as a $\mathcal{Z}$-module. It follows from the implication (iv) $\Rightarrow$ (i) of (5.2.4) that x is a $\mathcal{Z}$-integer, and so belongs to $\mathcal{O}$. $\qquad\square$

We next show that a ring satisfying the conditions of the previous theorem is a Dedekind domain. (The converse is also true; its proof is left as an exercise.) We divide the argument into a series of lemmas.

5.2.12 Lemma

Let $\mathcal{O}$ be a commutative Noetherian ring. Then every nonzero proper ideal of $\mathcal{O}$ contains a product of nonzero prime ideals.

Proof

Suppose that some such ideal does not. Since $\mathcal{O}$ is Noetherian, there is an ideal $\mathfrak{a}$ that is maximal among such ideals (3.1.6). Clearly, $\mathfrak{a}$ itself cannot be prime, so there are elements r, s in $\mathcal{O}$ with $r, s \notin \mathfrak{a}$ but $rs \in \mathfrak{a}$. But then the ideals $\mathfrak{a} + r\mathcal{O}$ and $\mathfrak{a} + s\mathcal{O}$ must each contain a product of prime ideals, and $(\mathfrak{a} + r\mathcal{O})(\mathfrak{a} + s\mathcal{O}) \subseteq \mathfrak{a}$, a contradiction. $\qquad\square$

5.2.13 Lemma

Let $\mathcal{O}$ be a commutative Noetherian domain in which every nonzero prime ideal is maximal, and let $\mathfrak{a}$ be a proper ideal of $\mathcal{O}$. Then there is an element x of the field of fractions $\mathcal{K}$ of $\mathcal{O}$ with $x \notin \mathcal{O}$ but $x\mathfrak{a} \subseteq \mathcal{O}$.

Proof

Take a nonzero element a of $\mathfrak{a}$. The ideal $a\mathcal{O}$ contains some product of primes $\mathfrak{p}_1 \cdots \mathfrak{p}_k$ by the previous lemma; choose such a product with k minimal. Since $\mathcal{O}$ is Noetherian, $\mathfrak{a}$ is contained in some maximal ideal $\mathfrak{p}$ which is therefore prime (5.1.2). If, for each i, there is an element $x_i \in \mathfrak{p}_i \setminus \mathfrak{p}$, then $x_1 \cdots x_k \in \mathfrak{p}$, a contradiction. Thus we must have, say, $\mathfrak{p}_1 \subseteq \mathfrak{p}$. Since every prime ideal is maximal, we obtain $\mathfrak{p}_1 = \mathfrak{p}$. By the minimality of k, we may choose $b \in \mathfrak{p}_2 \cdots \mathfrak{p}_k \setminus a\mathcal{O}$. Then $ba^{-1} \notin \mathcal{O}$, but $ba^{-1}\mathfrak{a} \subseteq ba^{-1}\mathfrak{p}_1 \subseteq \mathcal{O}$. $\qquad\square$

We can now prove that the conditions listed in Theorem 5.2.11 suffice to show that a ring is a Dedekind domain.

5.2.14 Theorem

Suppose that a commutative domain $\mathcal{O}$ satisfies the conditions below.

(i) *$\mathcal{O}$ is Noetherian.*

(ii) *Every nonzero prime ideal of $\mathcal{O}$ is a maximal ideal.*

(iii) *$\mathcal{O}$ is integrally closed.*

Then $\mathcal{O}$ is a Dedekind domain.

Proof

We have to show that, if $\mathfrak{a}$ is a nonzero fractional ideal of $\mathcal{O}$, then $\mathfrak{a}$ is invertible. By (5.1.8), there is some element y in $\mathcal{O}$ with $y\mathfrak{a}$ integral. Since $y\mathcal{O}$ is evidently invertible, it is enough to show that an integral ideal is invertible.

Suppose then that $\mathfrak{a}$ is integral and choose any nonzero $a \in \mathfrak{a}$. Define $\mathfrak{b} = \{r \in \mathcal{K} \mid r\mathfrak{a} \subseteq a\mathcal{O}\}$, where $\mathcal{K}$ is the field of fractions of $\mathcal{O}$. For any r in $\mathfrak{b}$, we have $ra\mathcal{O} \subseteq a\mathcal{O}$ and so $r \in \mathcal{O}$. Thus $\mathfrak{b}$ is an integral ideal of $\mathcal{O}$. We need only verify that $\mathfrak{a}\mathfrak{b} = a\mathcal{O}$. Suppose not. Then $a^{-1}\mathfrak{a}\mathfrak{b}$ is a proper ideal of $\mathcal{O}$, and, by the preceding lemma, there is some $x \in \mathcal{K} \setminus \mathcal{O}$ with $xa^{-1}\mathfrak{a}\mathfrak{b} \subseteq \mathcal{O}$. But then $x\mathfrak{b} \subseteq \mathfrak{b}$; hence x is an $\mathcal{O}$-integer by the implication (iv) $\Rightarrow$ (i) of (5.2.4), and so x is in $\mathcal{O}$ because $\mathcal{O}$ is assumed to be integrally closed. This gives a contradiction. $\square$

Combining (5.2.11) and (5.2.14) gives the result that we have been seeking.

5.2.15 Theorem

Let $\mathcal{Z}$ be a principal ideal domain with field of fractions $\mathcal{Q}$, and let $\mathcal{O}$ be the ring of $\mathcal{Z}$-integers in a separable finite extension field $\mathcal{K}$ of $\mathcal{Q}$.

Then $\mathcal{O}$ is a Dedekind domain. $\square$

This result also holds without the hypothesis of separability; proofs are given in [Cohn 1979], §11.5, Exercise 5 and [Zariski & Samuel 1963], Chapter V, §8.

Exercises

5.2.1   Show that

   (a) a Dedekind domain $\mathcal{O}$ is integrally closed,
   (b) if R is a domain and R is also a finite-dimensional $\mathcal{K}$-algebra, with $\mathcal{K}$ a field, then R is already a field.

   Deduce that the conditions of (5.2.14) provide an alternative definition of a Dedekind domain.

5.2.2   Let $\mathcal{O}'$ be a proper subring of the ring of integers $\mathcal{O}$ in a quadratic field $\mathcal{Q}(\sqrt{d})$. Show that $\mathcal{O}'$ satisfies two of the conditions of Theorem 5.2.14 but not the third. Find an ideal of $\mathcal{O}'$ which is not invertible – see Exercise 5.1.10.

5.2.3   **Splitting fields and separability**

   (a) Let f be an irreducible polynomial of degree n over the field $\mathcal{F}$. Verify that $\mathcal{K} = \mathcal{F}[X]/f(X)\mathcal{F}[X]$ is a field, which is an extension field of $\mathcal{F}$ of degree n (after identifying $\mathcal{F}$ with its image in $\mathcal{K}$).

(b) Let f be any (nonconstant) polynomial over $\mathcal{F}$. A *splitting field* for f is any extension $\mathcal{K}$ of $\mathcal{F}$ such that

$$f = (X - a_1)^{e(1)} \cdots (X - a_k)^{e(k)}$$

where $a_1, \ldots, a_k$ in $\mathcal{K}$ are all distinct. (In other words, the irreducible factors of f in $\mathcal{K}[X]$ are all linear.)
Show that f has a splitting field, of finite degree over $\mathcal{F}$.

(c) Let f, g be (nonconstant) polynomials over $\mathcal{F}$ and let (f, g) be their greatest common divisor. Show that $(f, g) = 1$ in $\mathcal{F}[X]$ if and only if $(f, g) = 1$ in $\mathcal{K}[X]$ for any extension field $\mathcal{K}$ of $\mathcal{F}$.

(d) The *derivative* is the function $D : \mathcal{F}[X] \to \mathcal{F}[X]$ given by

$$D(f_0 + f_1 X + f_2 X^2 + \cdots + f_n X^n) = f_1 + 2f_2 X + \cdots + n f_n X^{n-1}.$$

Verify that $D(f+g) = Df + Dg$ and $D(fg) = fDg + gDf$ always. Show that, if there is a nonconstant polynomial f with $Df = 0$, then $\mathcal{F}$ must have characteristic $p > 0$, and

$$f = f_0 + f_1 X^p + f_2 X^{2p} + \cdots + f_k X^{kp}.$$

(e) We say that f has a *repeated root* if $e(i) > 1$ for some i. Prove that f has a repeated root in a splitting field if and only if $(f, Df) \neq 1$.

(f) Deduce that an irreducible polynomial f has a repeated root if and only if $Df = 0$.

(g) Conclude that when $\mathcal{F}$ has characteristic 0, any finite extension of $\mathcal{F}$ is separable.

(h) Show further that, if the irreducible polymomial f does have a repeated root, then it has only one distinct root in any splitting field.

(i) Suppose that $\mathcal{F} = \mathcal{E}(Y)$, where $\mathcal{E}$ is a field of characteristic p (and Y is a polynomial variable). Verify that $X^p - Y$ is not separable.

5.2.4 Orders

Let $\mathcal{Z}$ be a principal ideal domain with field of fractions $\mathcal{Q}$ and let R be a $\mathcal{Z}$-order in the $\mathcal{Q}$-algebra $A = \mathcal{Q}R$; R need not be commutative. Such an order is said to be *maximal* if it is not properly contained in any $\mathcal{Z}$-order S with $A = \mathcal{Q}S$.

(a) Show that any element of R is a $\mathcal{Z}$-integer.

(b) Suppose that $A = \mathcal{K}_1 \times \cdots \times \mathcal{K}_s$ where each $\mathcal{K}_i$ is a field and each $\mathcal{K}_i$ is a separable extension of $\mathcal{Q}$ with ring of $\mathcal{Z}$-integers $\mathcal{O}_i$. Show that $\mathcal{M} = \mathcal{O}_1 \times \cdots \times \mathcal{O}_s$ is a $\mathcal{Z}$-order in A.

(c) Deduce that $\mathcal{M}$ is the unique maximal $\mathcal{Z}$-order in A.

(d) Let $S = M_2(\mathcal{Z})$ and $B = M_2(\mathcal{Q})$. Put $e = \begin{pmatrix} 0 & 1 \\ 0 & 0 \end{pmatrix}$ and $t = \begin{pmatrix} 0 & p \\ 1 & 0 \end{pmatrix}$ where p is a prime element of $\mathcal{Z}$. Show that tSt^{-1} is also a $\mathcal{Z}$-order in B. By considering $e + tet^{-1}$, show that there is no $\mathcal{Z}$-order containing both S and tSt^{-1}.

Remark. Both S and tSt^{-1} can be shown to be maximal $\mathcal{Z}$-orders in B. In general, if the noncommutative $\mathcal{Q}$-algebra A satisfies an extended definition of separability, then A always contains maximal $\mathcal{Z}$-orders, but these maximal orders are not unique. Full details are given in [Reiner 1975].

5.2.5   Let $\mathcal{Z}$ and $\mathcal{Q}$ be as before, and let $A = \mathcal{Q}[\epsilon]$ be the ring of dual numbers over $\mathcal{Q}$. Show that $a = a_0 + a_1\epsilon \in A$ is a $\mathcal{Z}$-integer if and only if $a_0 \in \mathcal{Z}$. Choose a prime element p of $\mathcal{Z}$ and put $R_i = \mathcal{Z} \cdot 1 + \mathcal{Z} p^{-i}\epsilon$ for $i \geq 1$. Show that $\{R_i\}$ is an infinite ascending chain of $\mathcal{Z}$-orders in A, and hence that no R_i is contained in a maximal $\mathcal{Z}$-order.

5.3 QUADRATIC FIELDS

We next take a closer look at the ideal theory of a quadratic extension field $\mathcal{Q}(\sqrt{d})$ of the field of fractions of a principal ideal domain $\mathcal{Z}$. The prime ideals of the ring of integers $\mathcal{O}$ in $\mathcal{Q}(\sqrt{d})$ can be described very explicitly in terms of the factorizations of the prime (that is, irreducible) elements of $\mathcal{Z}$. We are also able to give some concrete computations of class groups $\mathrm{Cl}(\mathcal{O})$.

As before, we consider two cases only: the number field case where $\mathcal{Z}$ is $\mathbb{Z}$, and the function field case, $\mathcal{Z} = \mathcal{K}[X]$, with $\mathcal{K}$ a field of characteristic not equal to 2. However, we are able to give details of class group computations only in the number field case. (There is a short summary of some results for function fields.)

5.3.1 Factorization in general

We start with some general observations about factorization in Dedekind domains. Let $\mathcal{Z}$ be a principal ideal domain and let the Dedekind domain $\mathcal{O}$ be the integral closure of $\mathcal{Z}$ in some finite extension $\mathcal{K}/\mathcal{Q}$ of the field of fractions of $\mathcal{Z}$.

Given a prime (or irreducible) element p in $\mathcal{Z}$, our general results (5.1.19) and (5.1.21) tell us that the ideal $p\mathcal{O}$ has a unique prime factorization

$$p\mathcal{O} = \mathfrak{p}_1^{e(1)} \cdots \mathfrak{p}_k^{e(k)},$$

where $\mathfrak{p}_1, \ldots, \mathfrak{p}_k$ are distinct prime ideals of $\mathcal{O}$ and $e(1), \ldots, e(k)$ are unique positive integers, the *ramification indices* of p.

Also, for any (nonzero) prime ideal $\mathfrak{p}$ of $\mathcal{O}$, there is some prime element p of $\mathcal{Z}$ in whose factorization $\mathfrak{p}$ occurs. To see this, we first note that $\mathfrak{p} \cap \mathcal{Z}$ is a nonzero ideal of $\mathcal{Z}$, since the minimal polynomial of any nonzero element of $\mathfrak{p}$ has a nonzero constant term a, and clearly $a \in \mathfrak{p} \cap \mathcal{Z}$. Thus $\mathfrak{p} \cap \mathcal{Z} = p\mathcal{Z}$ for some prime p. But then $p\mathcal{O} \subseteq \mathfrak{p}$, so $\mathfrak{p}$ must be one of the ideals $\mathfrak{p}_i$.

We often say that $\mathfrak{p}$ is *above* p or that p is *below* $\mathfrak{p}$ to indicate that $\mathfrak{p}$ occurs in the factorization of $p\mathcal{O}$.

5.3.2 Factorization in the quadratic case

Historically, much of the investigation of quadratic fields stems from the attempt to find all the integers that can be represented in the form $x^2 - dy^2$ where x, y and d are again integers. The case $d = -1$, that is, the two squares problem, was solved by Gauss in the early 19th century by using factorization in the ring $\mathbb{Z}[i]$, which ring is now named the *Gaussian integers* in his honour. Thus there was an intensive investigation of factorizations in rings of the form $\mathbb{Z}[\sqrt{d}]$ in the mid 19th century.

It was soon realized that unique factorization of elements into irreducibles was no longer valid in such a ring. One reason is the technicality that this ring is sometimes too small (Theorem 5.2.8), but, even when it is the correct ring, unique factorization may still fail, as the example in $\mathbb{Z}[\sqrt{-5}]$ shows. The essential ingredient that enabled unique factorization to be restored was the introduction of *ideal divisors* by Kummer, which were then given a concrete manifestation as *ideal numbers* by Dedekind. A survey of the development of these ideas can be found in [Dedekind 1996].

In essence, the use of an ideal (number) overcomes the absence of a greatest common divisor. If elements a, b of $\mathbb{Z}[\sqrt{d}]$ do have a greatest common divisor c, it can be written as $c = sa + tb$ for some s, t in $\mathbb{Z}[\sqrt{d}]$. If c does not exist, then a substitute for it is $Ra + Rb$, the ideal generated by a and b. Thus our exposition, which starts with abstract ideals and proceeds to concrete examples, is historically back to front.

Let $\mathcal{O}$ be the ring of integers in a quadratic extension of $\mathcal{Q}$. There is an obvious bijective correspondence between the ideals of $\mathcal{O}$ that contain $p\mathcal{O}$ and the ideals of the residue ring $\mathcal{O}/p\mathcal{O}$, which is a two-dimensional space over the field $\mathcal{Z}/p\mathcal{Z}$ (and so Artinian). Since an ideal of $\mathcal{O}/p\mathcal{O}$ is necessarily a subspace, any proper ideal of $\mathcal{O}$ which contains $p\mathcal{O}$ properly must already be a maximal ideal of $\mathcal{O}$. Thus $p\mathcal{O}$ cannot have more than two prime factors (allowing repetitions) and there are only three possible configurations for the set of ideals in $\mathcal{O}/p\mathcal{O}$. These are described by the following diagrams.

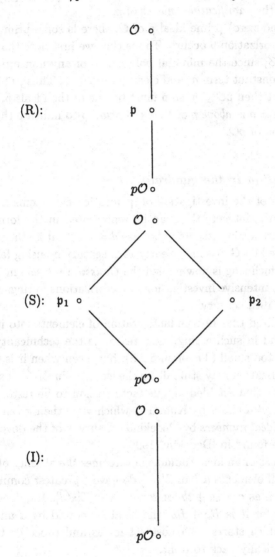

The labelling is explained by the corresponding factorizations of $p\mathcal{O}$.

(R): $p\mathcal{O} = \mathfrak{p}^2$. Then p is *ramified*. The image of $\mathfrak{p}$ in $\mathcal{O}/p\mathcal{O}$ is the Jacobson radical of $\mathcal{O}/p\mathcal{O}$.

(S): $p\mathcal{O} = \mathfrak{p}_1\mathfrak{p}_2$, with $\mathfrak{p}_1, \mathfrak{p}_2$ distinct primes of $\mathcal{O}$. Then p is said to be *unramified* and *split* (or *decomposed*) in $\mathcal{O}$. Invoking the Chinese Remainder

Theorem in its ideal-theoretic formulation (5.1.5), we see that the residue ring $\mathcal{O}/p\mathcal{O}$ has the form $\mathcal{O}/p\mathcal{O} \cong \mathcal{F}_1 \times \mathcal{F}_2$, where $\mathcal{F}_1$ and $\mathcal{F}_2$ are fields which are both isomorphic to $\mathcal{Z}/p\mathcal{Z}$.

(I): $p\mathcal{O}$ is a prime ideal of $\mathcal{O}$. In this case, p is said to be *inert* since it remains prime (or irreducible). The residue ring $\mathcal{O}/p\mathcal{O}$ is a field extension of $\mathcal{Z}/p\mathcal{Z}$, of degree 2.

We introduce some standard notation for the primes above p. In case (R), we write $\hat{p}$ for the unique prime ideal above p, and, in case (S), we write p_1, p_2 for the two distinct primes above p. (In case (I), we do not need any extra notation.)

5.3.3 Explicit factorizations

Now that we have described the possible factorizations of $p\mathcal{O}$, we investigate how these can be determined explicitly in terms of p itself. We restrict our attention to the two fundamental cases: either $\mathcal{Z}$ is $\mathbb{Z}$ (the number field case), or $\mathcal{Z} = \mathcal{K}[X]$, with $\mathcal{K}$ a field of characteristic not equal to 2 (the function field case).

Let $d \in \mathcal{Z}$ be squarefree. By (5.2.8), the ring of $\mathcal{Z}$-integers in $\mathcal{K} = \mathcal{Q}(\sqrt{d})$ is $\mathcal{O} = \mathcal{Z}[\delta]$, where

$$\delta = \begin{cases} \sqrt{d} & \mathcal{Z} = \mathcal{K}[X], \text{ or } \mathcal{Z} = \mathbb{Z}, \ d \equiv 2, 3 \pmod{4}, \\ (1 + \sqrt{d})/2 & \mathcal{Z} = \mathbb{Z}, \ d \equiv 1 \pmod{4}. \end{cases}$$

The decomposition of $p\mathcal{O}$ is governed by the relationship between p and d, and also the value of the residue of $d \pmod 4$ in the number field case. We make a case-by-case analysis, in the course of which we also find generators for the prime ideals of $\mathcal{O}$ (thus confirming the fact that at most two generators are needed, as promised by (5.1.23)).

(A)   Suppose first that $\mathcal{O} = \mathcal{Z}[\sqrt{d}]$. Then

$$\mathcal{O}/p\mathcal{O} \cong (\mathcal{Z}/p\mathcal{Z})[Y]/(Y^2 - \bar{d}),$$

where Y is a polynomial variable and, for any $r \in \mathbb{Z}$, $\bar{r}$ denotes the residue of $r \pmod p$. We subdivide the cases according to the possible factorizations of $Y^2 - \bar{d}$.

(i) Suppose that $\bar{d}$ is not a square in $\mathcal{Z}/p\mathcal{Z}$. Then $Y^2 - \bar{d}$ is irreducible, so that $\mathcal{O}/p\mathcal{O}$ is a field. In this case, p is inert.

(ii) $Y^2 - \bar{d}$ has two distinct roots in $\mathcal{Z}/p\mathcal{Z}$. These must be expressible as ϵ and $-\epsilon$, and so $p \neq 2$. We have $Y^2 - \bar{d} = (Y - \epsilon)(Y + \epsilon)$, and, by the

Chinese Remainder Theorem (5.1.5), $\mathcal{O}/p\mathcal{O} \cong \mathbb{Z}/p\mathbb{Z} \times \mathbb{Z}/p\mathbb{Z}$. Thus p is split.

Choose any element e of $\mathbb{Z}$ with $\bar{e} = \epsilon$. Then $p_1 = (p, \sqrt{d} - e)$ and $p_2 = (p, \sqrt{d} + e)$ are the primes over p.

Suppose next that $Y^2 - \bar{d}$ has a repeated root ϵ in $\mathbb{Z}/p\mathbb{Z}$. This can happen only if $2\epsilon = 0$, which means that $p|2d$. Thus $p = 2$ or $p|d$.

(iii) Suppose that $p = 2$. Then, whatever the value of d, the polynomial $Y^2 - \bar{d}$ is either Y^2 or $Y^2 - \bar{1} = (Y - \bar{1})^2$, so that 2 is ramified.

If $d \equiv 2 \pmod 4$, then $\widehat{2} = (2, \sqrt{d})$, and if $d \equiv 3 \pmod 4$, $\widehat{2} = (2, \sqrt{d} + 1)$.

(iv) Suppose that p divides d. Then p is ramified, and $\widehat{p} = (p, \sqrt{d})$.

(B)   We are left to consider the number field case with $d \equiv 1 \pmod 4$, so that $\mathcal{O} = \mathbb{Z}[(1 + \sqrt{d})/2]$. If p is odd, then 2 is invertible in $\mathbb{Z}/p\mathbb{Z}$, and $\mathcal{O}/p\mathcal{O} \cong (\mathbb{Z}/p\mathbb{Z})[Y]/(Y^2 - \bar{d})$ again. Thus we are in one of the situations covered by (i), (ii) and (iv) above.

Finally, we have $d \equiv 1 \pmod 4$ and $p = 2$. Since $(1 + \sqrt{d})/2$ has minimal polynomial $Y^2 - Y + (1 - d)/4$ by (5.2.8), we have

$$\mathcal{O}/2\mathcal{O} = (\mathbb{Z}/2\mathbb{Z})[Y]/(Y^2 - Y + \overline{(1 - d)/4}).$$

This gives two further cases.

(v) If $d \equiv 1 \pmod 8$, then the above polynomial is simply $Y^2 + Y$, which has two distinct irreducible factors, Y and $Y + \bar{1}$. Then 2 is split, with factors

$$2_1 = (2, (1 + \sqrt{d})/2)$$

and

$$2_2 = (2, 1 + (1 + \sqrt{d})/2) = (2, (1 - \sqrt{d})/2).$$

(vi) If $d \equiv 5 \pmod 8$, the polynomial reduces to $Y^2 - Y - \bar{1}$, which is irreducible over $\mathbb{Z}/2\mathbb{Z}$. Thus 2 is inert.

5.3.4 A summary

In the number field case, we can summarize these results in a table as follows. Here, the symbol (d/p) is the *Legendre symbol*, which is defined whenever $p \nmid d$, taking the value 1 when $\bar{d}$ is a square in $\mathbb{Z}/p\mathbb{Z}$ and -1 otherwise.

	$p = 2$	$p > 2$ $p \mid d$	$p > 2$ $(d/p) = 1$	$p > 2$ $(d/p) = -1$
$d \equiv 2, 3 \pmod 4$	R	R	S	I
$d \equiv 1 \pmod 8$	S	R	S	I
$d \equiv 5 \pmod 8$	I	R	S	I

In the number field case, the question of whether or not d is a square modulo p can be resolved by the law of quadratic reciprocity, which is discussed in most texts on algebraic number theory. We refer the reader to [Marcus 1977], Chapter 4 or [Fröhlich & Taylor 1991], Theorem 40, for accounts. Both texts also give the decompositions of primes in many types of extension of $\mathbb{Z}$, often in the form of exercises.

5.3.5 The norm

The norm is a useful measure of 'size' for elements and ideals of number fields. We define it here only for quadratic extensions in the number field case.

Let $\mathcal{O}$ be the ring of integers in $\mathbb{Q}(\sqrt{d})$ and let $\mathfrak{a}$ be an integral ideal of $\mathcal{O}$. The *norm* of $\mathfrak{a}$ is defined to be

$$\|\mathfrak{a}\| = |\mathcal{O}/\mathfrak{a}|,$$

the number of elements in the residue ring $\mathcal{O}/\mathfrak{a}$. (Sometimes, this norm is called the *index*, to avoid confusion with the closely related norm of an element.) Obviously, $\|\mathfrak{a}\| = 1$ precisely when $\mathfrak{a} = \mathcal{O}$. For an element a of $\mathcal{O}$, we write $\|a\|$ for $\|a\mathcal{O}\|$. The fact that the norm is always finite comes from the first basic result, which is as follows.

5.3.6 Proposition

(a) *Let* $\mathfrak{p}$ *be a nonzero prime ideal* $\mathfrak{p}$ *of* $\mathcal{O}$*, with* $\mathfrak{p} \cap \mathbb{Z} = p\mathbb{Z}$*,* p *a prime in* $\mathbb{Z}$*. Then*

$$\|\mathfrak{p}\| = \begin{cases} p & \text{if } p \text{ is either ramified or split,} \\ p^2 & \text{if } p \text{ is inert.} \end{cases}$$

(b) *If* $\mathfrak{b}, \mathfrak{c}$ *are any integral ideals of* $\mathcal{O}$*, then*

$$\|\mathfrak{b}\| \cdot \|\mathfrak{c}\| = \|\mathfrak{b}\mathfrak{c}\|.$$

Proof

(a) $\mathcal{O}/\mathfrak{p}$ is an extension field of $\mathbb{Z}/p\mathbb{Z}$, of degree 1 or 2, and it is also the

residue ring of $\mathcal{O}/p\mathcal{O}$ modulo its radical, so the result is immediate from the
listing of the possible ideal structures in $\mathcal{O}/p\mathcal{O}$ (5.3.1).

(b)   There are an exact sequence

$$0 \longrightarrow \mathfrak{b}/\mathfrak{bc} \longrightarrow \mathcal{O}/\mathfrak{bc} \longrightarrow \mathcal{O}/\mathfrak{b} \longrightarrow 0$$

and an isomorphism $\mathfrak{b}/\mathfrak{bc} \cong \mathcal{O}/\mathfrak{c}$ (5.1.24), which shows that the norm is
multiplicative. $\square$

By combining this result with the Unique Factorization Theorem for Ideals
(5.1.19), we obtain the following useful application.

5.3.7 Corollary

*For any nonnegative integer n, there are only finitely many integral ideals
of $\mathcal{O}$ whose norm is n.* $\square$

We have already defined the norm of an element a of $\mathcal{O}$ by $Na = a_0^2 - a_1^2 d$,
where $a = a_0 + a_1\sqrt{d}$. The connection between the two norms is given by the
next result.

5.3.8 Lemma

*Let a be a nonzero element of $\mathcal{O}$. Then $\|a\mathcal{O}\| = |Na|$, where $|Na|$ is the
absolute value of Na.*

Proof

Recall that $Na = a \cdot \gamma a$, where $\gamma a = a_0 - a_1\sqrt{d}$. For any ideal $\mathfrak{a}$ of $\mathcal{O}, \gamma \mathfrak{a}$
is also an ideal of $\mathcal{O}$. Further, if $\mathfrak{p}$ is a prime above p, so also is $\gamma\mathfrak{p}$. Thus
$\|\gamma\mathfrak{a}\| = \|\mathfrak{a}\|$ for all ideals $\mathfrak{a}$, and so

$$\|a\mathcal{O}\|^2 = \|Na \cdot \mathcal{O}\|.$$

However, $\|p\mathcal{O}\| = p^2$ for a prime p of $\mathbb{Z}$, hence $\|z\mathcal{O}\| = z^2$ for any element z
of $\mathbb{Z}$. $\square$

5.3.9 The Euclidean property

For some small values of d, the norm makes $\mathcal{O}$ into a Euclidean domain
as defined in (3.2.7). In detail, put $\varphi(a) = \|a\|$ for $a \in \mathcal{O}$. Then φ is a
multiplicative function from $\mathcal{O}$ to $\{0\} \cup \mathbb{N}$ (so axiom (ED2′) holds), with
$\varphi(a) = 0$ only if $a = 0$, and the division algorithm is satisfied:

for any $a, b \in \mathcal{O} \setminus 0$, we have $a = qb + r$, with $q, r \in \mathcal{O}$ and $\varphi(r) < \varphi(b)$.

As we observed in (3.2.10), the ring $\mathcal{O}$ is then a principal ideal domain, so that $\mathrm{Cl}(\mathcal{O}) = 1$ (5.1.13), and the structure of finitely generated $\mathcal{O}$-modules is given by (3.3.6).

The reader should have little difficulty in verifying that $\mathbb{Z}[i], \mathbb{Z}[\sqrt{2}], \mathbb{Z}[\sqrt{-2}]$ and $\mathbb{Z}[(1 + \sqrt{-3})/2]$ are all Euclidean under their norms; the essential point is to note that it is enough to show that, for any $ab^{-1} \in \mathbb{Q}(\sqrt{d})$, there is some $q \in \mathcal{O}$ with $\|ab^{-1} - q\| < 1$.

A complete determination of the quadratic number fields which are Euclidean in this way is given by [Hardy & Wright 1979], §§14.7ff. The topic is also discussed in [Eggleton, Lacampagne & Selfridge 1992].

Note that it is possible for $\mathcal{O}$ to be a principal ideal domain without being Euclidean; for example, the ring of integers in $\mathbb{Q}(\sqrt{-19})$ is not Euclidean, but it has trivial ideal class group – see (5.3.18) below.

5.3.10 The class group again

Our aim now is to make some calculations of ideal class groups. Recall that, for an arbitrary Dedekind domain $\mathcal{O}$, its class group $\mathrm{Cl}(\mathcal{O})$ consists of classes $\{a\}$, with a a fractional ideal of $\mathcal{O}$, where $\{a\} = \{b\}$ if $a = xb$ for some nonzero x in the field of fractions $\mathcal{K}$ of $\mathcal{O}$. We refer to a as a *representative* of the ideal class $\{a\}$. By (5.1.14), any class has an integral representative. In the cases of interest to us, it is not too hard to determine whether or not a particular ideal is principal.

Much more difficult is the problem of determining a set of ideals that contains at least one representative from each possible class. We are content to quote the results that we need, since the proofs would take us too far afield.

For number fields in general, we need only look at a finite set of integral ideals of $\mathcal{O}$ because of the following fact, proofs of which are given in [Marcus 1977], Chapter 5, [Hasse 1980], Chapter 29, §1 and Chapter 30, §3, and [Fröhlich & Taylor 1991], Theorem 31, among others.

5.3.11 Theorem

Let $\mathcal{K}$ be a finite extension of $\mathbb{Q}$, with ring of integers $\mathcal{O}$. Then there is a positive integer $B = B(\mathcal{K})$ such that each element of the ideal class group $\mathrm{Cl}(\mathcal{O})$ has the form $\{a\}$ for some integral ideal a with $\|a\| < B$. ○

Combining this fact with (5.3.7), we obtain a key result.

5.3.12 Theorem

Let $\mathcal{K}$ be a finite extension of $\mathbb{Q}$, with ring of integers $\mathcal{O}$. Then the class group $\mathrm{Cl}(\mathcal{O})$ is finite. □

5.3.13 The class number

The order of the class group is called the *class number* of $\mathcal{O}$ (or of $\mathcal{K}$) and usually denoted by h. The determination of this number remains a fundamental problem in algebraic number theory. For an account of the connection between the nature of h and Fermat's Last Theorem, see [Ireland & Rosen 1982] and [Fröhlich & Taylor 1991], which give many results about class numbers of explicit fields.

In the quadratic case, computations are facilitated by a rather sharp bound for the constant $B(\mathcal{K})$ due to Kneser; a proof is given by [Hasse 1980], Chapter 29, §2.

5.3.14 Theorem

Let $\mathcal{K} = \mathbb{Q}(\sqrt{d})$ be a quadratic number field, with ring of integers $\mathcal{O}$. Then each ideal class in $\mathrm{Cl}(\mathcal{O})$ has a representative integral ideal $\mathfrak{a}$ with

   (i) $\|\mathfrak{a}\| \le \sqrt{d}$ if $d > 0$, $d \equiv 2, 3$ (mod 4),

  (ii) $\|\mathfrak{a}\| \le (\sqrt{d})/2$ if $d > 0$, $d \equiv 1$ (mod 4),

 (iii) $\|\mathfrak{a}\| \le 2(\sqrt{|d|/3})$ if $d < 0$, $d \equiv 2, 3$ (mod 4),

 (iv) $\|\mathfrak{a}\| \le \sqrt{|d|/3}$ if $d < 0$, $d \equiv 1$ (mod 4). $\bigcirc$

5.3.15 Computations of class groups

These bounds permit the calculation of the class group of $\mathbb{Q}(\sqrt{d})$ by hand for small values of d. First, we show how one may determine whether or not a given ideal $\mathfrak{a}$ is principal.

Suppose that $\mathfrak{a}$ is principal, with $\|\mathfrak{a}\| = n$. Then there must be an element $a \in \mathcal{O}$ with $\|a\| = n$. Write $a = a_0 + a_1\delta$, where $a_0, a_1 \in \mathbb{Z}$ and $\delta = \sqrt{d}$ if $d \equiv 2, 3$ (mod 4), or $\delta = (1 + \sqrt{d})/2$ if $d \equiv 1$ (mod 4).

Then, by (5.3.8), either

$$\|a\| = |a_0^2 - a_1^2 d|$$

or

$$\|a\| = |(a_0 + a_1/2)^2 - d(a_1/2)^2|,$$

so we must be able to solve the integer equation

$$|a_0^2 - a_1^2 d| = n, \quad \text{if } d \equiv 2, 3 \text{ (mod 4)},$$

or the equation

$$|(2a_0 + a_1)^2 - da_1^2| = 4n, \quad \text{if } d \equiv 1 \text{ (mod 4)}.$$

For small values of n, it is often easy to show that there are no solutions of the relevant equation, or that the solutions do not in any case give a generator of the ideal $\mathfrak{a}$.

In the calculations that occupy the remainder of this section, we make repeated use of the following immediate consequence of the discussion in (5.3.2).

5.3.16 Lemma

(i) *If the prime p of $\mathbb{Z}$ ramifies in $\mathbb{Q}(\sqrt{d})$, then $\{\widehat{p}\}$ has order at most 2 in $\mathrm{Cl}(\mathcal{O})$.*

(ii) *If p is inert in $\mathbb{Q}(\sqrt{d})$, then $\{p\}$ is trivial.*

(iii) *If p splits in $\mathbb{Q}(\sqrt{d})$, then $\{p_1\} = \{p_2\}^{-1}$, and p_1 and p_2 both have norm p.*   □

For example, take $d = -5$. In this case, the ring of integers is $\mathbb{Z}[\sqrt{-5}]$. By part (iii) of the preceding result, every ideal class in $\mathbb{Z}[\sqrt{-5}]$ has representative with norm at most 2. We know from case (iii) of (5.3.3) that the prime 2 ramifies, being the square of the ideal $\widehat{2} = (2, 1 + \sqrt{-5})$. Clearly, $\|\widehat{2}\| = 2$. If $\mathfrak{p}$ is a prime ideal of $\mathbb{Z}[\sqrt{-5}]$ above an odd prime p of $\mathbb{Z}$, then $\|\mathfrak{p}\| \geq p$ (5.3.6), and so $\widehat{2}$ is the only possible representative of a nontrivial class. However, $\widehat{2}$ cannot be principal since there is no solution of $a_0^2 + 5a_1^2 = 2$ in $\mathbb{Z}$.

We summarize this as follows.

5.3.17 Theorem

The class group of the ring of integers $\mathbb{Z}[\sqrt{-5}]$ of $\mathbb{Q}(\sqrt{-5})$ has order 2. □

Now, a result promised in (5.3.9).

5.3.18 Theorem

The ring $\mathbb{Z}[(1 + \sqrt{-19})/2]$ of integers in $\mathbb{Q}(\sqrt{-19})$ is a principal ideal domain.

Proof

Since $-19 \equiv 1 \pmod 4$, the ring of integers is as claimed (5.2.8). By part (iv) of (5.3.14), each ideal class has a representative with norm at most 2. Since $-19 \equiv 5 \pmod 8$, the table in (5.3.4) shows that 2 is inert in $\mathcal{O}$. Thus $\|2\| = 4$ (5.3.6) and there is no ideal of norm 2. Furthermore any other prime ideal must have norm at least p for some odd prime p of $\mathbb{Z}$, and so the only permissible representative is the ring itself.   □

Next, a noncyclic example.

5.3.19 Theorem

The class group of the ring of integers $\mathbb{Z}[\sqrt{-21}]$ of $\mathbb{Q}(\sqrt{-21})$ is the Klein four group (that is, the direct product of two cyclic groups of order 2).

Proof

The bound on the norm of a representative integral ideal is $2\sqrt{21/3}$, effectively 5. Drawing on (5.3.3) again, we see that the primes 2 and 3 ramify, the primes above them being $\widehat{2} = (2, 1 + \sqrt{-21})$ and $\widehat{3} = (3, \sqrt{-21})$ respectively. Also, 5 splits since $-21 \equiv 2^2 \pmod 5$, the factors being $5_1 = (5, 2 + \sqrt{-21})$ and $5_2 = (5, 2 - \sqrt{-21})$. Any other prime ideal of $\mathbb{Z}[\sqrt{-21}]$ must have norm greater than 5, so the possible nontrivial ideal classes are $\{\widehat{2}\}$, $\{\widehat{3}\}$, $\{5_1\}$ and $\{5_2\} = \{5_1\}^{-1}$.

The classes $\{\widehat{2}\}$, $\{\widehat{3}\}$ and $\{\widehat{2}\} \cdot \{\widehat{3}\}$ are all nontrivial since the corresponding norms are 2, 3 and 6 respectively, and the equations

$$a_0^2 + 21a_1^2 = 2, 3, 6$$

have no solution in $\mathbb{Z}$. Since each of these classes has order 2 (5.3.16), they must therefore all be different.

It follows that $\{\widehat{2}\} \cdot \{\widehat{3}\}$ must be the remaining ideal class, that of $\{5_i\}$ for $i = 1$ or 2. But it matters not which, since then $\{5_1\} = \{5_2\}$ has order 2 also. □

Finally, a result which can be used to provide an interesting example in K-theory.

5.3.20 Theorem

The ring of integers $\mathbb{Z}[(1 + \sqrt{-71})/2]$ of $\mathbb{Q}(\sqrt{-71})$ has class group cyclic of order 7. Further, conjugation induces the homomorphism $x \mapsto x^{-1}$ on the class group.

Proof

Let $\delta = (1 + \sqrt{-71})/2$, which is a root of $Y^2 - Y + 18$. By (5.2.8), the ring of integers in $\mathbb{Q}(\sqrt{-71})$ is $\mathbb{Z}[\delta]$ because $-71 \equiv 1 \pmod 4$. Also, the bound on the norm of a representative integral ideal given by (5.3.13) is $\sqrt{71/3} = 4.86\ldots$

Since also $-71 \equiv 1 \pmod 8$, we see from (5.3.2) that the prime 2 splits into $2_1 = (2, \delta)$ and $2_2 = (2, \delta - 1)$, both of which have norm 2.

We have $-71 \equiv 1 \pmod 3$, so 3 also splits, the factors being

$$3_1 = (3, 1 + \sqrt{-71}) \quad \text{and} \quad 3_2 = (3, 1 - \sqrt{-71}),$$

which have norm 3. No other prime of $\mathbb{Z}$ need be considered, and so, listing all ideals with norm less than 5, we see that there are at most six nontrivial ideal classes:

$$\{2_1\}, \{2_1\}^2,$$
$$\{2_2\} = \{2_1\}^{-1}, \{2_2\}^2,$$
$$\{3_1\} \text{ and } \{3_2\} = \{3_1\}^{-1}.$$

We show directly that the powers $\{2_1\}^k, k = 1, \ldots, 6$, are all nontrivial, which proves that the class group is cyclic of order 7.

Suppose that $a = a_0 + a_1\delta$ is a generator of 2_1^k, where $1 \le k \le 6$. Then, computing the norm of 2_1^k in two ways, we find that the coefficients of a are integers which satisfy the equation

$$(2a_0 + a_1)^2 + 71a_1^2 = 4 \cdot 2^k.$$

It is therefore enough to show that this equation cannot be solved when $k \le 6$. The argument for $k = 6$ is typical.

We have $4 \cdot 2^6 = 256$. If $a_1 = 0$, then $a = \pm 2^3$; but $(2^3) = 2_1^3 2_2^3$, so $2^3 \notin 2_1^6$ cannot be a generator. If $a_1 = \pm 1$, then $(2a_0 \pm 1)^2 = 185$, which is impossible in $\mathbb{Z}$. No other value of a_1 need be considered, so 2_1^6 is not principal.

The conjugation homomorphism γ sends $\sqrt{-71}$ to $-\sqrt{-71}$ and so sends δ to $1 - \delta$. Thus the ideal 2_1 is sent to 2_2, and so the induced homomorphism on the class group sends the generator $\{2_1\}$ to its inverse. ◻

It is instructive to see what happens when $k = 7$ in the above example. Take $a_1 = 1$; then $(2a_0 + 1)^2 = 441$, so we have a solution $a = 10 + \delta$. Since $10 + \delta$ belongs to 2_1 but not 2_2, it generates 2_1^7. In fact, -10 is a solution of $Y^2 - Y + 18$ modulo 128, and $2_1^k = (2^k, 10 + \delta)$ for $k = 1, \ldots, 7$.

5.3.21 Function fields

The calculation of the class group for a quadratic extension of $\mathcal{K}(X)$ depends very much on the nature of the coefficient field $\mathcal{K}$, and requires techniques that we cannot develop in this text. We therefore summarize some results. (We thank Vernon Armitage and Robin Hartshorne for some very helpful correspondence about this topic.)

If the field $\mathcal{K}$ is finite then $\mathcal{K}(X)$ enjoys most of the properties that make number theory work over $\mathbb{Q}$. Properly speaking, both $\mathcal{K}(X)$ and $\mathbb{Q}$ are global fields, the theory of which is given in [Weiss 1963], Chapter 5, for example. The essential point is that each residue field $\mathcal{K}[X]/p(X)\mathcal{K}[X]$, with $p(X)$ an irreducible polynomial, is finite, as are all residue rings of $\mathcal{K}[X]$. Thus each nonzero ideal has a finite norm, which is defined as in (5.3.5). It is then possible to give estimates similar to (5.3.11) that use this norm. In particular, the class group of an extension of $\mathcal{K}(X)$ is finite.

Next, we look at some possibilities when $d = d(X)$ has small degree. First, suppose that $d(X)$ has degree 1. After a change of variable, we may as well suppose that $d(X) = X$. Then the quadratic extension $\mathcal{K}(X)(\sqrt{X}) = \mathcal{K}(Y)$ has ring of integers $\mathcal{K}[X][\sqrt{X}] = \mathcal{K}[Y]$, a polynomial ring again, whose class group is trivial.

Suppose that $d(X)$ has degree 2. After a change of variable, we can take $d(X) = aX^2 + b$. The quadratic extension $\mathcal{L} = \mathcal{K}(X)(\sqrt{d})$ is then the coordinate ring of the conic $Y^2 = aX^2 + b$, and there are two cases. If $Y^2 - aX^2$ is irreducible (as a polynomial in two variables), then the conic is an 'ellipse' and the associated class group is cyclic of order 2. A particular case is given by $d(X) = 1 - X^2$ over $\mathbb{R}[X]$; here, $\mathcal{L} = \mathcal{K}(X,Y)/(X^2 + Y^2 - 1)$ is the coordinate ring of the circle.

When $Y^2 - aX^2$ is reducible, we have a 'hyperbola', and the class group is trivial. Computations of this type are given in [Fossum 1973], §11.

Finally, suppose that $d(X)$ has degree 3. The curve $Y^2 = d(X)$ is called an elliptic curve, the theory of which is rich and deep. The graph of the curve in the plane $\mathcal{K}^2$ (or, more accurately, in the projective plane over $\mathcal{K}$) is itself a group under a geometrically defined rule of composition, and it turns out that $\mathrm{Cl}(\mathcal{O})$ can be identified with this group. In the language of algebraic geometry, $\mathrm{Cl}(\mathcal{O})$ is isomorphic to the group of 'divisors of degree 0' on the curve. Accessible accounts are given by [Fulton 1989], Chapter 8, and [Cohn 1991].

Exercises

5.3.1 Class groups

The reader is now in a position to compute the ideal class groups for many other quadratic fields $\mathbb{Q}(\sqrt{d})$ with d small, subject to the limitations of time and energy. Here is a sample of further results that one might wish to confirm:

- $d = \pm 2, \pm 3, 5, \pm 7$ – trivial;
- $d = \pm 10$ – cyclic, of order 2;
- $d = -14$ – cyclic, of order 4;
- $d = -23$ – cyclic, of order 3.

[Marcus 1977], Chapter 5, and [Fröhlich & Taylor 1991] give many more computations of class groups, some in the form of exercises.

Warning. The computation of class groups for quadratic fields in general is far from being a routine matter. For example, it is not known whether or not there is an infinite number of fields $\mathbb{Q}(\sqrt{d})$ with $d > 0$ and class number 1. Lists of some results are given in [Encyclopaedic Dictionary of Mathematics 1987] and [Pohst & Zassenhaus 1989]

5.3.2   Conjugation

Let $\mathcal{O}$ be the ring of integers in the quadratic field $\mathbb{Q}(\sqrt{d})$ and let γ be the conjugation automorphism of $\mathbb{Q}(\sqrt{d})$, which is given by

$$\gamma(x_0 + x_1\sqrt{d}) = x_0 - x_1\sqrt{d}$$

as in (5.2.6).

Verify the following assertions, in which we use the notation of (5.3.2):

(a) γ is an automorphism of $\mathcal{O}$;
(b) if the prime p of $\mathbb{Z}$ is ramified, then $\gamma(\hat{p}) = \hat{p}$;
(c) if p splits, then $\gamma(p_1) = p_2$ and $\gamma(p_2) = p_1$.

Deduce that γ induces an automorphism γ' of the class group $\mathrm{Cl}(\mathcal{O})$, and that γ' has order 1 or 2.

By considering (5.3.19) and (5.3.20), show that both possibilities occur.

Note. This exercise gives a very special case of an area of investigation of great interest. The group $G = \{1, \gamma\}$ is the Galois group of $\mathbb{Q}(\sqrt{d})$ over $\mathbb{Q}$ (see [Cohn 1979], §5.5 for an introduction to Galois theory). We have shown that the class group $\mathrm{Cl}(\mathcal{O})$ is a $\mathbb{Z}G$-module, albeit written multiplicatively. The ring of integers itself is also a $\mathbb{Z}G$-module.

Given an arbitrary Galois extension of a number field or a function field with Galois group G, it is possible to define many natural modules over $\mathbb{Z}G$, among which are the class group and the ring of integers in the extension. Such modules are called *Galois modules*, which are the subject of much current research – see [Fröhlich 1983] and [Snaith 1994].

5.3.3   Primes versus irreducibles: the struggle continues

Show that 3 is an irreducible element of the Dedekind domain $\mathbb{Z}[\sqrt{-5}]$, but that $3\mathbb{Z}[\sqrt{-5}]$ is not a prime ideal of $\mathbb{Z}[\sqrt{-5}]$. (This example was promised in (5.1.28).)

6

MODULES OVER DEDEKIND DOMAINS

Now that we have established the basic ring-theoretic properties of Dedekind domains, we turn to the problem of classifying their finitely generated modules. We attack this problem in three steps. In the first step, we obtain a structure theorem for the projective modules over an arbitrary Dedekind domain. Next, we specialize to the case that the Dedekind domain is a valuation ring, that is, it has only one nonzero prime ideal. Given a general Dedekind domain $\mathcal{O}$ and a prime $\mathfrak{p}$ of $\mathcal{O}$, there is a canonical valuation ring $\mathcal{O}_{\mathfrak{p}}$, the localization of $\mathcal{O}$ at $\mathfrak{p}$, whose prime ideal corresponds to the chosen prime ideal of $\mathcal{O}$. Since a valuation ring is a Euclidean domain, we can apply the results of section 3.3 to describe its modules. Finally, we piece together the structure of a module over a general Dedekind domain from our knowledge of the modules over its various localizations.

In passing, we obtain some results on modules over commutative principal ideal domains that were promised in Chapter 3, but which cannot be conveniently derived from a discussion of noncommutative Euclidean domains.

Two methods of argument in this chapter foreshadow techniques that we consider in greater depth in [BK: CM]. The construction of a localization in section 6.2 is a special case of a more general construction that we consider in section 6.1 of [BK: CM], and the arguments in section 6.3 are a first glimpse of the 'local–global' methods which we discuss in section 7.3 of [BK: CM].

As is customary in this text, we deal with right modules unless the contrary is stated. However, save for some examples, all rings mentioned in this chapter are commutative, and so all modules can be regarded as balanced bimodules (1.2.7). We take advantage of this observation to switch sides on occasion when it seems more natural to work with left scalar multiplication rather than right. This happens mostly when we deal with the ideals of a ring.

6.1 PROJECTIVE MODULES OVER DEDEKIND DOMAINS

Our aim in this section is to classify the finitely generated projective modules over a Dedekind domain $\mathcal{O}$. We show that every ideal of $\mathcal{O}$ is projective and that every projective $\mathcal{O}$-module is a direct sum of ideals. Furthermore, we can write a projective module in the form $P \cong \mathcal{O}^{r-1} \oplus \mathfrak{a}$ for some ideal $\mathfrak{a}$ of $\mathcal{O}$. The integer r and the ideal class $\{\mathfrak{a}\}$ in $\mathrm{Cl}(\mathcal{O})$ are uniquely determined by the module P (and vice versa), so we can see that there will be non-free projective $\mathcal{O}$-modules provided that the class group $\mathrm{Cl}(\mathcal{O})$ of $\mathcal{O}$ is nontrivial.

We begin with a fact that was first made explicit with the introduction of the techniques of homological algebra, [Cartan & Eilenberg 1956], VII, §§3, 5.

6.1.1 Lemma

Let $\mathfrak{a}$ be a fractional ideal of a Dedekind domain $\mathcal{O}$. Then $\mathfrak{a}$ is a finitely generated projective $\mathcal{O}$-module.

In particular, any integral ideal of $\mathcal{O}$ is a finitely generated projective $\mathcal{O}$-module.

Proof

By (5.1.8), $d\mathfrak{a} \subseteq \mathcal{O}$ for some nonzero element d of $\mathcal{O}$. Evidently, $d\mathfrak{a} \cong \mathfrak{a}$ as an $\mathcal{O}$-module, so we may assume that $\mathfrak{a}$ is integral.

By (5.1.23), $\mathfrak{a}$ is generated by two elements, a_1, a_2, say. Choose any nonzero element x in $\mathfrak{a}$ and define, using invertibility, a fractional ideal $\mathfrak{b}$ by $x\mathcal{O} = \mathfrak{a}\mathfrak{b}$. Then $x = a_1 b_1 + a_2 b_2$, with b_1, b_2 in $\mathfrak{b}$. There is a surjection $\pi : \mathcal{O}^2 \to \mathfrak{a}$ given by

$$\pi(y_1, y_2) = y_1 a_1 + y_2 a_2,$$

which is split by the map $z \mapsto (z b_1 / x, z b_2 / x)$. The assertion now follows by (2.5.8). □

This result leads to a first description of finitely generated projective modules.

6.1.2 Theorem

Let M be a module over a Dedekind domain $\mathcal{O}$. Then the following statements are equivalent.

(i) *$M \cong \mathfrak{a}_1 \oplus \cdots \oplus \mathfrak{a}_s$ for a finite set $\{\mathfrak{a}_1, \ldots, \mathfrak{a}_s\}$ of integral ideals of $\mathcal{O}$.*

(ii) *M is finitely generated and projective as an $\mathcal{O}$-module.*

(iii) *There is an injective homomorphism $\sigma : M \to \mathcal{O}^t$ for some integer t.*

Proof

(i) $\Rightarrow$ (ii):   Immediate from the preceding lemma and (2.5.5).

(ii) $\Rightarrow$ (iii):   Since M is finitely generated, there is a surjection from some finitely generated free module $\mathcal{O}^t$ to M; since M is projective, this homomorphism must be split (see (2.5.8)).

(iii) $\Rightarrow$ (i):   Argue by induction on t. If $t = 1$ (or 0), M is (isomorphic to) an integral ideal of $\mathcal{O}$, and thus M is projective by the preceding result. For $t > 1$, let $\epsilon : \mathcal{O}^t \to \mathcal{O}$ be given by projection to the t th component. Then $\epsilon\sigma M$ is an ideal of $\mathcal{O}$, so projective, and

$$M \cong (M \cap \operatorname{Ker} \epsilon\sigma) \oplus \epsilon\sigma M$$

with $\sigma(M \cap \operatorname{Ker} \epsilon\sigma) \subseteq \mathcal{O}^{t-1}$.   □

A more precise description of projective modules requires two preliminary results on ideals.

6.1.3 Proposition

Let $\mathfrak{a}$ and $\mathfrak{b}$ be integral ideals in a Dedekind domain $\mathcal{O}$. Then there is an integral ideal $\mathfrak{a}'$ in the ideal class of $\mathfrak{a}$ such that $\mathfrak{a}'$ and $\mathfrak{b}$ are coprime.

Proof

Take a nonzero element a of $\mathfrak{a}$ and let $\mathfrak{c} = (a\mathcal{O})\mathfrak{a}^{-1}$. Since $\mathcal{O}/\mathfrak{bc}$ is a principal ideal ring (5.1.22), we have $\mathfrak{c} = \mathfrak{bc} + c\mathcal{O}$ for some c in $\mathfrak{c}$. Hence $\mathfrak{ac} = \mathfrak{abc} + ac$ and so $\mathcal{O} = \mathfrak{b} + \mathfrak{a}(ca^{-1})$.   □

6.1.4 Lemma

Let $\mathfrak{a}$ and $\mathfrak{b}$ be integral ideals of the Dedekind domain $\mathcal{O}$. Then there is an $\mathcal{O}$-module isomorphism

$$\mathfrak{a} \oplus \mathfrak{b} \cong \mathcal{O} \oplus \mathfrak{ab}.$$

Proof

First, suppose that $\mathfrak{a}$ and $\mathfrak{b}$ are coprime. Define a homomorphism $\alpha : \mathfrak{a}\oplus\mathfrak{b} \to \mathcal{O}$ by $\alpha(a, b) = a - b$. Then α is surjective, so split, and $\operatorname{Ker} \alpha = \mathfrak{a} \cap \mathfrak{b} = \mathfrak{ab}$ by (5.1.5).

In general, the preceding result together with (5.1.14) shows that there is an ideal $\mathfrak{a}' \cong \mathfrak{a}$ with $\mathfrak{a}'$ and $\mathfrak{b}$ coprime; then $\mathfrak{a}'\mathfrak{b} \cong \mathfrak{ab}$.   □

We also need an enhanced version of the process of 'clearing denominators' introduced in (5.1.7).

6.1.5 Lemma

Let $\mathcal{O}$ be a Dedekind domain with field of fractions $\mathcal{K}$ and let $\mathfrak{a}_1, \ldots, \mathfrak{a}_r$ be integral ideals of $\mathcal{O}$. Then

(i) *any element of the vector space $\mathcal{K}^r$ can be written in the form*

$$\begin{pmatrix} x_1 \\ \vdots \\ x_r \end{pmatrix} = \begin{pmatrix} a_1 d^{-1} \\ \vdots \\ a_r d^{-1} \end{pmatrix}$$

with $d \in \mathcal{O}$ and $a_i \in \mathfrak{a}_i$ for each i,

(ii) $\mathcal{K}^r = (\mathfrak{a}_1 \oplus \cdots \oplus \mathfrak{a}_r)\mathcal{K}$.

Proof

Using (5.1.8), it is easily seen that for each i we can write $x_i = c_i/d_i$ with c_i in $\mathfrak{a}_i$ and d_i in $\mathcal{O}$. Taking $d = d_1 \cdots d_r$, we obtain (i), and (ii) is immediate. $\qquad\square$

We can now present the main result [Steinitz 1912].

6.1.6 Steinitz' Theorem

Let $\mathcal{O}$ be a Dedekind domain and let $\mathfrak{a}_1, \ldots, \mathfrak{a}_r$ and $\mathfrak{b}_1, \ldots, \mathfrak{b}_s$ be integral ideals of $\mathcal{O}$. Then the following assertions are equivalent:

(i) *there is an $\mathcal{O}$-module isomorphism*

$$\phi : \mathfrak{a}_1 \oplus \cdots \oplus \mathfrak{a}_r \longrightarrow \mathfrak{b}_1 \oplus \cdots \oplus \mathfrak{b}_s;$$

(ii) $\qquad\qquad r = s \text{ and } \{\mathfrak{a}_1 \cdots \mathfrak{a}_r\} = \{\mathfrak{b}_1 \cdots \mathfrak{b}_r\} \text{ in } \mathrm{Cl}(\mathcal{O}).$

Proof

(i) $\Rightarrow$ (ii):   Let $\mathcal{K}$ be the field of fractions of $\mathcal{O}$ and let $\phi : \mathfrak{a}_1 \oplus \cdots \oplus \mathfrak{a}_r \to \mathfrak{b}_1 \oplus \cdots \oplus \mathfrak{b}_s$ be the given isomorphism. By the preceding lemma, an element x of the space $\mathcal{K}^r$ can be expressed in the form $x = a/d$ with $a \in \mathfrak{a}_1 \oplus \cdots \oplus \mathfrak{a}_r$ and $d \in \mathcal{O}$. Extend ϕ to a map ϕ' from $\mathcal{K}^r$ to $\mathcal{K}^s$ by $\phi'(x) = \phi(a)/d$. It is straightforward to verify that ϕ' is a well-defined $\mathcal{K}$-linear map and an isomorphism of $\mathcal{K}$-spaces. Thus $r = s$, and both ϕ' and ϕ are represented by left multiplication by a matrix $Q = (q_{ij})$ over $\mathcal{K}$. (Recall that we work with right modules and therefore regard the vector space $\mathcal{K}^r$ as a 'column-space'; see (2.2.9) for a discussion of the relation between endomorphisms of a free module and matrices. In particular, note that the matrix Q is determined to within conjugacy, so that its determinant $\det Q$ is uniquely defined.)

Let

$$a = \begin{pmatrix} a_1 \\ \vdots \\ a_r \end{pmatrix} \in \mathfrak{a}_1 \oplus \cdots \oplus \mathfrak{a}_r.$$

Since $Qa \in \mathfrak{b}_1 \oplus \cdots \oplus \mathfrak{b}_r$, we have $\sum_j q_{ij} a_j \in \mathfrak{b}_i$ for all i. In particular, taking $a_h = 0$ for $h \neq j$, we find that $q_{ij} a_j \in \mathfrak{b}_i$ for all i, j. So

$$\det(Q) \cdot a_1 \cdots a_r = \det(Q \cdot \mathrm{diag}(a_1, \ldots, a_r))$$

$$= \det \begin{pmatrix} q_{11} a_1 & \cdots & q_{1r} a_r \\ \vdots & \ddots & \vdots \\ q_{r1} a_1 & \cdots & q_{rr} a_r \end{pmatrix}$$

$$\in \mathfrak{b}_1 \cdots \mathfrak{b}_r.$$

Since $\mathfrak{a}_1 \cdots \mathfrak{a}_r$ is generated by all products $a_1 \cdots a_r$, we find that

$$\det(Q) \mathfrak{a}_1 \cdots \mathfrak{a}_r \subseteq \mathfrak{b}_1 \cdots \mathfrak{b}_r.$$

Similarly, $\det(Q^{-1}) \mathfrak{b}_1 \cdots \mathfrak{b}_r \subseteq \mathfrak{a}_1 \cdots \mathfrak{a}_r$, which gives

$$\{\mathfrak{a}_1 \cdots \mathfrak{a}_r\} = \{\mathfrak{b}_1 \cdots \mathfrak{b}_r\}.$$

(ii) $\Rightarrow$ (i):   By Lemma (6.1.4) and induction,

$$\mathfrak{a}_1 \oplus \cdots \oplus \mathfrak{a}_r \cong \mathcal{O}^{r-1} \oplus \mathfrak{a}_1 \cdots \mathfrak{a}_r$$

and

$$\mathfrak{b}_1 \oplus \cdots \oplus \mathfrak{b}_r \cong \mathcal{O}^{r-1} \oplus \mathfrak{b}_1 \cdots \mathfrak{b}_r.$$

However, $\{\mathfrak{a}_1 \cdots \mathfrak{a}_r\} = \{\mathfrak{b}_1 \cdots \mathfrak{b}_r\}$ implies that $\mathfrak{a}_1 \cdots \mathfrak{a}_r \cong \mathfrak{b}_1 \cdots \mathfrak{b}_r$ by (5.1.14). $\square$

6.1.7 The standard form

Steinitz' Theorem tells us that the isomorphism type of a finitely generated projective $\mathcal{O}$-module $M \cong \mathfrak{a}_1 \oplus \cdots \oplus \mathfrak{a}_r$ is characterized by two invariants, namely, its *rank*, which is the integer r, and its *ideal class* $\{M\} = \{\mathfrak{a}_1 \cdots \mathfrak{a}_r\}$ in $\mathrm{Cl}(\mathcal{O})$.

Note in particular that M is isomorphic to a projective module in *standard form* $\mathcal{O}^{r-1} \oplus \mathfrak{a}$ with $\mathfrak{a}$ an integral ideal, since by (5.1.14) every ideal class has an integral representative.

The next consequence of Steinitz' Theorem should be contrasted with the phenomenon of non-cancellation which we saw in (3.3.10).

6.1.8 Corollary

Cancellation holds for projective modules over a Dedekind domain: if

$$M \oplus \mathcal{O} \cong N \oplus \mathcal{O}$$

with M and N finitely generated projective, then

$$M \cong N. \qquad \qquad \square$$

On the other hand, if we allow both components of the direct sum to vary, we obtain the following observation.

6.1.9 Corollary

If $\mathrm{Cl}(\mathcal{O})$ is nontrivial, with $\{\mathfrak{a}\} \neq 1$, then

$$\mathfrak{a} \oplus \mathfrak{a}^{-1} \cong \mathcal{O}^2,$$

but

$$\mathfrak{a} \not\cong \mathcal{O} \ \text{ and } \ \mathfrak{a}^{-1} \not\cong \mathcal{O}. \qquad \qquad \square$$

We also obtain a result which is more often proved without invoking the theory of Dedekind domains.

6.1.10 Corollary

Let $\mathcal{O}$ be a commutative principal ideal domain. Then every finitely generated projective $\mathcal{O}$-module is free. $\qquad \qquad \square$

6.1.11 *The noncommutative case*

The definition of a Dedekind domain can be extended to allow the possibility of noncommutative Dedekind rings, which include the noncommutative Euclidean rings that we discussed in Chapter 4. Such rings share many of the properties of commutative Dedekind domains; comprehensive details can be found in Chapter 5 of [McConnell & Robson 1987].

Exercises

6.1.1   A converse to (6.1.1); see also Exercise 5.1.2.

Let $\mathcal{O}$ be a commutative domain with field of fractions $\mathcal{K}$, and let $\mathfrak{a}$ be a (nonzero) integral ideal of $\mathcal{O}$. Suppose that there is an integral ideal $\mathfrak{a}'$ with $\mathfrak{a}\mathfrak{a}' = x\mathcal{O}$ for some nonzero element x in $\mathcal{O}$. Show that $\mathfrak{a}$ is finitely generated and projective.

6.1.2   Let $\mathcal{K}$ be a field. Show that the ideal (X, Y) in the polynomial ring $\mathcal{K}[X, Y]$ is not projective

(a) by invoking Exercise 5.1.6,

(b) by showing directly that the surjection $\pi : (\mathcal{K}[X, Y])^2 \to (X, Y)$, $\pi(f, g) = Xf - Yg$, cannot be split.

More generally, show that the ideal $(X_1, \ldots, X_k)$ of the polynomial ring $\mathcal{K}[X_1, \ldots, X_n]$ is not projective for $n \geq k \geq 2$.

6.1.3   **Tiled orders**

Let $\mathcal{O}$ be a Dedekind domain with field of fractions $\mathcal{K}$. Given a collection $X = \{\mathfrak{a}_{ij} \mid i, j = 1, \ldots, r\}$ of fractional ideals of $\mathcal{O}$, the *set of tiled matrices* associated to X is the set $T(X)$ of matrices $a = (a_{ij}) \in M_r(\mathcal{K})$ such that $a_{ij} \in \mathfrak{a}_{ij}$ for all i, j.

Show that $T(X)$ is a subring of $M_r(\mathcal{K})$ if and only if $\mathfrak{a}_{ii} = \mathcal{O}$ for all i and $\mathfrak{a}_{ij} \mathfrak{a}_{jk} \subseteq \mathfrak{a}_{ik}$ for all i, j, k.

Show also that, if $T(X)$ is a ring, it is an $\mathcal{O}$-order. Such an order is known as a *tiled order*.

Using Exercises 5.1.8 and 2.1.6, show that the endomorphism ring $\mathrm{End}(M_\mathcal{O})$ of a projective (right) $\mathcal{O}$-module $M = \mathfrak{a}_1 \oplus \cdots \oplus \mathfrak{a}_r$ is the tiled order associated to the set

$$\{\mathfrak{a}_i \mathfrak{a}_j^{-1} \mid i, j = 1, \ldots, r\}.$$

Write down this order when M is in standard form $\mathcal{O}^{r-1} \oplus \mathfrak{a}$.

6.1.4   Suppose that $M = \mathfrak{a}_1 \oplus \cdots \oplus \mathfrak{a}_r$ and $N = \mathfrak{b}_1 \oplus \cdots \oplus \mathfrak{b}_s$ are projective (right) $\mathcal{O}$-modules. Generalize the previous exercise by showing that $\mathrm{Hom}(M, N)$ can be described as the set of $s \times r$ tiled matrices associated with the set of ideals

$$\{\mathfrak{b}_i \mathfrak{a}_j^{-1} \mid i = 1, \ldots, s, \ j = 1, \ldots, r\}.$$

Verify that $\mathrm{Hom}(M, N)$ is an $\mathrm{End}(N)$-$\mathrm{End}(M)$-bimodule by checking the matrix multiplications.

6.2 VALUATION RINGS

In this section, we introduce the valuation associated to a nonzero prime ideal $\mathfrak{p}$ of a Dedekind domain $\mathcal{O}$ and the corresponding valuation ring $\mathcal{O}_\mathfrak{p}$. Such a ring $\mathcal{O}_\mathfrak{p}$ has a very transparent internal structure. It is a Euclidean domain and hence a principal ideal domain; further, it has only one nonzero prime ideal, and any nonzero ideal is a power of this prime. Thus the module theory of $\mathcal{O}_\mathfrak{p}$ is known since it is a special case of that given in section 3.3 for

Euclidean domains. In the next section we show how to combine the results for each separate prime $\mathfrak{p}$ to complete the determination of the structure of $\mathcal{O}$-modules.

6.2.1 Valuations

Let $\mathcal{O}$ be a Dedekind domain with field of fractions $\mathcal{K}$ and let $\mathfrak{p}$ be a nonzero prime ideal of $\mathcal{O}$. By (5.1.19), any fractional ideal $\mathfrak{a}$ of $\mathcal{O}$ has a factorization $\mathfrak{a} = \mathfrak{p}^{v(\mathfrak{a})}\mathfrak{a}'$, where $v(\mathfrak{a}) = v(\mathfrak{p}, \mathfrak{a})$ is an integer uniquely determined by $\mathfrak{a}$, and $\mathfrak{a}'$ is a product of prime ideals other than $\mathfrak{p}$. Of course, we may have $\mathfrak{a}' = \mathcal{O}$, and, by definition, $v(\mathfrak{p}, \mathfrak{a}') = 0$. The integer $v(\mathfrak{a})$ is called the $\mathfrak{p}$-adic valuation of $\mathfrak{a}$.

If $\mathfrak{p} = p\mathcal{O}$ is a principal ideal, we may speak instead of the p-adic valuation.

If x is a nonzero element of $\mathcal{K}$, we define the $\mathfrak{p}$-adic valuation of x to be $v(x) = v(x\mathcal{O})$. For convenience, we put $v(0) = \infty$, with 0 being either the ideal or the element.

Thus we have functions

$$v : \mathrm{Frac}(\mathcal{O}) \cup \{0\} \longrightarrow \mathbb{Z} \cup \{\infty\}$$

and

$$v : \mathcal{K} \longrightarrow \mathbb{Z} \cup \{\infty\};$$

the context should make it clear which is intended. It is helpful to extend the usual ordering on $\mathbb{Z}$ to $\mathbb{Z} \cup \{\infty\}$ by setting $n < \infty$ for any integer n.

The elementary properties of these functions are easy to check. For example, to check surjectivity, note as in Exercise 5.1.4 that a proper invertible ideal $\mathfrak{p}$ can never have $\mathfrak{p}^k = \mathfrak{p}^{k+1}$. We record the results for ideals and for elements separately.

6.2.2 Lemma

Let $\mathfrak{a}$ and $\mathfrak{b}$ be fractional ideals of a Dedekind domain $\mathcal{O}$, and let v be the $\mathfrak{p}$-adic valuation for some prime $\mathfrak{p}$ of $\mathcal{O}$. Then

$$v : \mathrm{Frac}(\mathcal{O}) \cup \{0\} \longrightarrow \mathbb{Z} \cup \{\infty\}$$

is a surjective function with the following properties:

(i) $v(\mathfrak{a}) = \infty$ *if and only if* $\mathfrak{a} = 0$;
(ii) $v(\mathfrak{a} + \mathfrak{b}) \geq \min(v(\mathfrak{a}), v(\mathfrak{b}))$, *with equality if* $v(\mathfrak{a}) \neq v(\mathfrak{b})$;
(iii) $v(\mathfrak{a}\mathfrak{b}) = v(\mathfrak{a}) + v(\mathfrak{b})$;
(iv) $v(\mathfrak{p}) = 1$;
(v) $v(\mathcal{O}) = 0$;

(vi) $v(\mathfrak{a}^{-1}) = -v(\mathfrak{a})$. $\square$

6.2.3 Lemma

Let x and y be nonzero elements of the field of fractions $\mathcal{K}$ of a Dedekind domain $\mathcal{O}$, with v as before. Then

$$v : \mathcal{K} \longrightarrow \mathbb{Z} \cup \{\infty\}$$

and its restriction

$$v : \mathcal{O} \longrightarrow \{0\} \cup \mathbb{N} \cup \{\infty\}$$

are surjective functions with the following properties:

(i) $v(x) = \infty$ *if and only if* $x = 0$;
(ii) $v(x + y) \geq \min(v(x), v(y))$, *with equality if* $v(x) \neq v(y)$;
(iii) $v(xy) = v(x) + v(y)$;
(iv) $v(p) = 1$ *for any element* $p \in \mathfrak{p} \backslash \mathfrak{p}^2$, *and such elements exist*;
(v) $v(1) = 0$;
(vi) $v(x^{-1}) = -v(x)$. $\square$

Observe that in each of the lemmas above, properties (v) and (vi) are of secondary importance, since they can be deduced from (iii).

The function v is more properly described as a *discrete rank one valuation*, since there is a wider theory of valuations which may have values in ordered groups other than $\mathbb{Z}$, for example $\mathbb{R}$ (non-discrete) or $\mathbb{Z}^n$ (rank n). Such matters are discussed in [Cohn 1991]. However, we have no cause to consider these more general valuations, so we dispense with the qualifying adjectives.

6.2.4 Localization

The *valuation ring* or *localization* of $\mathcal{O}$ at $\mathfrak{p}$ is defined to be the subring $\mathcal{O}_\mathfrak{p}$ of $\mathcal{K}$ consisting of all fractions which can be written in the form a/b with $a, b \in \mathcal{O}$ and $b \notin \mathfrak{p}$. It is not hard to verify that $\mathcal{O}_\mathfrak{p}$ is a subring of $\mathcal{K}$, the crucial point being that, if $b_1, b_2 \in \mathcal{O}$ and $b_1, b_2 \notin \mathfrak{p}$, then $b_1 b_2 \notin \mathfrak{p}$ since $\mathfrak{p}$ is a prime ideal.

To identify the localization in terms of the corresponding valuation, we need a useful lemma.

6.2.5 Lemma

Let $\mathfrak{p}$ be a nonzero prime ideal of a Dedekind domain $\mathcal{O}$ and choose any $p \in \mathfrak{p} \backslash \mathfrak{p}^2$. Let $a \in \mathcal{O}$ and put $v = v(a)$. Then there exist

$$a', a'' \in \mathcal{O} \backslash \mathfrak{p}$$

with

$$aa' = p^v a''.$$

Proof

By (6.2.3), $v(ap^{-v}) = 0$, and so (5.1.19) shows that $ap^{-v}\mathcal{O} = \mathfrak{a}\mathfrak{b}^{-1}$ for integral ideals $\mathfrak{a}, \mathfrak{b}$, neither of which is divisible by $\mathfrak{p}$. Take any $a' \in \mathfrak{b} \backslash \mathfrak{p}$; then $aa'p^{-v} = a'' \in \mathfrak{a} \backslash \mathfrak{p}$. □

We obtain an alternative characterization of the localization.

6.2.6 Theorem

Let $\mathfrak{p}$ be a nonzero prime ideal of the Dedekind domain $\mathcal{O}$. Then

$$\mathcal{O}_\mathfrak{p} = \{x \in \mathcal{K} \mid v(x) \geq 0\}.$$

Proof

It is clear that $\mathcal{O}_\mathfrak{p}$ is contained in the right-hand set. For the converse, take x in $\mathcal{K}$ with $v(x) \geq 0$, and write $x = a/b$ with a and b in $\mathcal{O}$. Put $y = v(a)$ and $z = v(b)$, so that $y \geq z$.

Choose some element p in $\mathfrak{p} \backslash \mathfrak{p}^2$. By the lemma, $aa' = p^y a''$ and $bb' = p^z b''$ with $a', a'', b', b'' \in \mathcal{O} \backslash \mathfrak{p}$. Then

$$x = aa'b'/bb'a' = p^{y-z}a''b'/b''a' \in \mathcal{O}_\mathfrak{p}.$$ □

Consider an element x of $\mathcal{K}$. Clearly, x belongs to $\mathcal{O}$ precisely when the ideal $x\mathcal{O}$ is integral, which by (5.1.20) is the same as requiring that $v_\mathfrak{p}(x) \geq 0$ for all prime ideals $\mathfrak{p}$. This observation gives an important property of Dedekind domains.

6.2.7 Theorem

Let $\mathcal{O}$ be a Dedekind domain. Then $\mathcal{O} = \bigcap_\mathfrak{p} \mathcal{O}_\mathfrak{p}$, where the intersection is taken over all the (nonzero) prime ideals of $\mathcal{O}$. □

The following result is a straightforward application of the properties of valuations that are listed in (6.2.2) and (6.2.3). It also relies on our discussion of local rings in (4.3.24).

6.2.8 Proposition

(i) *Let $x, y \in \mathcal{O}_{\mathfrak{p}}$ be nonzero. Then $x \mid y$ if and only if $v(x) \leq v(y)$.*

(ii) *Let $x \in \mathcal{K}$, $\neq 0$. Then either x or x^{-1} is in $\mathcal{O}_{\mathfrak{p}}$.*

(iii) *The group of units in $\mathcal{O}_{\mathfrak{p}}$ is $U(\mathcal{O}_{\mathfrak{p}}) = \{x \in \mathcal{K} \mid v(x) = 0\}$.*

(iv) *$\mathcal{O}_{\mathfrak{p}}$ is a local ring with maximal ideal*

$$\mathfrak{p}\mathcal{O}_{\mathfrak{p}} = \{x \in \mathcal{K} \mid v(x) \geq 1\}.$$

(v) *The ideal $\mathfrak{p}\mathcal{O}_{\mathfrak{p}}$ is principal, generated by any element $p \in \mathcal{K}$ with $v(p) = 1$.*

(vi) *The fractional $\mathcal{O}_{\mathfrak{p}}$-ideals in $\mathcal{K}$ are the powers*

$$(\mathfrak{p}\mathcal{O}_{\mathfrak{p}})^i = \{x \in \mathcal{K} \mid v(x) \geq i\}, \; i \in \mathbb{Z}.$$

(vii) *$\mathcal{O}_{\mathfrak{p}}$ is a principal ideal domain.*   □

6.2.9 Uniformizing parameters

An element p of $\mathcal{O}_{\mathfrak{p}}$ with $v(p) = 1$ is sometimes called a *uniformizing parameter*, or *uniformizer* or *prime element*, because of property (v) above. Note that p can always be chosen to be an element of $\mathcal{O}$ itself. In view of (vii), the ring $\mathcal{O}_{\mathfrak{p}}$ may also be referred to as a *principal valuation ring* (nonprincipal valuation rings appear in geometric contexts that we do not consider in this text).

6.2.10 The localization as a Euclidean domain

We obtain the structure theory for modules over the valuation ring $\mathcal{O}_{\mathfrak{p}}$ as a special case of that for Euclidean domains. Recall from (3.2.7) that we need to define a function

$$\varphi : \mathcal{O}_{\mathfrak{p}} \longrightarrow \{0\} \cup \mathbb{N}$$

with the following properties:

(ED1) $\varphi(a) = 0 \Leftrightarrow a = 0$, where a is in $\mathcal{O}_{\mathfrak{p}}$;

(ED2) $\varphi(ab) \geq \varphi(a)$ for all nonzero a and b in $\mathcal{O}_{\mathfrak{p}}$;

(ED3) given a and b in $\mathcal{O}_{\mathfrak{p}}$, we have $a = bq + r$ for some q and r in $\mathcal{O}_{\mathfrak{p}}$ with $0 \leq \varphi(r) < \varphi(b)$.

The simplest choice is to put $\varphi(0) = 0$ and $\varphi(a) = 1 + v(a)$ otherwise. By definition, (ED1) holds, and (ED2) is immediate from part (ii) of (6.2.3). The division algorithm is rather trivially satisfied, since part (i) of (6.2.8) shows that we can take $r = a$ if $b \nmid a$.

Thus the Diagonal Reduction Theorem (3.3.2) holds for $\mathcal{O}_\mathfrak{p}$, from which we obtain, as in section 3.3, the structure of $\mathcal{O}_\mathfrak{p}$-modules. For convenience, we record this special case of (3.3.6) separately.

6.2.11 The Invariant Factor Theorem

Let M be a finitely generated (right) $\mathcal{O}_\mathfrak{p}$-module. Then

$$M \cong \mathcal{O}_\mathfrak{p}/d_1\mathcal{O}_\mathfrak{p} \oplus \cdots \oplus \mathcal{O}_\mathfrak{p}/d_\ell\mathcal{O}_\mathfrak{p} \oplus (\mathcal{O}_\mathfrak{p})^s,$$

where $d_1, \ldots, d_\ell$ are nonunits in $\mathcal{O}_\mathfrak{p}$, $d_1 \mid d_2 \mid \cdots \mid d_\ell$ and $s \geq 0$. $\qquad\square$

It is convenient to rephrase the Invariant Factor Theorem in terms of ideals. To do this, put $\delta(i) = v(d_i)$ for $i = 1, \ldots, \ell$.

6.2.12 Corollary

Let M be a finitely generated (right) $\mathcal{O}_\mathfrak{p}$-module. Then

$$M \cong \mathcal{O}_\mathfrak{p}/(\mathfrak{p}\mathcal{O}_\mathfrak{p})^{\delta(1)} \oplus \cdots \oplus \mathcal{O}_\mathfrak{p}/(\mathfrak{p}\mathcal{O}_\mathfrak{p})^{\delta(\ell)} \oplus (\mathcal{O}_\mathfrak{p})^s,$$

where $\delta(1) \leq \cdots \leq \delta(\ell)$ are positive integers. $\qquad\square$

6.2.13 Rank and invariant factors

As defined in (3.3.7), the elements $d_1, \ldots, d_\ell$ of $\mathcal{O}_\mathfrak{p}$ are the invariant factors of M, and the integer s is the rank of M. We can equally refer to the ideals $(\mathfrak{p}\mathcal{O}_\mathfrak{p})^{\delta(1)}, \ldots, (\mathfrak{p}\mathcal{O}_\mathfrak{p})^{\delta(\ell)}$ as being the invariant factors of M.

In the next section, we show that the invariant factors are uniquely determined by the module M (6.3.12), that is, the elements $d_1, \ldots, d_\ell$ are determined up to multiplication by units of $\mathcal{O}_\mathfrak{p}$. The rank of M is also unique, which we confirm in (6.3.6).

Exercises

6.2.1   Let $\mathcal{O}$ be a Dedekind domain and let $\mathbf{Q}$ be a subset of the set $\mathbf{P}$ of nonzero prime ideals of $\mathcal{O}$. Show that

$$\mathcal{O}_\mathbf{Q} = \bigcap_{\mathfrak{p} \in \mathbf{Q}} \mathcal{O}_\mathfrak{p}$$

is also a Dedekind domain and find its prime ideals.

Show that $\mathcal{O}_\mathbf{Q}$ is not a finitely generated $\mathcal{O}$-module unless $\mathbf{P} = \mathbf{Q}$. (Exercise 5.1.10 helps.)

6.2.2   Let $\mathcal{O} = \mathcal{O}_\mathfrak{p}$ be a valuation ring with (unique) maximal ideal $\mathfrak{p}$, and suppose that R is an $\mathcal{O}$-order.

Show that any finitely generated right R-module is also finitely generated as an $\mathcal{O}$-module.

Deduce that

(i) $\mathfrak{p}R \subseteq \text{rad}(R)$, the Jacobson radical of R as a ring,

(ii) R is a semilocal ring,

(iii) $\text{rad}(R) = \pi^{-1}(\text{rad}(R/\mathfrak{p}R))$, the inverse image with respect to the natural homomorphism π from R to $R/\mathfrak{p}R$,

(iv) $(\text{rad}(R))^h \subseteq \mathfrak{p}R$ for some integer h.

Hint. Nakayama's Lemma (4.3.10), together with (4.3.12), may be useful.

6.2.3   Let $\mathcal{O}$ be a valuation ring with maximal ideal $\mathfrak{p}$ and suppose that $\mathcal{O}$ has characteristic 0, so that $\mathbb{Z}$ can be regarded as a subring of $\mathcal{O}$ (1.1.10). Show that $\mathfrak{p} \cap \mathbb{Z} = p\mathbb{Z}$, where the prime p is the characteristic of the field $\mathcal{O}/\mathfrak{p}$.

Let G be a finite group, and let $\mathcal{O}G$ be its group ring over $\mathcal{O}$.

(i) Show that, if p does not divide the order $|G|$ of G, then

$$\text{rad}(\mathcal{O}G) = \mathfrak{p}G.$$

(ii) Suppose that $G = \langle \gamma \rangle$ is cyclic of order p. Show that

$$\text{rad}(\mathcal{O}G) = (1 - \gamma)\mathcal{O}G.$$

[Exercises 4.3.9 and 4.3.11 are relevant.]

6.2.4   Let $\mathcal{O}$ be a valuation ring with maximal ideal $\mathfrak{p}$, and let $R = \begin{pmatrix} \mathcal{O} & \mathfrak{p}^h \\ \mathcal{O} & \mathcal{O} \end{pmatrix}$ be the tiled order consisting of those matrices in $M_2(\mathcal{O})$ with $(1,2)$th entry belonging to $\mathfrak{p}^h$ where h is an integer, $h \geq 0$.

Show that

$$\text{rad}(R) = \begin{cases} \mathfrak{p}R & h = 0 \\ \begin{pmatrix} \mathfrak{p} & \mathfrak{p}^h \\ \mathcal{O} & \mathfrak{p} \end{pmatrix} & h \geq 1 \end{cases}$$

and that

$$\text{rad}(R)^2 = \begin{cases} \mathfrak{p}R & h = 1 \\ \mathfrak{p}\,\text{rad}(R) & h \neq 1 \end{cases}$$

Let $C(i) = \begin{pmatrix} \mathfrak{p}^i \\ \mathcal{O} \end{pmatrix}$ for $i = 0, \ldots, h$. Prove that each $C(i)$ is a right R-module, and that $C(i) \cong C(j)$ (as an R-module) if and only if $i = j$.

Hint. Any such isomorphism extends to an $M_2(\mathcal{K})$-homomorphism of the irreducible right module $\begin{pmatrix} \mathcal{K} \\ \mathcal{K} \end{pmatrix}$, where $\mathcal{K}$ is the field of fractions of $\mathcal{O}$, and so must be given by left multiplication by an element of $\mathcal{K}$.

In Exercise 7.2.5 of [BK: CM], we show that $C(i)$ is projective as an R-module if and only if $i = 0$ or h, and that any finitely generated projective right R-module is isomorphic to a direct sum of copies of $C(0)$ and $C(h)$. *Assuming* this result, deduce that $\mathrm{rad}(R)$ is projective as a right R-module if and only if $h = 0, 1$.

Remark. For $h = 0, 1$, R is an example of a hereditary order, the theory of which is discussed in detail in [Reiner 1975], Chapter 9; for $h = 0$, the order is maximal.

6.2.5 Repeat the previous exercise with

$$R = \begin{pmatrix} \mathcal{O} & \mathfrak{p}^h & \cdots & \mathfrak{p}^h & \mathfrak{p}^h \\ \mathcal{O} & \mathcal{O} & \cdots & \mathfrak{p}^h & \mathfrak{p}^h \\ \vdots & \vdots & \ddots & \vdots & \vdots \\ \mathcal{O} & \mathcal{O} & \cdots & \mathcal{O} & \mathfrak{p}^h \\ \mathcal{O} & \mathcal{O} & \cdots & \mathcal{O} & \mathcal{O} \end{pmatrix},$$

the $\mathcal{O}$-order consisting of all $n \times n$ matrices with entries above the diagonal belonging to $\mathfrak{p}^h$.

(*Warning.* The same ideas work but are trickier to implement; the generalization of the module $C(i)$ requires multiple indices.)

6.3 TORSION MODULES OVER DEDEKIND DOMAINS

We now combine the results of the preceding two sections to obtain a complete description of the finitely generated modules over an arbitrary Dedekind domain $\mathcal{O}$. We show that an $\mathcal{O}$-module can be decomposed into a direct sum of a torsion-free component and a torsion component. The results of section 6.1 show that the torsion-free component is projective and so its structure is known. The torsion component further decomposes into $\mathfrak{p}$-primary components, one for each nonzero prime ideal $\mathfrak{p}$ of $\mathcal{O}$. Almost all of these primary components are zero, and each is a torsion module over the corresponding valuation ring $\mathcal{O}_\mathfrak{p}$, so that its structure is known by the results of the previous section.

As a special case, we obtain a description of the structure of the finitely generated modules over an arbitrary commutative principal ideal domain.

The systematic use of the local rings $\mathcal{O}_\mathfrak{p}$ to obtain results on $\mathcal{O}$-modules is comparatively recent, originating with [Krull 1938].

6.3.1 Torsion modules

Let M be a right module over a commutative domain $\mathcal{O}$. An element m in M is said to be a *torsion element* if $mr = 0$ for some $r \neq 0$ in $\mathcal{O}$. The set of all torsion elements in M is denoted by $T(M)$ and called the *torsion submodule* of M. An easy verification reveals that $T(M)$ is indeed a submodule of M.

We say that M is a *torsion module* if $M = T(M)$ and, as in (1.2.23), that M is torsion-free if $T(M) = 0$. (The zero module is therefore both a torsion module and torsion-free; to avoid a proliferation of qualifying phrases, we agree to ignore trivial modifications that result from inserting or deleting zero modules.)

The following is routine.

6.3.2 Lemma

Let M be an $\mathcal{O}$-module, where $\mathcal{O}$ is a commutative domain. Then $T(M)$ is a torsion module and $M/T(M)$ is torsion-free.   □

It is clear that the free module $\mathcal{O}^k$ is torsion-free, as are all its submodules. The next result tells us that, in essence, any finitely generated torsion-free module arises as a submodule of a free module.

6.3.3 Lemma

Let M be a finitely generated torsion-free $\mathcal{O}$-module, where $\mathcal{O}$ is a commutative domain. Then there is an injective $\mathcal{O}$-module homomorphism from M to $\mathcal{O}^k$ for some integer k.

Proof

By (1.2.23), the $\mathcal{O}$-module M spans a vector space V over the field of fractions $\mathcal{K}$ of $\mathcal{O}$. Since M is finitely generated, V is finite-dimensional. Choose a set of generators $\{m_1, \ldots, m_\ell\}$ of M and a basis $\{e_1, \ldots, e_k\}$ of V, and write

$$m_j = \sum_i e_i x_{ij} \text{ with } x_{ij} \in \mathcal{K}.$$

The coefficients can be put over a common denominator d, and then

$$M \cong Md \subseteq e_1 \mathcal{O} \oplus \cdots \oplus e_k \mathcal{O}.   □$$

We combine the above result with (6.1.2).

6.3.4 Corollary

Let M be a finitely generated torsion-free module over a Dedekind domain $\mathcal{O}$. Then M is projective. ☐

Since $M/T(M)$ is torsion-free, the next result is immediate from the definition of a projective module (2.5.1).

6.3.5 Theorem

Let M be a finitely generated module over a Dedekind domain $\mathcal{O}$. Then there is an $\mathcal{O}$-module isomorphism

$$M \cong T(M) \oplus M/T(M).$$ ☐

6.3.6 The rank

We can also extend the definition of the *rank* to any finitely generated module M over a commutative domain $\mathcal{O}$. If M is torsion-free, we put $\text{rank}(M) = \dim(M\mathcal{K})$, where $M\mathcal{K}$ is the space spanned by M, and, in general, we set $\text{rank}(M) = \text{rank}(M/T(M))$. It is clear that this definition coincides with our previous definition when $\mathcal{O}$ is a Euclidean domain (3.3.7), (6.2.13). It is also obvious that the rank of a module is uniquely defined, and that $\text{rank}(M) = 0$ precisely when M is a torsion module.

6.3.7 Primary modules

Recall that, given an $\mathcal{O}$-module M, the annihilator of M is defined to be the ideal $\text{Ann}(M)$ of $\mathcal{O}$ given by

$$\text{Ann}(M) = \{r \in \mathcal{O} \mid mr = 0 \text{ for all } m \in M\}.$$

It is clear that a finitely generated $\mathcal{O}$-module has a nonzero annihilator if and only if it is a torsion module.

Let $\mathfrak{p}$ be a nonzero prime ideal of $\mathcal{O}$. An $\mathcal{O}$-module M is called $\mathfrak{p}$-*primary* if $\text{Ann}(M) = \mathfrak{p}^\delta$ for some natural number δ. (The zero module is admitted as a $\mathfrak{p}$-primary module.) If $\mathfrak{p} = p\mathcal{O}$ is a principal ideal, we sometimes prefer to say that a module is p-*primary*. This terminology is more natural if the coefficient ring is a principal ideal domain, such as $\mathbb{Z}$.

Clearly, the factor module $\mathcal{O}/\mathfrak{p}^\delta$ is $\mathfrak{p}$-primary, as is any finite direct sum of such factor modules (with possibly differing exponents). Our aim is to show that any finitely generated $\mathfrak{p}$-primary module can be described in this way, in an essentially unique fashion.

The argument exploits the Invariant Factor Theorem (6.2.11) for the localization $\mathcal{O}_\mathfrak{p}$, the next result providing the connection. Note that any finitely generated torsion $\mathcal{O}_\mathfrak{p}$-module is necessarily $\mathfrak{p}\mathcal{O}_\mathfrak{p}$-primary, since, by (6.2.8), $\mathfrak{p}\mathcal{O}_\mathfrak{p}$ is the unique nonzero prime ideal of the principal ideal domain $\mathcal{O}_\mathfrak{p}$.

6.3.8 Proposition

For any $\delta > 0$, $\mathcal{O}/\mathfrak{p}^\delta \cong \mathcal{O}_\mathfrak{p}/(\mathfrak{p}\mathcal{O}_\mathfrak{p})^\delta$ both as rings and as $\mathcal{O}$-modules.

Proof

The natural ring homomorphism from $\mathcal{O}$ to $\mathcal{O}_\mathfrak{p}/(\mathfrak{p}\mathcal{O}_\mathfrak{p})^\delta$ clearly has kernel $\mathfrak{p}^\delta$, so there is an induced injection ι from $\mathcal{O}/\mathfrak{p}^\delta$ to $\mathcal{O}_\mathfrak{p}/(\mathfrak{p}\mathcal{O}_\mathfrak{p})^\delta$.

To see that ι is surjective, take an element $a/b \in \mathcal{O}_\mathfrak{p}$, where $a \in \mathcal{O}$ and $b \in \mathcal{O} \setminus \mathfrak{p}$, and consider the ideal $b\mathcal{O} + \mathfrak{p}^\delta$ of the Dedekind domain $\mathcal{O}$. If this ideal is proper, then, by (5.1.19), it has a unique factorization in terms of prime ideals. Combining (5.1.17) and (5.1.21), we see that the only prime ideal which could possibly occur as a factor of $b\mathcal{O} + \mathfrak{p}^\delta$ is $\mathfrak{p}$ itself. Since $b \notin \mathfrak{p}$, we have $b\mathcal{O} + \mathfrak{p}^\delta = \mathcal{O}$.

Thus $1 = bc + z$ for some c in $\mathcal{O}$ and z in $\mathfrak{p}^\delta$, so $a/b \equiv ac \pmod{(\mathfrak{p}\mathcal{O}_\mathfrak{p})^\delta}$, that is, $\overline{(a/b)} = \iota(\overline{ac})$, which establishes surjectivity.

Finally, observe that the $\mathcal{O}$-module structure on $\mathcal{O}_\mathfrak{p}/(\mathfrak{p}\mathcal{O}_\mathfrak{p})^\delta$ arises from this isomorphism – by 'change of rings' (1.2.14). $\qquad\qquad\square$

We can now give a structure theorem for primary modules (but not yet a uniqueness theorem).

6.3.9 Theorem

Let $\mathfrak{p}$ be a nonzero prime ideal of the Dedekind domain $\mathcal{O}$, and let M be a finitely generated $\mathfrak{p}$-primary $\mathcal{O}$-module.

Then M is also a finitely generated $\mathcal{O}_\mathfrak{p}$-module, and there is a direct decomposition of M, both as an $\mathcal{O}$-module and as an $\mathcal{O}_\mathfrak{p}$-module,

$$M \cong \mathcal{O}/\mathfrak{p}^{\delta(1)} \oplus \cdots \oplus \mathcal{O}/\mathfrak{p}^{\delta(\ell)},$$

where $\delta(1) \leq \cdots \leq \delta(\ell)$ are positive integers.

Proof

By definition, the annihilator of M is $\mathfrak{p}^\delta$ for some $\delta > 0$. Thus M is naturally an $(\mathcal{O}/\mathfrak{p}^\delta)$-module, hence an $(\mathcal{O}_\mathfrak{p}/(\mathfrak{p}\mathcal{O}_\mathfrak{p})^\delta)$-module, and therefore an $\mathcal{O}_\mathfrak{p}$-module, from the previous result. By (6.2.12), M has a direct decomposition as an $\mathcal{O}_\mathfrak{p}$-module as claimed, and this decomposition is clearly also an $\mathcal{O}$-module decomposition. $\qquad\qquad\square$

6.3.10 Elementary divisors

When the Dedekind domain $\mathcal{O}$ has more than one nonzero prime ideal, the ideals $\mathfrak{p}^{\delta(1)}, \ldots, \mathfrak{p}^{\delta(\ell)}$ occurring in the decomposition of a $\mathfrak{p}$-primary module M are called the *elementary divisors* of M.

If $\mathfrak{p} = p\mathcal{O}$ is principal, in particular, if $\mathcal{O}$ is a principal ideal domain, the corresponding powers $p^{\delta(\mathfrak{p},1)}, \ldots, p^{\delta(\mathfrak{p},\ell(\mathfrak{p}))}$ of the irreducible element p are often called the elementary divisors of M.

This definition anticipates the definition of elementary divisors for an arbitrary torsion module. The elementary divisors of a $\mathfrak{p}$-primary module M as an $\mathcal{O}$-module are, by definition, the same as its invariant factors when M is regarded as an $\mathcal{O}_\mathfrak{p}$-module. The extension of the notion of invariant factors to modules over an arbitrary Dedekind domain is outlined in Exercise 6.3.6 below.

We associate with each direct sum decomposition of a $\mathfrak{p}$-primary module into cyclic modules a sequence

$$\text{edt}_\mathfrak{p}(M) = (\alpha_1, \alpha_2, \ldots)$$

of natural numbers in which α_i is the number of times that the term $\mathcal{O}/\mathfrak{p}^i$ occurs in the given direct sum. Once uniqueness of the decomposition has been established, as in the next theorem, this sequence may be called the $\mathfrak{p}$-*elementary divisor type* of M.

Note that $\alpha_i = 0$ for all but a finite set of indices; the maximum of the indices for which $\alpha_i \neq 0$ is the *length* of $\text{edt}_\mathfrak{p}(M)$. Clearly, the length is the exponent k occurring in the annihilator: $\text{Ann}(M) = \mathfrak{p}^k$.

The zero module corresponds to the zero sequence $(0, 0, \ldots)$, which has length 0, and $\mathcal{O}/\mathfrak{p}^k$ has type $(0, \ldots, 0, 1, 0, \ldots)$ of length k. This notion of length is not the same as the length of a composition series introduced in (4.1.8). Since the terms are used in different contexts, there should be no confusion.

Our main result shows that the elementary divisor type $\text{edt}_\mathfrak{p}(M)$ describes a $\mathfrak{p}$-primary module completely.

6.3.11 Theorem

Let M and N be finitely generated $\mathfrak{p}$-primary $\mathcal{O}$-modules. Then $M \cong N$ if and only if $\text{edt}_\mathfrak{p}(M) = \text{edt}_\mathfrak{p}(N)$.

Proof

It is obvious that if $\text{edt}_\mathfrak{p}(M) = \text{edt}_\mathfrak{p}(N)$ then $M \cong N$. Suppose conversely that $M \cong N$, and write

$$\mathrm{edt}_\mathfrak{p}(M) = (\alpha_1, \alpha_2, \ldots, \alpha_k, 0, \ldots)$$

and

$$\mathrm{edt}_\mathfrak{p}(N) = (\beta_1, \beta_2, \ldots, \beta_\ell, 0, \ldots),$$

where k and ℓ are the lengths of the sequences. We argue by induction on k.

If $k = 1$, then M is in effect a vector space over the field $\mathcal{O}/\mathfrak{p}$ and α_1 is its dimension; since $\mathrm{Ann}(N) = \mathfrak{p}$ also, we must have $\ell = 1$ and $\alpha_1 = \beta_1$.

Now suppose $k > 1$. First, we note that the quotient modules $M/\mathfrak{p}M$ and $N/\mathfrak{p}N$ are isomorphic and have types

$$\mathrm{edt}_\mathfrak{p}(M/\mathfrak{p}M) = (\alpha_1 + \alpha_2 + \cdots + \alpha_k, 0, \ldots)$$

and

$$\mathrm{edt}_\mathfrak{p}(N/\mathfrak{p}N) = (\beta_1 + \beta_2 + \cdots + \beta_\ell, 0, \ldots)$$

respectively, which shows that

$$\alpha_1 + \alpha_2 + \cdots + \alpha_k = \beta_1 + \beta_2 + \cdots + \beta_\ell.$$

Next we note that, since $\mathfrak{p}/\mathfrak{p}^i \cong \mathcal{O}/\mathfrak{p}^{i-1}$ for $i > 1$ by (5.1.24), the isomorphic modules $\mathfrak{p}M$ and $\mathfrak{p}N$ have types

$$\mathrm{edt}_\mathfrak{p}(\mathfrak{p}M) = (\alpha_2, \ldots, \alpha_k, 0, \ldots)$$

and

$$\mathrm{edt}_\mathfrak{p}(\mathfrak{p}N) = (\beta_2, \ldots, \beta_\ell, 0, \ldots)$$

of lengths $k - 1$ and $\ell - 1$; the result now follows by induction.   □

For reference, we state an immediate consequence of the preceding result.

6.3.12 Corollary

Let $\mathfrak{p}$ be a nonzero prime ideal of a Dedekind domain $\mathcal{O}$, and let M be a finitely generated torsion $\mathcal{O}_\mathfrak{p}$-module. Then the invariant factors

$$(\mathfrak{p}\mathcal{O}_\mathfrak{p})^{\delta(1)}, \ldots, (\mathfrak{p}\mathcal{O}_\mathfrak{p})^{\delta(\ell)}, \quad \delta(1) \leq \cdots \leq \delta(\ell),$$

of M are unique.   □

6.3.13 Primary decomposition

We now consider the decomposition of a finitely generated torsion $\mathcal{O}$-module M into a direct sum of $\mathfrak{p}$-primary submodules as $\mathfrak{p}$ ranges over the set $\mathbf{P} = \mathbf{P}(\mathcal{O})$ of nonzero primes of the Dedekind domain $\mathcal{O}$. This gives a characterization of M in terms of the collection $\{\mathrm{edt}_{\mathfrak{p}}(M) \mid \mathfrak{p} \in \mathbf{P}\}$ of elementary divisor types arising from its summands.

We first show that, for each $\mathfrak{p}$, M has a maximal $\mathfrak{p}$-primary submodule. To see this, note that, if M' and M'' are both $\mathfrak{p}$-primary submodules of M (that is, both have annihilators that are powers of $\mathfrak{p}$), then so also is $M' + M''$. Thus, if there were no maximal $\mathfrak{p}$-primary submodule, we could construct an infinite ascending chain in M, contrary to the fact that M is Noetherian, since it is a finitely generated module over the Noetherian ring $\mathcal{O}$ and so itself Noetherian (3.1.4).

We can therefore define the $\mathfrak{p}$-*component* $T_{\mathfrak{p}}(M)$ of M to be the maximal $\mathfrak{p}$-primary submodule of M, that is, the maximal submodule with the property that $\mathrm{Ann}(T_{\mathfrak{p}}(M)) = \mathfrak{p}^k$ for some natural number k. Alternative terms for $T_{\mathfrak{p}}(M)$ are the $\mathfrak{p}$-*torsion part* and the $\mathfrak{p}$-*primary part* of M. Since M is Noetherian, $T_{\mathfrak{p}}(M)$ is finitely generated.

The next result is the key to the existence of the primary decomposition.

6.3.14 Lemma

Let $\mathfrak{a}$ and $\mathfrak{b}$ be coprime ideals of $\mathcal{O}$, and let M be an $\mathcal{O}$-module such that $\mathrm{Ann}(M) = \mathfrak{a}\mathfrak{b}$.

Then $M = \mathfrak{a}M \oplus \mathfrak{b}M$, and $\mathrm{Ann}(\mathfrak{a}M) = \mathfrak{b}$ and $\mathrm{Ann}(\mathfrak{b}M) = \mathfrak{a}$.

Proof

Since $\mathfrak{a}$ and $\mathfrak{b}$ are coprime, $\mathcal{O} = \mathfrak{a} + \mathfrak{b}$. Thus $1 = a + b$ for some elements a, b of $\mathfrak{a}, \mathfrak{b}$ respectively, and so $M = \mathfrak{a}M + \mathfrak{b}M$.

Suppose that $m \in \mathfrak{a}M \cap \mathfrak{b}M$. Then $am = bm = 0$ because $\mathfrak{a}\mathfrak{b}M = 0$, hence $m = (a + b)m = 0$, which gives the direct sum decomposition.

Clearly $\mathfrak{b} \subseteq \mathrm{Ann}(\mathfrak{a}M)$. Let x be in $\mathrm{Ann}(\mathfrak{a}M)$. Then $x\mathcal{O} \cdot \mathfrak{a} \subseteq \mathrm{Ann}(M) = \mathfrak{a}\mathfrak{b}$ and hence $x\mathcal{O} \subseteq \mathfrak{b}$, since $\mathfrak{a}$ is an invertible ideal. $\qquad\square$

Given the module M and a nonzero prime ideal $\mathfrak{p}$, we write the factorization of the annihilator in the form

$$\mathrm{Ann}(M) = \mathfrak{p}^k \mathfrak{c}(\mathfrak{p}),$$

where $k \geq 0$ is the $\mathfrak{p}$-adic valuation of $\mathrm{Ann}(M)$ and $\mathfrak{p}$ and $\mathfrak{c}(\mathfrak{p})$ are coprime (see (6.2.1)). Note that $k = 0$ and $\mathfrak{c}(\mathfrak{p}) = \mathrm{Ann}(M)$ except for the finite set of primes that actually occur as factors of $\mathrm{Ann}(M)$.

6.3.15 Proposition

Let M be a finitely generated torsion module over a Dedekind domain $\mathcal{O}$. Then the following hold.

(i) $T_{\mathfrak{p}}(M) = \mathfrak{c}(\mathfrak{p})M$.

(ii) $T_{\mathfrak{p}}(M) = 0$ *for all but finitely many primes* $\mathfrak{p}$.

(iii) $M = \bigoplus_{\mathfrak{p}} T_{\mathfrak{p}}(M)$, *where the sum is taken over the set* $\mathbf{P}$ *of all nonzero primes* $\mathfrak{p}$ *of* $\mathcal{O}$.

Proof

Clearly $\mathfrak{c}(\mathfrak{p})M \subseteq T_{\mathfrak{p}}(M)$. By the lemma, we have $M = \mathfrak{p}^k M \oplus \mathfrak{c}(\mathfrak{p})M$ and $\mathrm{Ann}(\mathfrak{p}^k M) = \mathfrak{c}(\mathfrak{p})$. Thus if $m = x + n$ is in $T_{\mathfrak{p}}(M)$, where x and n belong to the respective summands of M, we find that x is annihilated by both the coprime ideals $\mathfrak{p}^k$ and $\mathfrak{c}(\mathfrak{p})$, so must be 0. Thus (i) holds, and (ii) follows since $\mathfrak{c}(\mathfrak{p}) = \mathrm{Ann}(M)$ for almost all primes.

The last part follows by induction on the number of distinct prime factors of $\mathrm{Ann}(M)$. Note that, if $\mathfrak{p}$ is a factor of $\mathrm{Ann}(M)$, then, by the preceding lemma,

$$M = \mathfrak{c}(\mathfrak{p})M \oplus \mathfrak{p}^k M = T_{\mathfrak{p}}(M) \oplus \mathfrak{p}^k M.$$

Now $\mathrm{Ann}(\mathfrak{p}^k M) = \mathfrak{c}(\mathfrak{p})$ has fewer prime factors than $\mathrm{Ann}(M)$, and $T_{\mathfrak{q}}(\mathfrak{p}^k M) = T_{\mathfrak{q}}(M)$ if $\mathfrak{q}$ is a prime different from $\mathfrak{p}$.   □

We note that all possible primary decompositions actually occur.

6.3.16 Proposition

Given any set of finitely generated $\mathfrak{p}$-primary $\mathcal{O}$-modules $M_{\mathfrak{p}}$, $\mathfrak{p} \in \mathbf{P}$, only finitely many of which are nonzero, the external direct sum $M = \bigoplus M_{\mathfrak{p}}$ has $T_{\mathfrak{p}}(M) \cong M_{\mathfrak{p}}$ for all $\mathfrak{p}$.   □

6.3.17 Elementary divisors again

Given an arbitrary finitely generated torsion $\mathcal{O}$-module M, the *elementary divisors* of M are the ideals

$$\mathfrak{p}^{\delta(\mathfrak{p},1)}, \ldots, \mathfrak{p}^{\delta(\mathfrak{p},\ell(\mathfrak{p}))}$$

that occur in the nontrivial $\mathfrak{p}$-primary components

$$T_{\mathfrak{p}}(M) \cong \mathcal{O}/\mathfrak{p}^{\delta(\mathfrak{p},1)} \oplus \cdots \oplus \mathcal{O}/\mathfrak{p}^{\delta(\mathfrak{p},\ell(\mathfrak{p}))}$$

of M, where $\mathfrak{p}$ varies through the set $\mathbf{P}$ of nonzero prime ideals of $\mathcal{O}$. So that we can list the elementary divisors unambiguously, we must choose some convenient ordering of **P**; then, for a given $\mathfrak{p}$, we take the exponents in non-decreasing order.

When $\mathcal{O}$ is a principal ideal domain, the powers

$$p^{\delta(\mathfrak{p},1)}, \ldots, p^{\delta(\mathfrak{p},\ell(\mathfrak{p}))}$$

of the irreducible elements p are more often called the elementary divisors of M. Now p will run through a representative set **P**e of irreducible elements of $\mathcal{O}$, that is, each prime ideal $\mathfrak{p} \in \mathbf{P}$ is generated by exactly one member p of **P**e. We usually write $T_p(M)$ rather than $T_{p\mathcal{O}}(M)$ and call it the *p-primary submodule* or *p-component* of M.

The *elementary divisor type* of M is defined as follows. For each nonzero prime ideal $\mathfrak{p}$ of $\mathcal{O}$, put

$$\text{edt}_{\mathfrak{p}}(M) = \text{edt}_{\mathfrak{p}}(T_{\mathfrak{p}}(M))$$

which is a sequence of finite length. Then the elementary divisor type of M is defined to be

$$\text{edt}(M) = (\text{edt}_{\mathfrak{p}}(M) \mid \mathfrak{p} \in \mathbf{P}),$$

a sequence of such sequences, where the set **P** of primes is again given some convenient ordering. For all except a finite set of primes, $\text{edt}_{\mathfrak{p}}(M) = (0, 0, \ldots)$.

It is clear that we can construct the elementary divisors of a module from its elementary divisor type, and vice versa. Before we can show that its elementary divisor type gives a complete description of a module, we need to discuss how homomorphisms affect the primary components of modules.

6.3.18 Homomorphisms

Suppose that $\lambda : M \to N$ is a homomorphism between finitely generated torsion $\mathcal{O}$-modules, where $\mathcal{O}$ is a Dedekind domain. It is clear from the description of the $\mathfrak{p}$-components of M and N that λ induces a family of $\mathcal{O}$-module homomorphisms

$$T_{\mathfrak{p}}(\lambda) : T_{\mathfrak{p}}(M) \longrightarrow T_{\mathfrak{p}}(N).$$

For all except a finite set of $\mathfrak{p}$, $T_{\mathfrak{p}}(\lambda)$ is the zero map between zero modules.

Conversely, given a family $\{\lambda(\mathfrak{p}) : T_{\mathfrak{p}}(M) \to T_{\mathfrak{p}}(N)\}$ of $\mathcal{O}$-module homomorphisms, the direct sum $\lambda = \bigoplus_{\mathfrak{p}} \lambda(\mathfrak{p})$ is a homomorphism from M to N with $T_{\mathfrak{p}}(\lambda) = \lambda(\mathfrak{p})$ for all $\mathfrak{p}$.

Our discussion should have made the following result obvious.

6.3.19 Proposition

Let M and N be finitely generated torsion $\mathcal{O}$-modules, where $\mathcal{O}$ is a Dedekind domain. Then $M \cong N$ if and only if $T_{\mathfrak{p}}(M) \cong T_{\mathfrak{p}}(N)$ for all primes $\mathfrak{p}$ of $\mathcal{O}$.

$\square$

Combining the preceding proposition with the description of primary modules in (6.3.9), together with the fact that the elementary divisor type characterizes primary modules (6.3.11), we obtain the classical description of torsion modules over a Dedekind domain.

6.3.20 The Primary Decomposition Theorem

Let M and N be finitely generated torsion modules over the Dedekind domain $\mathcal{O}$. Then

(i)

$$ M \cong \bigoplus\nolimits_{\mathfrak{p}} \left(\mathcal{O}/\mathfrak{p}^{\delta(\mathfrak{p},1)} \oplus \cdots \oplus \mathcal{O}/\mathfrak{p}^{\delta(\mathfrak{p},\ell(\mathfrak{p}))} \right), $$

where $\delta(\mathfrak{p},1) \leq \cdots \leq \delta(\mathfrak{p},\ell(\mathfrak{p}))$ are positive integers and $\ell(\mathfrak{p}) = 0$ for all except a finite set of primes $\mathfrak{p}$,

(ii) *$M \cong N$ if and only if $\mathrm{edt}(M) = \mathrm{edt}(N)$, that is, $\mathrm{edt}_{\mathfrak{p}}(M) = \mathrm{edt}_{\mathfrak{p}}(N)$ for all (nonzero) primes $\mathfrak{p}$ of $\mathcal{O}$,*

(iii) *the set of prime ideals $\mathfrak{p}$ of $\mathcal{O}$ with $\ell(\mathfrak{p}) \neq 0$ and the positive integers $\delta(\mathfrak{p},1) \leq \cdots \leq \delta(\mathfrak{p},\ell(\mathfrak{p}))$ are uniquely determined by M, and vice versa – informally, M is determined by its elementary divisors.* $\square$

6.3.21 Alternative decompositions

The Primary Decomposition Theorem shows that a finitely generated torsion module M is a direct sum of cyclic modules. In general, there are many ways in which a torsion module can be written as a direct sum of cyclic modules, because, by the Chinese Remainder Theorem (5.1.5), there is an isomorphism $\mathcal{O}/\mathfrak{ab} \cong \mathcal{O}/\mathfrak{a} \oplus \mathcal{O}/\mathfrak{b}$ for any pair of coprime ideals $\mathfrak{a}$ and $\mathfrak{b}$ of $\mathcal{O}$.

Thus, any uniqueness assertion for a direct decomposition of M into cyclic summands requires the imposition of some extra condition. In the Primary Decomposition Theorem, we in effect require that there are as many cyclic summands as possible. At the opposite extreme, if we ask for as few cyclic summands as possible, we are led to the Invariant Factor Theorem, which is sketched in Exercise 6.3.6 below.

In the case that $\mathcal{O}$ has only one nonzero prime ideal, that is, $\mathcal{O}$ is already local, the Invariant Factor Theorem is the same as the Primary Decomposition Theorem.

6.3.22 Homomorphisms again

We need some further remarks on homomorphisms as a preliminary to summarizing our results on modules over a Dedekind domain $\mathcal{O}$.

Let M and N be arbitrary finitely generated $\mathcal{O}$-modules and suppose that $\lambda : M \to N$ is a homomorphism between them. It is easy to see that λ induces a homomorphism

$$T(\lambda) : T(M) \longrightarrow T(N)$$

between the torsion submodules of M and N, and hence a homomorphism

$$F(\lambda) : M/T(M) \longrightarrow N/T(N)$$

between their torsion-free quotient modules.

By (6.3.4), $M/T(M)$ and $N/T(N)$ are projective, and so there are internal direct decompositions

$$M = T(M) \oplus M' \text{ and } N = T(N) \oplus N'$$

with

$$M/T(M) \cong M' \text{ and } N/T(N) \cong M',$$

as noted in (6.3.5). It is easy to check that λ is an isomorphism precisely when both $T(\lambda)$ and $F(\lambda)$ are isomorphisms.

Note that there is usually no canonical choice for the submodules M' and N' – see Exercise 6.3.9. In the language of categories, the methods by which $T(M)$ and $M/T(M)$ are constructed from M are both functorial, but the construction of M' is not.

We summarize all our results in one compendium, which completely classifies finitely generated modules for Dedekind domains.

6.3.23 Theorem

Let M be a finitely generated module over a Dedekind domain $\mathcal{O}$. Then the following assertions hold.

(i) $M = P \oplus T$ *where P is a finitely generated projective $\mathcal{O}$-module and $T = T(M)$ is a finitely generated torsion $\mathcal{O}$-module.*

(ii) $P \cong \mathcal{O}^{r-1} \oplus \mathfrak{a}$ *where $\mathfrak{a}$ is an ideal of $\mathcal{O}$.*

(iii) $T \cong \bigoplus_{\mathfrak{p}} T(\mathfrak{p})$, *where $\mathfrak{p}$ runs through the nonzero prime ideals of $\mathcal{O}$, each $T(\mathfrak{p}) = T_{\mathfrak{p}}(M)$ is a finitely generated $\mathfrak{p}$-primary $\mathcal{O}$-module, and $T(\mathfrak{p}) = 0$ for almost all $\mathfrak{p}$.*

(iv) *If $T(\mathfrak{p}) \neq 0$, then*

$$T(\mathfrak{p}) \cong \mathcal{O}/\mathfrak{p}^{\delta(\mathfrak{p},1)} \oplus \cdots \oplus \mathcal{O}/\mathfrak{p}^{\delta(\mathfrak{p},\ell(\mathfrak{p}))},$$

where $0 < \delta(\mathfrak{p}, 1) \leq \cdots \leq \delta(\mathfrak{p}, \ell(\mathfrak{p}))$ (and if $T(\mathfrak{p}) = 0$, then $\ell(\mathfrak{p}) = 0$).

Furthermore, the module M determines, and in turn is determined to within isomorphism by, the following information:

- *the integer $r = \text{rank}(M)$,*
- *the integers $\ell(\mathfrak{p})$ for all $\mathfrak{p}$,*
- *the integers $\delta(\mathfrak{p}, i)$ for all $1 \leq i \leq \ell(\mathfrak{p})$ and all $\mathfrak{p}$,*
- *the class $\{\mathfrak{a}\}$ of $\mathfrak{a}$ in the ideal class group $\text{Cl}(\mathcal{O})$.*

(That is, a module N is isomorphic to M if and only if the set of integers and the ideal class attached to N are the same as those for M.)

In particular, the torsion part of M is determined up to isomorphism by its set of elementary divisors

$$\{\mathfrak{p}^{\delta(\mathfrak{p},1)}, \ldots, \mathfrak{p}^{\delta(\mathfrak{p},\ell(\mathfrak{p}))} \mid \mathfrak{p} \in \mathbf{P}\}.$$

Proof

Assertion (i) follows by (6.3.4) and (6.3.5). For (ii), see (6.1.7), and (iii) is in (6.3.15), noting that here $T(\mathfrak{p}) = T_{\mathfrak{p}}(T)$. Statement (iv) is given in (6.3.20).

The claim about isomorphism follows from (6.3.20) again, combined with Steinitz' Theorem (6.1.6), using the discussion of homomorphisms above. □

Finally, it is useful to have a reinterpretation of the above result when $\mathcal{O}$ is a (commutative) principal ideal domain. In this case, any projective module is free, and it is more usual to describe torsion modules in terms of their p-primary components (6.3.17), where p runs through a representative set **Pe** of irreducible elements of $\mathcal{O}$, that is, **Pe** has exactly one member p for each nonzero prime ideal $\mathfrak{p}$ of $\mathcal{O}$. We also regard the elementary divisors of a module as powers of irreducible elements rather than ideals.

The next result is an immediate translation of its predecessor into the changed vocabulary.

6.3.24 Theorem

Let M be a finitely generated module over a commutative principal ideal domain $\mathcal{O}$. Then the following assertions hold.

(i) *$M = F \oplus T$ where $F \cong \mathcal{O}^r$ is a finitely generated free $\mathcal{O}$-module and T is a finitely generated torsion $\mathcal{O}$-module.*

(ii) *$T \cong \bigoplus_p T(p)$, where $p \in \mathbf{Pe}$, each $T(p)$ is a finitely generated p-primary $\mathcal{O}$-module, and $T(p) = 0$ for almost all p.*

(iii) *If $T(p) \neq 0$, then*

$$T(p) \cong \mathcal{O}/p^{\delta(p,1)}\mathcal{O} \oplus \cdots \oplus \mathcal{O}/p^{\delta(p,\ell(p))}\mathcal{O},$$

where $0 < \delta(p, 1) \leq \cdots \leq \delta(p, \ell(p))$ (and if $T(p) = 0$, $\ell(p) = 0$).

Furthermore, the module M determines, and in turn is determined to within isomorphism by, the following information:

- the integer $r = \text{rank}(M)$,
- the integers $\ell(p)$ for all p in **Pe**,
- the integers $\delta(p, i)$ for all i with $1 \leq i \leq \ell(p)$ and all p in **Pe**.

In particular, the torsion part of M is determined up to isomorphism by its set of elementary divisors

$$\{p^{\delta(p,1)}, \ldots, p^{\delta(p,\ell(p))} \mid p \in \textbf{Pe}\}. \qquad \square$$

Exercises

In these exercises, the ring $\mathcal{O}$ is a Dedekind domain. For a pair of (right) $\mathcal{O}$-modules M and N we abbreviate $\text{Hom}_{\mathcal{O}}(M, N)$ to $\text{Hom}(M, N)$.

6.3.1   Suppose that $\mathcal{O}$ has infinitely many distinct prime ideals, and put $M = \bigoplus_{\mathfrak{p}} (\mathcal{O}/\mathfrak{p})$. Show that M is a torsion module, but $\text{Ann}(M) = 0$.
Let $N = \bigoplus_{i>0} (\mathcal{O}/\mathfrak{p}^i)$ for any fixed $\mathfrak{p}$. Show that $\text{Ann}(N) = 0$ but that any finitely generated submodule of N is $\mathfrak{p}$-primary.

(This shows the significance of restricting our attention to finitely generated modules, particularly in (6.3.7).)

6.3.2   We can extend the definition of $\mathfrak{p}$-primary to $\mathcal{O}$-modules which are not finitely generated by saying that an $\mathcal{O}$-module M is $\mathfrak{p}$-primary provided that all its finitely generated submodules are $\mathfrak{p}$-primary.

Using Zorn's Lemma, show that any $\mathcal{O}$-module has a maximal $\mathfrak{p}$-primary submodule.

Verify that parts (ii) and (iii) of (6.3.15) continue to hold for non-finitely-generated modules.

6.3.3   Find the composition series of $\mathcal{O}/\mathfrak{p}^\delta$. Hence find the composition factors of an arbitrary $\mathfrak{p}$-primary module M, and find a formula for the length of a composition series for M in terms of its elementary divisor type.

Extend the result to arbitrary finitely generated torsion modules.

6.3.4   Let $\mathfrak{a}$ be an integral ideal of $\mathcal{O}$. Find the primary decomposition of $\mathcal{O}/\mathfrak{a}$.

Suppose that $\mathfrak{a}_1, \ldots, \mathfrak{a}_r$ are ideals of $\mathcal{O}$ with $\mathfrak{a}_1 | \cdots | \mathfrak{a}_r$, and let $M = \mathcal{O}/\mathfrak{a}_1 \oplus \cdots \oplus \mathcal{O}/\mathfrak{a}_r$ (the external direct sum). Describe the primary decomposition of M and its elementary divisor type.

6.3.5   Uniqueness of invariant factors

Let $\mathcal{O}$ be a (commutative) Euclidean domain, in particular, a polynomial ring $\mathcal{K}[T]$, and let M be a finitely generated $\mathcal{O}$-module. By (3.3.6),

$$M \cong \mathcal{O}/e_1\mathcal{O} \oplus \cdots \oplus \mathcal{O}/e_\ell\mathcal{O} \oplus \mathcal{O}^s,$$

where $e_1 \mid e_2 \mid \cdots \mid e_\ell$ are the invariant factors of M.

Identify $T(M)$ and $M/T(M)$, and, using Exercise 6.3.4 above, show that the ideals $e_1\mathcal{O},\ldots,e_\ell\mathcal{O}$ are unique. Deduce that the invariant factors of M are unique up to multiplication by units.

6.3.6   Invariant Factor Theorem for Dedekind Domains

Let $\mathcal{O}$ be an arbitrary Dedekind domain, and let M be a finitely generated torsion module over $\mathcal{O}$. Using the primary decomposition of M, show that there are ideals $\mathfrak{a}_1,\ldots,\mathfrak{a}_\ell$ of $\mathcal{O}$ with $\mathfrak{a}_1|\cdots|\mathfrak{a}_\ell$ and $M \cong \mathcal{O}/\mathfrak{a}_1 \oplus \cdots \oplus \mathcal{O}/\mathfrak{a}_\ell$.

Deduce that if K is a submodule of a free $\mathcal{O}$-module $\mathcal{O}^n$, then there is a basis $(f_1,\ldots,f_n)$ of $\mathcal{O}^n$ such that $K = \mathfrak{a}_1 f_1 \oplus \cdots \oplus \mathfrak{a}_\ell f_\ell$.

(For a direct proof, see [Curtis & Reiner 1966], 22.12.)

6.3.7   Let $\mathfrak{a}$ and $\mathfrak{b}$ be fractional ideals of $\mathcal{O}$. Show that there is an exact sequence

$$0 \longrightarrow \mathrm{Hom}(\mathfrak{a},\mathfrak{b}) \longrightarrow \mathrm{Hom}(\mathfrak{a},\mathcal{O}) \longrightarrow \mathrm{Hom}(\mathfrak{a},\mathcal{O}/\mathfrak{b}) \longrightarrow 0,$$

and deduce that

$$\mathrm{Hom}(\mathfrak{a},\mathcal{O}/\mathfrak{b}) \cong \mathfrak{a}^{-1}/\mathfrak{a}^{-1}\mathfrak{b} \cong \mathcal{O}/\mathfrak{b}.$$

Hints. (6.1.1), (5.1.24) and Exercise 5.1.8.

6.3.8   Let M and N be (right) $\mathcal{O}$-modules. Show that

(i) $\mathrm{Hom}(M,N) = 0$ if $\mathrm{Ann}(M)$ and $\mathrm{Ann}(N)$ are coprime,

(ii) $\mathrm{Hom}(M,N) = 0$ if M is a torsion module and N is torsion-free.

6.3.9   Let M and N be finitely generated (right) $\mathcal{O}$-modules, and choose internal direct sum decompositions $M = T(M) \oplus M'$ and $N = T(N) \oplus N'$, so that $M' \cong M/T(M)$, etc.

Show that

$$\mathrm{Hom}(M,N) \cong \begin{pmatrix} \mathrm{Hom}(T(M),T(N)) & \mathrm{Hom}(M',T(N)) \\ 0 & \mathrm{Hom}(M',N') \end{pmatrix},$$

where the matrices act as left multipliers on the 'column' M – Exercises 2.1.6 and 2.1.7 are relevant.

Confirm that $\mathrm{End}(M)$ is a triangular ring of matrices.

Compute $\mathrm{Hom}(\mathcal{O}/\mathfrak{p} \oplus \mathcal{O}, \mathcal{O}/\mathfrak{q} \oplus \mathcal{O})$, where $\mathfrak{p}, \mathfrak{q}$ are prime ideals of $\mathcal{O}$, possibly the same.

Remark. The term $\mathrm{Hom}(M', N')$ is known, by Exercise 6.1.4, and $\mathrm{Hom}(M', T(N))$ is computable by Exercise 6.3.7 above. As noted in (6.3.18), an element λ in $\mathrm{Hom}(T(M), T(N))$ can be represented as a sequence $(\lambda_\mathfrak{p})$, where each $\lambda_\mathfrak{p}$ is in in $\mathrm{Hom}(T_\mathfrak{p}(M), T_\mathfrak{p}(N))$. The next exercise gives $\lambda_\mathfrak{p}$.

6.3.10 Let $\mathfrak{p}$ be a nonzero prime ideal of $\mathcal{O}$. Show that

$$\mathrm{Hom}_{\mathcal{O}}(\mathcal{O}/\mathfrak{p}^r, \mathcal{O}/\mathfrak{p}^s) = \{x \in \mathcal{O}/\mathfrak{p}^s \mid \mathfrak{p}^r x = 0\} \cong \begin{cases} \mathcal{O}/\mathfrak{p}^s & r \geq s, \\ \mathcal{O}/\mathfrak{p}^r & r < s. \end{cases}$$

Let $M = \mathcal{O}/\mathfrak{p}^{\delta(1)} \oplus \cdots \oplus \mathcal{O}/\mathfrak{p}^{\delta(k)}$ and $N = \mathcal{O}/\mathfrak{p}^{\epsilon(1)} \oplus \cdots \oplus \mathcal{O}/\mathfrak{p}^{\epsilon(\ell)}$ be $\mathfrak{p}$-primary modules. Describe $\mathrm{Hom}(M, N)$ as a set of $\ell \times k$ matrices.

References

Each item is accompanied by a list of the pages on which it is cited, with the exception of our companion volume on categories and modules, [BK: CM].

[Abrams 1987] G. D. Abrams, On dense subrings of RFM(R), *J. Algebra* **110** (1987), 243–248. [73]

[Allenby 1991] R. B. J. T. Allenby, *Rings, Fields and Groups*, 2nd edition (Edward Arnold, London, 1991). [1]

[Artin 1927] E. Artin, Zur Theorie der hypercomplexer Zahlen, *Abh. Math. Seminaire Univ. Hamburg* **5** (1927) 251–260. [146, 161]

[Auslander, Reiten & Smalø 1995] M. Auslander, I. Reiten & S. O. Smalø Representations of Artin Algebras, Cambridge Studies in Advanced Mathematics 36 (Cambridge University Press, Cambridge, 1995). [182]

[Baer 1940] R. Baer, Abelian groups that are direct summands of every containing abelian group, *Bull. Amer. Math. Soc.* **46** (1940), 800–806. [95]

[Bass 1968] H. Bass, *Algebraic K-Theory* (Benjamin, New York, 1968). [181]

[Berrick 1982] A. J. Berrick, *An Approach to Algebraic K-Theory*, Research Notes in Mathematics 56 (Pitman, Boston, Mass., 1982). [73]

[BK: CM] A. J. Berrick & M. E. Keating, *Categories and Modules* (Cambridge University Press, Cambridge, to appear).

[Berrick & Keating 1997] A. J. Berrick & M. E. Keating, Rectangular invertible matrices, *Amer. Math. Monthly* **104** (1997), 297–302. [73, 185]

[Bourbaki 1991] N. Bourbaki, *Elements of the History of Mathematics*, trans. J. Meldrum (Springer-Verlag, Berlin, 1991). [xv]

[Camillo 1984] V. Camillo, Morita equivalence and infinite matrix rings, *Proc. Amer. Math. Soc.* **90** (1984), 186–188. [73]

[Campbell 1978] P. J. Campbell, The origin of "Zorn's Lemma", *Historia Math.* **5** (1978), 77–89. [26]

[Cartan & Eilenberg 1956] H. Cartan & S. Eilenberg, *Homological Algebra* (Oxford University Press, London, 1956). [95, 225]

[Claborn 1966] L. Claborn, Every abelian group is a class group, *Pacific J. Math.* **18** (1966), 219–222. [191]

[Cohn 1966] P. M. Cohn, Some remarks on the invariant basis property, *Topology* **5** (1966), 215–228. [71]

[Cohn 1979] P. M. Cohn, *Algebra*, Volume 2 (John Wiley & Sons, Chichester, 1979). [163, 172, 182, 206, 208, 223]

[Cohn 1981] P. M. Cohn, *Universal Algebra* (Reidel, Dordrecht, Holland, 1981). [74]

[Cohn 1982] P. M. Cohn, *Algebra*, Volume 1, 2nd edition, (John Wiley & Sons, Chichester, 1982). [137]

[Cohn 1985] P. M. Cohn, *Free Rings and their Relations*, 2nd edition, London Math. Soc. Monograph 19 (Academic Press, London, 1985).
[128, 129, 135, 137, 140, 142]
[Cohn 1991] P. M. Cohn, *Algebraic Numbers and Algebraic Functions* (Chapman & Hall, London, 1991). [222, 232]
[Cohn 1995] P. M. Cohn, *Skew Fields: Theory of General Division Rings*, Encyclopedia of Mathematics and its Applications 57 (Cambridge University Press, Cambridge, 1995). [13]
[Curtis & Reiner 1966] C. W. Curtis & I. Reiner, *Representation Theory of Finite Groups and Associative Algebras* (John Wiley & Sons, New York, 1966). [250]
[Curtis & Reiner 1981] C. W. Curtis & I. Reiner, *Methods of Representation Theory with Applications to Finite Groups and Orders*, Volume I (John Wiley & Sons, New York, 1981). [184]
[Dedekind 1932] R. Dedekind, *Gesammmelte mathematische Werke* 3 vol. (Vieweg, Braunschweig, 1932). [109]
[Dedekind 1996] R. Dedekind, *Theory of Algebraic Numbers*, trans. J. Stillwell, Cambridge Mathematical Library (Cambridge University Press, Cambridge, 1996). [xv, 15, 109, 186, 211]
[Deuring 1937] M. Deuring, *Algebren*, Ergebnisse der Math. und ihrer Grenzgebiete, Band 4 (Springer-Verlag, Berlin, 1937). [113]
[Dickson 1938] L. E. Dickson, *Algebras and their Arithmetics* (Stechert & Co, New York, 1938). [130]
[Eggleton, Lacampagne & Selfridge 1992] R. B. Eggleton, C. B. Lacampagne & J. L. Selfridge, Euclidean quadratic fields, *Amer. Math. Monthly* **99** (1992), 829–837. [217]
[Encyclopaedic Dictionary of Mathematics 1987] *Encyclopaedic Dictionary of Mathematics, 3rd edition*, editor K. Ito (MIT Press, Cambridge, Mass., 1987). [223]
[Euclid 1956] *Euclid's Elements*, Volume 2, trans. T. L. Heath (Dover, New York, 1956). [122]
[Fossum 1973] R. M. Fossum *The Divisor Class Group of a Krull Domain*, Ergebnisse der Mathematik und ihrer Grenzgebiete, Band 74 (Springer-Verlag, Berlin, 1973). [191, 222]
[Fröhlich 1983] A. Fröhlich, *Galois Module Structure of Algebraic Integers*, Ergebnisse der Mathematik und ihrer Grenzgebiete, 3. Folge, Band 1 (Springer-Verlag, Berlin, 1983). [223]
[Fröhlich & Taylor 1991] A. Fröhlich & M. J. Taylor, *Algebraic Number Theory*, Cambridge Studies in Advanced Mathematics 27 (Cambridge University Press, Cambridge, 1991). [206, 215, 217, 218, 222]
[Fuchs 1967] L. Fuchs, *Abelian Groups* (Pergamon Press, Oxford, 1967). [72]
[Fulton 1989] W. Fulton, *Algebraic Curves* (Addison-Wesley, Reading, Mass., 1989). [222]
[Goodearl 1992] K. R. Goodearl, Prime ideals in skew polynomial rings and quantized Weyl algebras, *J. Algebra* **150** (1992), 324–377. [129]
[Guralnick & Levy 1988] R. M. Guralnick & L. S. Levy, Presentations of modules when ideals need not be principal, *Illinois J. Math.* **32** (1988), 593–653. [140]
[Guralnick, Levy & Odenthal 1988] R. M. Guralnick, L. S. Levy & C. Odenthal, Elementary divisor theorem for noncommutative PIDs, *Proc. Amer. Math. Soc.* **103** (1988), 1003–1012. [140]
[Hamilton 1853] Sir W. R. Hamilton, *Lectures on Quaternions* (Dublin, 1853). [131]

[Hardy & Wright 1979] G. H. Hardy & E. M. Wright, *An Introduction to the Theory of Numbers*, 5th edition (Oxford University Press, Oxford, 1979). [217]

[Hasse 1980] H. Hasse, *Number Theory* (Springer-Verlag, Berlin, 1980). [217, 218]

[Higgins 1975] P. J. Higgins, *A First Course in Abstract Algebra* (Van Nostrand Reinhold, London, 1975) [1]

[Hilbert 1932-35] D. Hilbert, *Gesammelte Abhandlung*, 3 vols (Springer, Berlin, 1932–35). [119]

[Howie 1976] J. M. Howie, *An Introduction to Semigroup Theory*, London Math. Soc. Monograph 7 (Academic Press, London, 1976). [74]

[Hurewicz 1941] W. Hurewicz, On duality theorems, *Bull. Amer. Math. Soc.* **47** (1941), 562–563. [76]

[Hurwitz 1896] A. Hurwitz, Uber die Zahlentheorie der Quaternion, *Nachrichten der Gesellschaft der Wissenshaft, Göttingen* (1896), 313–340. [130]

[Ireland & Rosen 1982] K. Ireland & M. Rosen, *A Classical Introduction to Number Theory*, Graduate Texts in Mathematics 84 (Springer-Verlag, New York, 1986). [218]

[Jacobson 1945] N. Jacobson, The radical and semisimplicity for arbitrary rings, *Amer. J. Math.* **67** (1945), 300–320. [173]

[Jacobson 1964] N. Jacobson, *Lectures in Abstract Algebra, Volume III - Theory of Fields and Galois Theory* (Van Nostrand, New York, 1964). [118]

[Jacobson 1968] N. Jacobson, *Theory of Rings*, 4th edition, Mathematical Surveys No. 2, (Amer. Math. Soc. Providence, R. I., 1968). [128]

[Jech 1973] T. J. Jech, *The Axiom of Choice*, Studies in Logic Vol. 75 (North-Holland, Amsterdam, 1973). [26]

[Jordan 1993] D. A. Jordan, Iterated skew polynomial rings and quantum groups, *J. Algebra*, **156** (1993), 194–218. [129]

[Keating 1998] M. E. Keating, *A First Course on Module Theory* (Imperial College Press, London, 1998) [15]

[Kelley & Pitcher 1947] J. L. Kelley & E. Pitcher, Exact homomorphism sequences in homology theory, *Annals of Math.* **48** (1947), 682–709. [76]

[Kleiner 1996] I. Kleiner, The genesis of the abstract ring concept, *Amer. Math. Monthly* **103** (1996), 417–423. [1]

[Krull 1938] W. Krull, Dimensiontheorie in Stellenringe, *J. reine angew. Math.* **169** (1938), 204–226. [238]

[Kuratowski 1922] C. Kuratowski, Une méthode d'élimination des nombres transfinis des raisonnements mathématiques, *Fund. Math.* **3** (1922), 76–108. [26]

[Lam & Ang 1992] L. Y. Lam & T. S. Ang, *Fleeting Footsteps: Tracing the Conception of Arithmetic and Algebra in Ancient China* (World Scientific, Singapore, 1992). [188]

[Lam 1978] T. Y. Lam, *Serre's Conjecture*, Lecture Notes in Mathematics 635 (Springer, Berlin, 1978). [140]

[Lam & Leroy 1992] T. Y. Lam & A. Leroy, *Homomorphisms between Ore extensions*, Contempory Math. 124 (Amer. Math. Soc. , Providence, R. I., 1992) 83–110. [129]

[Leavitt 1957] W. G. Leavitt, Rings without invariant basis number, *Proc. Amer. Math. Soc.* **8** (1957), 322–328. [71, 104]

[Leavitt 1962] W. G. Leavitt, The module type of a ring, *Trans. Amer. Math. Soc.* **103** (1962), 113–130. [71, 104]

[McConnell & Robson 1987] J. C. McConnell & J. C. Robson, *Noncommutative Noetherian Rings* (Wiley-Interscience, John Wiley, Chichester, 1987).
[123, 129, 132, 142, 162, 229]

[Mal'cev 1937] A. I. Mal'cev, On the immersion of an algebraic ring in a field, *Math. Ann.* **117** (1937), 686–691. [13]

[Marcus 1977] D. A. Marcus, *Number Fields*, Universitext (Springer-Verlag, Berlin, 1977). [215, 217, 222]

[Nagata 1978] M. Nagata, *On Euclid algorithm, C. P. Ramanujan – a Tribute*, Studies in Math., 8 Tata Institute of Fundamental Research, Bombay, 1978, 175–186. [130]

[Noether 1926] E. Noether, Abstrakter Aufbau der Idealtheorie in algebraischen Zahl- und Funktionenkörper, *Math. Ann.* **96** (1926), 26–61. [109, 186]

[Noether 1929] E. Noether, Hypercomplexe Grössen und Darstellungtheorie, *Math. Zeitschrift* **30** (1929), 641–692. [15, 104, 150]

[Noether & Schmeider 1920] E. Noether & W. Schmeidler, Moduln in nichtcommutativen Bereichen, inbesondere aus Differential- und Differenzenausdrucken, *Math. Zeitschrift* **8** (1920), 1–35. [15, 104]

[Ore 1933] O. Ore, Theory of noncommutative polynomials, *Annals of Math.* **34** (1933), 480–508. [129]

[Pierce 1881] B. Pierce, Linear associative algebra, *Amer. J. Math.* **IV** (1881) 97–221. [99]

[Pohst & Zassenhaus 1989] M. Pohst & H. Zassenhaus, *Algorithmic Algebraic Number Theory*, Encyclopedia of Mathematics and its Applications 30 (Cambridge University Press, Cambridge, 1989). [223]

[Reiner 1975] I. Reiner, *Maximal Orders*, London Math. Soc. Monograph 5 (Academic Press, London, 1975). [210, 237]

[Rotman 1979] J. J. Rotman, *An Introduction to Homological Algebra* (Academic Press, Boston, Mass., 1979). [95]

[Rowen 1988] L. H. Rowen, *Ring Theory*, Volume I, (Academic Press, Boston, Mass., 1988). [13, 129, 138, 182]

[Samuel 1948] P. Samuel, On universal mappings and free topological groups, *Bull. Amer. Math. Soc.* **54** (1948), 591–598. [104]

[Smith 1981] S. P. Smith, An example of a ring Morita equivalent to the Weyl algebra A_1, *J. Algebra* **73** (1981), 552–555. [64]

[Snaith 1994] V. P. Snaith, *Galois Module Structure*, Fields Institute Monograph No. 2, (Amer. Math. Soc., Providence, R. I., 1994). [223]

[Srinivasan & Sally 1983] B. Srinivasan & J. Sally (editors), *Emmy Noether in Bryn Mawr* (Springer-Verlag, Berlin, 1983). [xv]

[Steinitz 1912] E. Steinitz, Rechticke Systeme und Moduln in algebraischen Zahlkörpern, *Math. Ann.* **71** (1912), 328–354 and **72** (1912), 297–345. [227]

[van der Waerden 1980] B. L. van der Waerden, *A History of Algebra* (Springer-Verlag, Berlin, 1980). [xv]

[Wedderburn 1908] J. H. M. Wedderburn, On the hypercomplex numbers, *J. reine angew. Math.* **167** (1932), 129–141. [161]

[Wedderburn 1932] J. H. M. Wedderburn, Non-commutative domains of integrity, *Proc. London Math. Soc.* (2) **6** (1908), 77–117. [123]

[Weiss 1963] E. Weiss, *Algebraic Number Theory* (McGraw-Hill, New York, 1963). [221]

[Zariski & Samuel 1963] O. Zariski & P. Samuel, *Commutative Algebra I* (Van Nostrand, Princeton, N. J., 1963). [208]

[Zorn 1935] M. Zorn, A remark on method in transfinite algebra, *Bull. Amer. Math. Soc.* **41** (1935), 667–670.   [26, 27]

Index

Index of names

Index of terms

Printed in the United States
By Bookmasters